T0310405

AeroMACS

AeroMACS

An IEEE 802.16 Standard-Based Technology for the Next Generation of Air Transportation Systems

Behnam Kamali

Sam Nunn Eminent Scholar of Telecommunications and
Professor of Electrical and Computer Engineering
Mercer University
Macon, GA

Published by
Standards Information Network

IEEE PRESS

WILEY

Registered Office
John Wiley & Sons, Inc., 111 River Street, Hoboken, NJ 07030, USA

Editorial Office
111 River Street, Hoboken, NJ 07030, USA

For details of our global editorial offices, customer services, and more information about Wiley products visit us at www.wiley.com.

Wiley also publishes its books in a variety of electronic formats and by print-on-demand. Some content that appears in standard print versions of this book may not be available in other formats.

Library of Congress Cataloging-in-Publication Data

ISBN: 9781119281108

Set in 10/12 pt WarnockPro-Regular by Thomson Digital, Noida, India

Printed in the United States of America

C10004631_092718

This book is dedicated to the memory of my father,
Abdul Hossain Kamali (1915–1973),
who was taken away from me unexpectedly,
but his quest for knowledge, his enthusiasm for technology,
and his insistence on the independent search for truth
have remained with me and inspired me.

Table of Contents

Preface

Civil aviation plays a major role in driving sustainable global and national economic and social development. During the year 2015, civil aviation created 9.9 million jobs inside the industry, and directly and indirectly supported the employment of 62.7 million people around the world. The total global economic impact of civil aviation was $2.7 trillion (including the effects of tourism). In the same year, approximately 3.6 billion passengers were transported through air. The volume of freight carried via air reached 51.2 million tons. Today, the value of air-transported goods stands at $17.5 billion per day. Accordingly, in the year 2015, approximately 3.5% of global GDP was supported by civil aviation. Research conducted in the United States suggests that every $100 million dollars invested in aerospace yields an extra $70 million in GDP year after year[1]. In addition to economic prosperity, civil aviation brings about a number of social and human relation benefits, ranging from swift delivery of health care, emergency services, and humanitarian aid, to the promotion of peace and friendship among various groups of people through trade, leisure, and cultural experiences and exchanges.

The global air transportation system is a worldwide network, consisting of four components of airport and airport infrastructures, commercial aircraft operators, air navigation service providers, and the manufacturers of aircraft and associated components. The airport component plays a central role in air traffic management, air traffic control, and the management of national and global airspace systems. From the technical point of view air transportation operation is centered around three elements of communications, navigation, and surveillance. The safety of air transportation is critically linked to the availability of reliable aeronautical communication systems that support all aspects of air operations and air traffic management, including navigation and

---

1 IATA (International Air Transport Association) Fact Sheet Economic & Social Benefits of Air Transport, 2017.

surveillance. Owing to the fact that flight safety is the highest priority in aviation, extreme measures must be taken to protect the aeronautical communication systems against harmful interference, malfunction, and capacity limitation.

In the early days of commercial aviation, the 1940s, analog AM radio over VHF band was adopted for aeronautical communications. This selection was made mostly for the reason that analog AM was the only fully developed and proven radio communications technology at the time. However, by the late 1980s, spectrum congestion in aeronautical VHF band, due to rapid growth in both commercial and general sectors of civil aviation, became a concern for the aviation community in the United States and in Europe. The concerns about inability of the legacy system to safely manage future levels of air traffic, called for modernization of air transportation systems. This in turn led to the initiatives of Next Generation Air Transportation System Integrated Plan (NextGen) in the United States, and European Commission Single European Sky ATM Research (SESAR) in Europe. A joint FAA-EUROCONTROL technology assessment study on communications for future aviation systems had already come to the conclusion that no single communication technology could satisfy all physical, operational, and functional requirements of various aeronautical transmission domains. Based on recommendations made by that study, a broadband wireless mobile communications technology based on IEEE 802.16e (Mobile WiMAX) was selected for airport surface domain, leading to the advent of aeronautical mobile airport communications system, AeroMACS, the subject of focus in this book.

Over the past few years AeroMACS has evolved from a technology concept to a deployed operating communications network over a number of major U.S. airports. Projections are that AeroMACS will be deployed across the globe by the year 2020. It is worth noting that AeroMACS, as a new broadband data link able to support the ever-expanding air traffic management communications requirements, is emerging out of the modernization initiatives of NextGen and SESAR, and therefore should be considered to be an integral and enabling part of both NextGen and SESAR visions.

The main feature of this book is its pioneering focus on AeroMACS, representing, perhaps, the first text written entirely on the technology and how it relates to its parental standards (although book chapters on the subject have been published previously). The text is prepared, by and large, from a system engineering perspective, however, it also places emphasis on the description of IEEE 802.16e standards and how they can be tied up with communications requirements on the airport surface. A second contribution that this book aspires to make; when viewed on the whole, is to provide a complete picture of the overall process of how a new technology is developed based on an already established standard, in this case IEEE 802.16e standards. AeroMACS, like its parent standards, mobile WiMAX and IEEE 802.16-2009 WirelessMAN, is a complex technology that is impossible to fully describe in a few hundred pages. Nonetheless, it is hoped that this book will be able to provide an overall understanding of several facets of this fascinating technology that will be a key component of modern global air transportation systems. Another feature of this text is the simplicity of the language that is used for the description of complicated concepts. Efforts have also been made, to the extent possible and despite all the challenges, to make this book self-contained. To this end, review chapters are included and a large number of footnotes are provided in each chapter.

1 Synopsis of Chapters

This book, for the most part, reflects the results of the author's research activities in the field of aeronautical communications in conjunction with several summer research fellowships at NASA Glenn Research Center. The book consists of eight chapters. Chapter 1 presents an introduction to the applications of wireless communications in the airport environment. The chapter portrays a continuous picture of the evolution of airport surface communications techniques from the legacy VHF analog AM radio, to the appearance of digital communications schemes for various airport surface functionalities, and to the making of the AeroMACS concept. The rationales and the reasons behind the emergence of Aero-MACS technology are described. The large arenas over which AeroMACS will operate, that is, the National Airspace System (NAS) and the International Airspace System, are concisely overviewed. The Federal Aviation Administration's NextGen and European SESAR programs, planned to transform and modernize air transportation, are discussed as well. Auxiliary wireless and wireline systems for airport surface communications, including airport fiber optic cable loop system, are briefly covered in the conclusion.

In modern wireless communication theory, a formidable challenge is the integration of an astonishing breath of topics that are tied together to provide the necessary background for thorough understanding of a wireless technology such as AeroMACS. It is no longer possible to separate signal processing techniques, such as modulation and channel coding, from antenna systems (traditionally studied as a topic in electro-magnetic theory), and from networking issues involving physical layer and medium access control sublayer protocols. To this end, Chapter 2 is the first of the three review chapters in which two topics of cellular networking and wireless channel characterizations are addressed. The main objective for this and other review chapters is to ensure, as much as possible, that the text is self-contained. This approach is conducive to the understanding of the cellular architecture of the network and the challenges posed by airport surface radio channel in design, implementation, and deployment stages of AeroMACS systems.

Chapter 3, authored by Dr. David Matolak of the University of South Carolina, is dedicated to the airport surface radio channel characterization over the 5 GHz band. The chapter commences with describing the motivation and the need for this topic, followed by some background on wireless channels and modeling, and specific results for the airport surface channel. An extensive airport surface area channel measurement campaign is summarized. Example measurement results for RMS delay spread, coherence bandwidth, and small-scale fading Rician K-factors are provided. Detailed airport surface area channel models over the 5 GHz band, in the form of tapped-delay lines, are then presented.

Chapter 4 is the second review chapter, focusing on orthogonal frequency-division multiplexing (OFDM), coded OFDM, orthogonal frequency-division multiple access (OFDMA), and scalable OFDMA (SOFDMA). OFDMA is an access technology that offers significant advantages for broadband wireless transmission over its rival technologies such as CDMA. Accordingly, it is shared by a number of contemporary wireless telecommunication networks, including IEEE 802.16-Std-based networks such as WiMAX and AeroMACS. The primary advantage of OFDMA over rival access technologies is the ability of OFDM to convert a wideband frequency selective fading channel to a series of narrowband flat fading channels. This is the mechanism by which frequency selective fading effects of hostile multipath environments, such as the airport surface channel, are mitigated or eliminated altogether. Performance of channel coding in OFDM, that is, modulation–coding combination, is explored in this chapter, providing some background for understanding of adaptive modulation coding (AMC) scheme discussed in later chapters. Scalable OFDMA, which presents a key feature of mobile WiMAX networks, is covered in some detail.

Chapter 5 provides a brief review on IEEE 802.16-2009 and IEEE 802.16j-2009 standards as well as an overview on Worldwide Interoperability for Microwave Access (WiMAX); an IEEE 802.16-standard-based broadband access solution for wireless metropolitan area networks. AeroMACS mandatory and optional protocols are a subset of those inherited by mobile WiMAX from IEEE 802.16e standards. The main purpose of this review chapter is to provide technical background information on various algorithms and protocols that support Aero-MACS networks. A high point of WiMAX technology is the fact that only physical (PHY) layer and medium access control (MAC) sublayer protocols have been defined while the higher layer protocols and the core network architecture are left unspecified to be filled by other technologies such as IP network architecture. The backbone of WiMAX technology is formed by OFDMA, multiple-input multiple-output (MIMO) concept, and IP architecture, all inherited by AeroMACS networks.

Chapter 6 is entirely dedicated to AeroMACS, providing an introduction to information related to the creation, standardization, and test and evaluation (through test beds) of this aviation technology. The core of this chapter is the AeroMACS standardization process that starts with technology selection. In contrast with assembling a proprietary dedicated technology, AeroMACS is constructed based on an interoperable version of IEEE 802.16-2009 standards (mobile WiMAX). The advantages of using an established standard are listed in the chapter. The IEEE 802.16e standard brings with itself a large number of PHY layer and MAC sublayer optional and mandatory protocols to select from for any driven technology. The WiMAX Forum System Profile Version 1.09, which assembles a subset of the IEEE standard protocols together, is such a technology that was selected as the parent standard for AeroMACS. Based on this selection, RTCA has developed a profile for AeroMACS. An overview of AeroMACS profile is presented in Chapter 6. Standards and Recommended Practices (SARPS) was developed almost simultaneously with the AeroMACS Profile by RTCA and EUROCAE. The last pieces of standardization process for AeroMACS to follow, as the chapter explains, were Minimum Operation Performance Standards (MOPS) and Minimum Aviation System Standards (MASPS). Finally, the AeroMACS standardization documents became a source for developments of an AeroMACS technical manual and an installation guide document. Potential airport surface services and functionalities that may be carried by AeroMACS are also addressed in Chapter 6. The chapter elaborates on AeroMACS test bed configuration and summarizes the early test and evaluation results, as well.

Chapter 7 explores AeroMACS as a short-range high-aggregate-data-throughput broadband wireless communications system, and concentrates on the detailed characterization of AeroMACS PHY layer and MAC sublayer features. AeroMACS main PHY layer feature is its multipath resistant multiple access technology, OFDMA, which allows 5 MHz channels within the allocated ITU-regulated aeronautical C-band of 5091–5150 MHz. The duplexing method is TDD, which enables asymmetric signal transmission over uplink (UL) and downlink (DL) paths. Adaptive modulation and coding (AMC) is another key physical layer feature of AeroMACS network. AMC allows for a proper combination of a modulation and coding schemes commensurate with the channel conditions. Multiple-input multiple-output (MIMO) and smart antenna systems are another PHY layer feature of AeroMACS networks. The chapter also discusses AeroMACS MAC sublayer. In particular, scheduling, QoS, ARQ system, and handover (HO) procedure are described. AeroMACS network architecture and Network Reference Model (NRM) are discussed. It is explained that AeroMACS is planned to be an all-IP network that supports high-rate packet-switched air traffic control (ATC) and Aeronautical Operational Control (AOC) services for efficient and safe management of flights, while providing connectivity to aircraft, operational support vehicles, and personnel within the airport area. Finally, the chapter highlights the position and the role of the AeroMACS network within the larger contexts of the Airport Network and the global Aeronautical Telecommunications Network (ATN).

The core idea of Chapter 8 is the demonstration of the fact that the IEEE 802.16j Amendment is highly feasible to be utilized as the foundational standard upon which AeroMACS networks are developed. This amendment enables the network designer to use the multihop relay as yet another design option in their device arsenal set. The chapter contains a great deal of information regarding the applications and usage scenarios for multihop relays in AeroMACS networks. Since the C-band spectrum allocated for AeroMACS is shared by other applications, interapplication interference (IAI) becomes a critical issue. It is shown, through a preliminary simulation study, that deployment of IEEE 802.16j Aero-MACS poses no additional IAI to coallocated applications. An important consideration, given the AeroMACS constraints in both bandwidth and power, is how to increase AeroMACS capacity for accommodation of all assigned existing and potential future fixed and mobile services. This chapter demonstrates that gains that can be derived from the addition of IEEE 802.16j multihop relays to the AeroMACS standard can be exploited to improve capacity or to extend radio outreach of the network with no additional spectrum required. Hence, it is shown that it would make sense to allow the usage of relays, at least as an option, in AeroMACS networks. Furthermore, it is pointed out that it would always be possible to incorporate IEEE 802.16j standards into AeroMACS networks, even if the network is originally rolled out as an IEEE 802.16-2009-based network. The chapter introduces the key concept of "multihop gain" with a detailed analysis that quantifies this gain for a simple case. The chapter concludes with a strong case made in favor of IEEE 802.16j-based AeroMACS networks.

2 The Audience

This book can serve as a professional text assisting experts involved in research, development, deployment, and installation of AeroMACS systems. It can also be used as an academic textbook in wireless communications and networking, with case study application of WiMAX and AeroMACS, for a senior level undergraduate course or for a graduate level course in Electrical Engineering, Computer Engineering, and Computer Science programs.

The specific list of professional groups and individuals who may benefit from this text includes engineers and technical professionals involved in the R&D of AeroMACS systems, technical staff of government agencies working in aviation sectors, technical staff of private aviation firms all over the world involved in manufacturing of AeroMACS equipment, engineers and professionals who are interested or active in the design of standard-based wireless networks, and new researchers in wireless network design.

Acknowledgments

Although composed by a single author (or few authors), technical texts are drawn from the contributions of a large number of experts and the immense quantity of literature that they have created. I would like to acknowledge the groundbreaking research and development efforts of many researchers and engineers in the aviation industry, research institutions, academia, and national and international standardization bodies, whose contributions were instrumental in creating the groundwork for this book. In particular, I am appreciative to NASA Glenn Research Center's Communication, Control, and Instrumentation group.

I am deeply grateful to Robert J. Kerczewski of NASA Glenn Research Center for introducing me to AeroMACS technology and providing me with the opportunity to conduct research in AeroMACS area during my several summer research fellowships at NASA Glenn, and for being so generous with his time for discussion and exchange of ideas. Special thanks and appreciation is extended to Dr. David W. Matolak of the University of South Carolina for contributing Chapter 3 on the key topic of airport channel characterization over the 5 GHz band. I would also like to thank my NASA colleagues Rafael Apaza and Dr. Jeffery Wilson for sharing their insights on AeroMACS technology.

Special note of gratitude goes to John Wiley & Sons, Inc. publishing team, in particular to my editor, Mary Hatcher, for her continuous assistance and support for this book from proposal to production. I am also grateful to anonymous reviewers for their careful reading of the manuscript and their insightful comments and suggestions that have improved the quality of this book.

I would also like to recognize and appreciate the assistance that I have received from my former graduate student Laila Wise, who meticulously plotted some of the curves that I have included in Chapter 2. Last but not the least, I wish to express my appreciation to my life partner, Angela J. Manson, for her nonstop encouragement, patience, affection, and constructive editorial suggestions throughout the preparation of this book; without her support and love this book would not have been completed.

Behnam Kamali

Acronyms

A

AAA	Authentication, Authorization, and Accounting
A/A	Aircraft-to-Aircraft or Air-to-Air
AAS	Adaptive Array System
ABS	Advanced Base Station
ACARS	Aircraft Communications and Addressing Reporting System
ACAST	Advanced CNS Architectures and Systems Technologies
ACF	Area Control Facility
ACI	Adjacent Channel Interference
ACK	ARQ/HARQ positive acknowledgement
ACM	ATC Communications Management
ACP	Aeronautical Communications Panel
ACSP	Aeronautical Communication Service Provider
ADS	Automatic Dependent Surveillance
ADS-B	Automatic Dependent Surveillance-Broadcast
ADSL	Asymmetric Digital Subscriber Links
AeroMACS	Aeronautical Mobile Airport Communications System
AES	Advanced Encryption Standard
A/G	Air-to-Ground
AI	Aeronautical Information
AIP	Airport Improvement Program (Plan)
AIP	Aeronautical Information Publication
AIRMET	Airmen's Meteorological Information
AIS	Aeronautical Information Services
AM	Amplitude Modulation
AMC	Adaptive Modulation Coding
AMC	ATC Microphone Check
AMPS	Advanced Mobile Phone Services

AM(R)S	Aeronautical Mobile Route Services
AMS	Advanced Mobile Station
ANC	Air Navigation Conference
ANSP	Air Navigation Service Provider
AOC	Airline Operational Control
AP	Action Plan
APN	Airline Private Networks
APCO	Association of Public Safety Communications Officials-International
ARINC	Aeronautical Radio Incorporation
ARQ	Automatic Repeat Request
ARTCC	Air Route Traffic Control Center
ASA	Adjacent Subcarrier Allocation
ASA	Airport Surface Area
ASBU	Aviation System Block Upgrade
ASDE	Airport Surface Detection Equipment
ASN	Access Service Network
ASN-GW	Access Service Network Gateway
ASP	Application Service Provider
ASR	Airport Surveillance Radar
ASSC	Airport Surface Surveillance Capability
ATC	Air Traffic Control
ATCBI	Air Traffic Control Beacon Interrogator
ATCT	Air Traffic Control Tower
ATIS	Automatic Terminal Information Service
ATM	Air Traffic Management
ATN	Aeronautical Telecommunications Network
AWG	Aviation Working Group
AWGN	Additive White Gaussian Noise

B

BBC	British Broadcasting Company
BC	Boundary Coverage
BE	Best Effort
BER	Bit Error Rate
BFSK	Binary Frequency Shift Keying
BFWA	Broadband Fixed Wireless Applications
BGP	Border Gate Protocol
BPSK	Binary Phase Shift Keying
BR	Bandwidth Request
BS	Base Station
BSID	Base Station ID
BSN	Block Sequence Number

BTC	Block Turbo Code
BTS	Base Transceiver Station
B-VHF	Broadband VHF
BWA	Broadband Wireless Access

C

CC	Convolutional Code
CCI	Co-Channel Interference
CCM	Counter with Cipher-block chaining Message authentication code
CCRR	Co-Channel Reuse Ratio
CCTV	Close Circuit Television
CDM	Collaborative Decision Making
CDMA	Code Division Multiple Access
CE	Cyclic Extension
CFR	Code of Federal Regulation
CID	Connection Identifier
CINR	Carrier to Interference and Noise Ratio
CIR	Channel Impulse Response
CLCS	Cable Loop Communications Systems
CLE	Cleveland-Hopkins International Airport
CM	Context Management
CMAC	Cipher-based Message Authentication Code
CNR	Carrier-to-Noise Ratio
CNS	Communications, Navigation, and Surveillance
COCR	Communications Operating Concept and Requirements
COFDM	Coded Orthogonal Frequency Division Multiplexing
CO-MIMO	Cooperative MIMO
COST	European Cooperation for Scientific and Technical Research
COTS	Commercial Off of The Shelf
CP	Cyclic Prefix
CPDLC	Controller–Pilot Data Link Communications
CPE	Customer Premises Equipment
CPS	Common Part Sublayer
CQI	Channel Quality Indicator
CQICH	Channel Quality Indicator Channel
CRC	Cyclic Redundancy Check
CRD	Clearance Request and Delivery
CRSCC	Circular Recursive Systematic Convolutional Code

CS	Convergence Sublayer (Service Specific Convergence Layer)
CSA	Commercial Service Airports
C-SAP	Control-Service Access Point
CSI	Channel State Information
CSMA	Carrier Sense Multiple Access
CSN	Connectivity Service Network
CTC	Convolutional Turbo Code
CTF	Channel Transfer Function
CWG	Certification Working Group

D

DAB	Digital Audio Broadcasting
DAL	Design Assurance Levels
D-ATIS	Digital Automatic Terminal Information System
D-AUS	Data Link Aeronautical Update Service
DBFSK	Differential Binary Phase Shift Keying
DCL	Departure Clearance
DFF	D (Delay) Flip-Flop
D-FIS	Digital Flight Information Services
DFT	Discrete Fourier Transform
DHCP	Dynamic Host Configuration Protocol
DHS	Department of Homeland Security
DIUC	DL Interval Usage Code (DIUC)
D-LIGHTING	Active Runway Lighting Systems
DME	Distance Measuring Equipment
D-NOTAM	Digital Notice to Airmen
DOCSIS	Data Over Cable Service Interface Specification
DOC	Department of Commerce
DOD	Department of Defense
DOT	Departments of Transportation
D-OTIS	Downlink (DL) Operational Terminal Information Service
DPSK	Differential Phase Shift Keying
DRNP	Dynamic Required Navigation Performance
DRR	Deficit Round-Robin
D-RVR	Download Runway Visual Range
DSB	Double Side Band
DSB-TC	Double Sideband Transmitted Carrier
D-SIG	Digital (or DL) Surface Information Guidance
DSP	Digital Signal Processing
DSSS	Direct Sequence Spread Spectrum
D-TAXI	Data Link Taxi

4DTRAD	4-D Trajectory Data Link
D-WPDS	Data Link Weather Planning Decision Service

E

EAP	Extensible Authentication Protocol
ECC	Error Correction Coding
EDF	Earliest Deadline First
EDS	Evenly Distributed Subcarrier
EFB	Electronic Flight Bag
ERIP	Effective Isotropic Radiated Power
ertPS	Extended Real-Time Polling Services
ESMR	Enhanced Specialized Mobile Radio
EUROCAE	European Organization for Civil Aviation Equipment
EUROCONTROL	European Organization for the Safety of Air Navigation

F

FAA	Federal Aviation Administration
FAR	Federal Aviation Regulations
FBSS	Fast Base Station Switching
FCH	Frame Control Header
FCI	Future Communications Infrastructure
FCS	Future Communications Studies
FDD	Frequency Domain (Division) Duplexing
4DTRAD	4D Trajectory Data Link
FDM	Frequency Division Multiplexing
FDMA	Frequency Division Multiple Access
FEC	Forward Error Correction
FER	Frame Error Rate
FFR	Fractional Frequency Reuse
FFT	Fast Fourier Transform
FH	Frequency Hopping
FIFO	First-In-First-Out
FirstNet	First Responder Network Authority
FIS	Flight Information Services
FL	Forward Link
FM	Frequency Modulation
FMS	Flight Management System
FOM	Flight Operations Manual
FRF	Frequency Reuse Factor
FSS	Flight Service Stations
FTP	File Transfer Protocol

FUSC	Full Usage of Subchannels
FWA	Fixed Wireless Access

G

GA	General Aviation
G/A	Ground-to-Air
GANP	Global Air Navigation Plan
GF	Galois Field
G/G	Ground to Ground
GMH	Generic MAC Header
GoS	Grade of Service
GPS	Global Positioning System
GRE	Generic Routing Encapsulation
GRC	Glenn Research Center
GTG	Graphical Turbulence Guidance

H

HARQ	Hybrid Automatic Repeat reQuest
HDSL	High-bit-rate Digital Subscriber Links
HDTV	High Definition Television
HF	High Frequency
HFDD	Half Frequency Division Duplexing
HHO	Hard Handover
HMAC	Hash Message Authentication Code
HNSP	Home Network Service Provider
HO	Handover, Handoff
HTTP	Hypertext Transport Protocol

I

IAI	Inter-Application Interference
IAIP	Integrated Aeronautical Information Package
IATA	International Air Transport Association
ICAO	International Civil Aviation Organization
ICI	Inter Carrier Interference
ICIC	Inter-Cell Interference Coordination
IDFT	Inverse Discrete Fourier Transform
IDR	Inter Domain Routers
IEEE	The Institute of Electrical and Electronic Engineers
IER	Information Exchange and Reporting
IETF	Internet Engineering Task Force
IFFT	Inverse Fast Fourier Transform
IFR	Instrument Flight Rules
IMT	International Mobile Telecommunications

IP	Internet Protocols
IPS	Internet Protocol Suite
IPTV	Internet Protocol Television
IPv6	Internet Protocols Version 6
ISDN	Integrated Services Digital Network
ISG	Internet Service Gateway
ISI	Intersymbol Interference
ISL	Instrument Landing System
ISM	Industrial, Scientific, Medical
ITS	Intelligent Transportation System
ITT	International Telephone & Telegraph
ITU	International Telecommunication Union
ITU-R	International Telecommunication Union-Radiocommunication

J

JPDO	Joint Planning and Development Office

L

LAN	Local Area Network
LCR	Level Crossing Rate
LDL	L-band Data Link
LDPC	Low Density Parity Check
LEO	Low Earth Orbit
LMR	Land Mobile Radio
LOS	Line of Sight
LOS-O	LOS-Open
LSB	Least Significant Bit
LTE	Long Term Evolution

M

MAN	Metropolitan Area Network
MAP	Media Access Protocol
MASPS	Minimum Aviation System Performance Standards
MBR	Maximum Burst Rate
MBS	Multicast-Broadcast Service
MCBCS	Multicast and Broadcast Services
MCM	Multicarrier Modulation
MDHO	Micro Diversity Handover
MET	Meteorological Data
METARS	Meteorological Aerodrome Reports
MFD	Multifunction Display
MIMO	Multiple-Input-Multiple-Output

ML	Maximum Likelihood
MLS	Microwave Landing System
MLT	Maximum Latency Tolerance
MMR	Mobile Multihop Relay
MODEM	Modulation/Demodulation
MOPS	Minimum Operational Performance Standards
MPC	Multipath Component
MPSK	M-ary Phase Shift Keying
MR-BS	Multihop Relay-Base Station
MRS	Minimum Receiver Sensitivity
MRTR	Minimum Reserved Traffic Rate
MS	Mobile Station
M-SAP	Management-Service Access Point
MSB	Most Significant Bit
MSC	Mobile Switching Center
MSP	Master-Slave Protocol
MSS	Mobile Satellite Service
MSTR	Maximum Sustained Traffic Rate
MTSO	Mobile Telephone Switching Office
MU-MIMO	Multiple User MIMO

N

NACK	Negative ARQ/HARQ Acknowledgement
NAP	Network Access Provider
NAS	National Airspace System
NASA	National Aeronautics and Space Administration
NASP	National Airport System Plan
NAVAID	Navigation Aids
NCMS	Network Control and Management System
NextGen	Next Generation Air Transportation System
NLOS	None Line of Sight
NLOS-S	NLOS-Specular
NNEW	Network Enabled Weather
NOTAM	Notice to Airman
NPIAS	National Plan of Integrated Airport Systems
NRM	Network Reference Model
nrtPS	Non-Real-Time Polling Services
NRT-VR	Non-Real-Time Variable Rate
NSNRCC	Non-Systematic Non-Recursive Convolutional Code
NSP	Network Service Provider
NTIA	National Telecommunications and Information Administration

NTIS	National Traffic Information Service
NWG	Network Working Group

O

OCL	Oceanic Clearance Delivery
OFDM	Orthogonal Frequency Division Multiplexing
OFDMA	Orthogonal Frequency Division Multiple Access
OFUSC	Optional FUSC
OOOI	Out, Off, On, In (time)
OPUSC	Optional PUSC
OSI	Open System Interconnection
OTIS	Operational Traffic Information System

P

PAPR	Peak-to-Average Power Ratio
PBN	Performance Based Navigation
PCS	Personal Communications Systems
PDC	Pre-Departure Clearance
PDF	Probability Density Function
PDP	Power Delay Profile
PDU	Protocol Data Unit
PDV	Packet Delay Variation
PIB	Pre-flight Information Bulletins
PKM	Privacy Key Management
PKMv2	Privacy Key Management version 2
PMDR	Private Mobile Digital Radio
PMP	Point-to-Multipoint
PMR	Private/Professional Mobile Radio
PN	Pseudo Noise
PS	Public Safety
PSC	Public Safety Communications
PSD	Power Spectral Density
PSTN	Public Switched Telephone (Telecommunications) Networks
PUSC	Partial Usage of Subchannels

Q

QAM	Quadrature Amplitude Modulation
QoC	Quality of Communication
QoS	Quality of Service
QPSK	Quadrature Phase Shift Keying

R

RADIUS	Remote Authentication Dial-In User Service
RARA	Rate Adaptive Resource Allocation

R&O	Report and Order
RCPC	Rate Compatible Punctured Convolutional Code
RCF	Remote Communications Facility
RDS	Randomly Distributed Subcarrier
RFI	Radio Frequency Interference
RL	Reverse Link
R-MAC	Relay Media Access Control
RMM	Remote Maintenance and Monitoring
RMS-DS	Root-Mean Square Delay Spread
RP	Reference Point
RR	Round-Robin
RRA	Radio Resource Agent
RRC	Radio Resource Controller
RRM	Radio Resource Management
RS	Relay Station
RS	Reed Solomon
RSS	Received Signal Strength
RSSI	Received Signal Strength Indicator
RTCA	Radio Technical Commission for Aeronautics
RTG	Receive Time Gap
rtPS	Real-Time Polling Services
RTR	Remote Transmitter Receiver
RT-VR	Real-Time Variable Rate
RVR	Runway Visual Range
R_x	Receiver

S

SA	Security Association
SANDRA	Seamless Aeronautical Networking Through Integration of Data Links, Radios, and Antennas
SAP	Service Access Point
SARPS	Standards and Recommended Practices
SAS	Smart Antenna System
SBS	Surveillance Broadcast System
SBS	Serving Base Station
SC	Single Carrier
SC	Special Committee
SD	Stationarity Distance
SDU	Service Data Unit
SESAR	European Commission Single European Sky ATM Research
SF	Service Flow

SFID	Service Flow Identifier
SHO	Soft Handover
SIGMET	Significant Meteorological Information
SIM	Subscriber Identify Module
SINR	Signal-to-Interference-Plus-Noise Ratio
SIP	Session Initiation Protocol
SIR	Signal to Co-Channel Interference Ratio
SISO	Single-Input Single-Output
SLA	Service Level Agreements
SMR	Specialized Mobile Radio
SNR	Signal-to-Noise Ratio
SOFDMA	Scalable Orthogonal Frequency Division Multiple Access
SONET	Synchronous Optical Network
SPWG	Service Provider Working Group
SS	Stationary (Subscriber) Station
STBC	Space-Time Block Code
STC	Time Space Coding
Std.	Standard
STDMA	Self-Organized Time Division Multiple Access
STTC	Space-Time Trellis Code
STTD	Space-Time Transmit Diversity
SU-MIMO	Single User MIMO
SWIM	System Wide Information Management
T	
TBCC	Tail Biting Convolution Codes
TBS	Target Base Station
T-CID	Tunneling Connection Identifier
TCM	Trellis Coded Modulation
TCP	Transmission Control Protocol
TDD	Time Division (Domain) Duplexing
TDL	Tapped-Delay Line
TDLS	Tower Data Link System
TDM	Time Division Multiplexing
TDMA	Time Division Multiple Access
TDLS	Tower Data Link System
TETRA	Terrestrial Trunk Radio
3GPP	Third Generation Partnership Project
TIA	Telecommunications Industry Association
TLV	Type, Length, Value
TO	Transmission Opportunities
ToR	Terms of References

TR	Transmitter Receiver
TRACON	Terminal Radar Approach Control
TSO	Technical Standard Orders
TTG	Transmit Time Gap
TUSC1	Tile Usage of Subchannels 1
TUSC2	Tile Usage of Subcarrier 2
TWG	Technical Working Group
Tx	Transmitter

U

UA (γ)	Percentage of Useful Area Coverage when Receiver Sensitivity is γ dB
UAT	Universal Access Transceiver
UCA	Useful Coverage Area
UGS	Unsolicited Grant Services
UISC	UL Interval Usage Code
US	Uncorrelated Scattering
USAS	User Applications and Services Survey
USIM	Universal Subscriber Identify Module
UWB	Ultrawideband

V

VDL	VHF Data Link
VHF	Very High Frequency
VLSI	Very Large-Scale Integration
VNSP	Visited Network Service Provider
VoIP	Voice over Internet Protocols
VOLMET	French acronym of VOL (flight) and METEO (weather)

W

WAAS	Wide Area Augmentation System
WDM	Wavelength Division Multiplexing
WFQ	Weighted Fair Queue
Wi-Fi	Wireless Fidelity
WiMAX	Worldwide Interoperability for Microwave Access
WMAN	Wireless Metropolitan Area Network
WRC	World Radiocommunication Conference
WSS	Wide-Sense Stationarity
WSSUS	Wide Sense Stationary Uncorrelated Scattering
WWAN	Wireless Wide Area Network
WXGRAPH	Graphical weather information

1

Airport Communications from Analog AM to AeroMACS

1.1 Introduction

The safety of air travel and air operations is critically linked to the availability of reliable aeronautical communications and navigation systems. Owing to the fact that flight safety is the highest priority in aviation, extreme measures must be taken to protect the aeronautical communication systems against harmful interference, malfunction, and capacity limitation.

In the early days of commercial aviation in the 1940s, analog double-sideband transmitted-carrier (DSB-TC) amplitude modulation (AM) over VHF band was adopted for aeronautical radio. This selection was made mostly for the reason that analog AM was the only fully developed and proven radio communications technology at the time. The number of VHF radio channels increased over the decades subsequent to the end of the World War II. In the 1980s, the VHF band of 118–137 MHz was allocated to aeronautical radio. With channel spacing of 25 kHz, 760 VHF AM (25-AM) radio channels became available. During the same decade, the avionics community predicted that early in the next century growth in flight operations and air traffic volume would demand communication capacity[1] that would be well beyond what was available in those days.

The air-to-ground (A/G) and ground-to-air (G/A) VHF communications system for civil air traffic control consisted of AM voice networks, where each flight domain had its own dedicated network. These networks were not interconnected and actually operated independently; however, their architecture was roughly the same. The pilot-to-control tower

---

1 Communications capacity may be defined in several different ways. Communications capacity in the context of this chapter denotes the quantity of the required aeronautical radio channels.

AeroMACS: An IEEE 802.16 Standard-Based Technology for the Next Generation of Air Transportation Systems, First Edition. Behnam Kamali.
© 2019 the Institute of Electrical and Electronics Engineers, Inc. Published 2019 by John Wiley & Sons, Inc.

(uplink; UL, also known as reverse channel or reverse link; RL) and controller-to-pilot (downlink, DL also called forward channel or forward link, FL) radio voice links were half-duplex connections and operated on a "push-to-talk" basis. Backup radio channels were provided in the event of system malfunction, power failure, or other unexpected situations. The VHF radio equipment was digitally controlled with the total of 760 channels, of which 524 channels were dedicated to A/G and G/A communications for air traffic control (ATC) purposes. The remaining channels were used by airlines for airline operational control (AOC). The AOC predominantly used and still uses a data service called the aircraft communications and address reporting system (ACARS) to manage and track the aircraft. However, the radio link can also be used for voice communications between pilots and airline agents [1]. Currently, the bulk of ground-to-ground (G/G) communications on the surface of airports is supported by wired and guided transmission systems, primarily through buried copper and fiber-optic cable loops. The G/G communications is also supported by a number of wireless systems, among them are VHF AM radio, airport WiFi system, and even some airport radar facilities.

In addition to the allocated VHF spectrum, two other spectral bands were considered to become available for aviation on a shared basis with other applications. First is an L-band spectrum of 960–1024 MHz, originally allocated for distance measuring equipment (DME). The second one is a C-band spectrum over 5000–5150 MHz, traditionally earmarked for microwave landing system (MLS). This radio spectrum was later allocated as the frequency band to carry aeronautical mobile airport communications system (AeroMACS). AeroMACS technology is the main focus of this text and at the time of its preparation, AeroMACS was already standardized and globally harmonized as a broadband IP data communication link for safety and regularity of flight at the airport surface. Currently, AeroMACS is being tested over several major U.S. airports and, barring any unforeseen complications, it is expected to be deployed globally by the year 2020. For future airports, AeroMACS is envisioned to constitute the backbone of the communications system for the airport surface, whereas older airports can form a communications infrastructure in which AeroMACS is complimented with the airport fiber optic and cable loops that are already in place.

1.2 Conventional Aeronautical Communication Domains (Flight Domains)

Aeronautical signals pass through several wireless communication channels before they reach the destination. Four possible transmission links exist in aeronautical communications path: aircraft (air)-to-controller (ground), A/G; controller-to-aircraft, G/A; ground-to-ground, G/G, and aircraft-to-aircraft; A/A links. The aircraft continuously communicates with the NAS (National Airspace System), or the global airspace system, throughout the flight duration. There are several different domains (channels) through which the aircraft may be required to communicate with a ground station. Each one is a wireless channel with its own particular conditions, constraints, and characteristics. For an overall aeronautical communications system design or simulation, each of the channels listed below must be considered and characterized.

1) *Enroute Communication Channel:* This is the domain when the aircraft is airborne and A/G and G/A transmissions are required. This is essentially a high-speed mobile communication link in which the aircraft flying is at high altitude and close to its maximum speed. This link can be modeled as a simple double-ray wireless channel, or a Rayleigh fading channel. However, in the majority of cases the channel contains a line-of-sight (LOS) path and a ground reflection. When the aircraft elevation angle is high the ground reflection takes place at a point very close to the ground station, therefore, the path length between the two rays is very small and hence they cannot be resolved by the receiver [2].

2) *Flying Over a Ground Station:* This is a special case of enroute channel during which the Doppler effect changes its sign. For design and simulation of the aeronautical communications links, this mode must be considered separately from the enroute case [3].

3) *Landing and Takeoff Domain:* The aircraft is airborne at low altitudes and moving at its landing and takeoff speed, it is engaged in A/G and G/A communications and is close to the control tower. The channel is multipath with a strong LOS component.

4) *Surface (Taxiing) Channel:* In this domain the aircraft moves rather slowly toward or away from the terminal, it is therefore a low-speed low-range mobile communications affected by multipath and some Doppler effect.

5) *Parking Mode:* This mode is applicable when the aircraft is on the ground and close to a terminal and traveling at a very low speed or is parked. This requires essentially a stationary wireless transmission of low range.

6) *Air-to-Air:* This channel is used for the purpose of communications between two aircraft while they are in flight.

7) *Oceanic Domain:* This channel has its own characteristics in the sense that it is a long-range communications channel for the most parts. VHF LOS transmission is not feasible for this domain.

8) *Polar Domain:* This is also a channel in which long-range communications take place. This domain has a limited satellite access.

In some literature, communications in domain 3 is referred to as *terminal communications.* Communications over domains 3–5 together are what is referred to as *airport surface communication* in this chapter. For oceanic and remote areas, such as polar regions, since LOS transmission to ground stations is not possible, HF (high frequency) band and satellite systems are used.

1.3 VHF Spectrum Depletion

It was long accepted that as a rule of thumb, and baring any unexpected sudden traffic increase, the aviation traffic is anticipated to have an annual growth of at least 2%. However, the spectrum that was allocated for various functionalities of aerospace management system remained fixed, except for the abovementioned L-band and C-band that later became available on a spectrum sharing basis. The safety, security, growth, and efficient operation of national and global aviation systems are vitally dependent on reliable communications, navigation, and surveillance (CNS) services. Communications provides wireless and wireline connections for voice and data exchange between various entities involved in the aviation system, that is, aircraft, airports, terminals, runways, control towers, satellite transponders, and so on. The other essential component of aviation system is the air traffic management (ATM) system that heavily relies on communications and surveillance components of CNS [4].

In the late 1990s, the demand for aeronautical communications links surpassed what the existing VHF radio channels could supply without unacceptable level of interference. In the United States, rapid increase in air traffic due to commercial transportation and general aviation (GA) (private aircrafts) was the culprit. In Europe, owing to an almost exponential growth in commercial flights in the 1990s, the problem was more severe. Besides, many major European airports with large volume of air traffic are located at close geographical proximity of each other. In the early 2000s, the Europeans proposed a scheme in which 25 kHz spacing band is reduced to 8.33 kHz and thereby the number of available radio channels is tripled to 2280. This scheme that became known as "8.33-AM" ran into some standardization problems and was not implemented in the United States although it was accepted and deployed in Europe.

As the capacity of VHF aeronautical radio was reaching saturation in the United States and in Europe, the International Civil Aviation

Organization[2] (ICAO) at its 11th Air Navigation Conference held in September 2003 made a number of recommendations. One recommendation specifically called for exploration of new terrestrial and satellite-based technologies on the basis of their potential for standardization for aeronautical mobile communications use. A second recommendation asked for monitoring emerging communications technologies but undertaking standardization work only when the technologies can meet current and emerging ICAO ATM requirements. These requirements asked for technologies that are technically proven, meet the safety standards of aviation, are cost-effective, can be implemented without prejudice to global harmonization, and are consistent with Global Air Navigation Plan (GANP) for CNS/ATM.

The key functional objective for future aeronautical communications systems was deemed to provide relief to the congested VHF aeronautical band by either substantially increasing the number of voice channels or using the spectrum more efficiently or a combination thereof. In doing so, one could contemplate several options. A direct possibility was using the available VHF band more efficiently by introducing new communication technologies that save spectrum. The other option was utilizing the available VHF spectrum more efficiently by reducing the channel spacing and guard bands. Another approach was incorporating data communications links such that the majority of required voice messages can be transmitted more efficiently by data and text messages. Yet, another alternative was to take advantage of appropriate frequencies outside of the aeronautical VHF band that were available on a shared spectrum basis. One could also contemplate applying technologies such as GPS and other satellite-based technologies that have their own allocated spectrum and are suitable for carrying some components of aeronautical communications [4].

1.4 The ACAST Project

In 2003, NASA initiated an R&D project for future CNS/ATM infrastructure that was termed as "Advanced CNS Architectures and Systems Technologies"; ACAST. The main objective of the ACAST project was to define a transitional architecture to support the transformation of the present day patched-together CNS infrastructure into an integrated

---

2 "The International Civil Aviation Organization is the global forum of States for international civil aviation. ICAO develops policies, standards, undertakes compliance audits, performs studies and analyses, provides assistance, and builds aviation capacity through the cooperation of Member States and stakeholders" [5].

high-performance digital network-centric system. This was to take place, perhaps, through technologies that can be implemented in near-term and midterm to address the airspace urgent needs, while they can simultaneously be a part of the long-term solution. It was suggested that one long-term solution that is most cost-effective and can support present and potential future requirements is a network-oriented hybrid of satellite and ground-based communication systems. It was further recommended that all ATM and nonpassenger enroute communications be handled by the satellite-based technology, and all terminal and surface communications be placed on the ground-based system [6]. The ATM communications consists of several components: ATC that includes CPDLC (controller–pilot data link communications) – a method by which control tower can communicate with pilots via data and text (to be discussed in Section 1.5.4)-, automatic dependent surveillance (ADS), National Traffic Information Service (NTIS), AOC, and advisory service; such as flight information services (FIS) and weather sensor data downlink.

There were 10 partially overlapping subprojects envisioned in the ACAST project. The first three subprojects were considered foundation or "guiding frameworks" for other technology development in the ACAST project. The first was *Transitional CNS Architecture* philosophy in which the key requirements for CNS transitional architecture were increased integration of data transmission, full A/G network connectivity, high capacity, global coverage, efficient use of spectrum, and capability to evolve into a long-term CNS architecture. The second subproject was *Global A/G Network*. This formed the backbone of the CNS infrastructure. The major feature of this network was full CNS information sharing with all network users. The required protocols that were gradually emerging indicated that the Internet techniques are likely to be applied in A/G network as well. The third subproject was related to *Spectrum Research*. There was and is an ever-increasing demand for spectrum for aviation, thus efficient usage of the spectrum and development of new CNS technologies that would use the available spectrum to meet the future needs of aeronautical applications was deemed to be a key component of the ACAST project.

Another ACAST subproject was "VHF systems Optimization." This subproject investigated the methods and techniques that optimize the performance of the then VHF aeronautical band [6].

In meeting the key functional objectives of VHF aeronautical communications, one should not lose the sight of the strategic objectives of the global airspace system; that the change must be cost justified, it should be globally applicable and interoperable, and it should allow a smooth transition for service providers and users, and should avoid needless avionics [7]. In providing short-term or midterm resolution to congestion problems, it would be prudent and desirable to ensure that the technology under consideration has the potential of becoming a part of the long-term solution, and is able to furnish a smooth transition from present to near-term and to long-term aeronautical communication system.

1.5 Early Digital Communication Technologies for Aeronautics

For over three decades, analog VHF DSB-AM system represented the dominant radio technology for aeronautics. In the late 1970s and early 1980s, data communications techniques gradually permeated into aeronautical information exchange systems; following the general trend in the then telecommunications industry, morphing into computer communication era. In this section, pre-AeroMACS digital communications schemes applied and implemented for aeronautics, as well as technologies that were considered for this application but were never implemented, are briefly reviewed in a historical context.

1.5.1 ACARS

The application of digital communications in civil aviation began with the introduction of Aircraft Communications Addressing and Reporting System (ACARS) technology in 1978. ACARS is a data communications scheme designed and commercialized by Aeronautical Radio, Inc. (ARINC) for short burst message communications between the aircraft crew and control tower, national aviation authorities, and airline operation centers. In other words, ACARS transmission link carrying operational information for AOC. ACARS is a packet radio system with a data rate of 2400 bps using DBFSK (differential binary frequency shift keying) modulation, operating over VHF AM channels. Burst messages in ACARS are limited to contain no more than 220 characters and transmission often lasts less than one second. To provide data integrity over ACARS link, CRC codes and automatic repeat request (ARQ) protocol are applied to each packet. Information carried by ACARS, often automatically sent without requiring any pilot action; ranging from departure and arrival time to reports on engine parameters that alert the ground maintenance crew that a fault requires attention upon the arrival of the aircraft. Over the past two decade, ACARS has been extended to provide assistance in air traffic control by transmitting text in lieu of voice messages. This reduces the need for voice channels since spoken English is converted to written messages, a feature that is most desirable for foreign pilots. However, ACARS has many limitations that prevented its consideration as a contender for next generation of

VHF data communication link. The primary shortcoming of ACARS is that it is a message link and as such it allows the transmission of a maximum number of textual characters, thus it is constrained by presentation and the length of the message that it can transmit.

1.5.2 VHF Data Link (VDL) Systems

In the late 1990s, ICAO endorsed the concept of supplementing VHF voice communications with data links. It has been shown that air traffic controller workload is directly proportional to the ongoing amount of voice communications. With the introduction of data communications through ACARS, voice traffic was shown to have dropped dramatically [8], that is, data transmission offers a more efficient use of the spectrum. This lead to the development of VHF Data Link Systems.

1.5.2.1 Aeronautical Telecommunications Network (ATN)

The Aeronautical Telecommunications Network (ATN), which is the international infrastructure that provides support for digital data transport, was originally envisioned to provide features such as network mobility and multiple data link availability [9]. Four A/G applications of ATN were originally standardized; Controller Pilot Data Link Communications (CPDLCs) that replaced most of the functions of the controller–pilot voice interaction with text messages, Automatic Dependent Surveillance (ADS) data was designed to provide position information to the ground station, digital flight information services (D-FIS) that allows the pilot to continuously receive information about flight conditions, and context management (CM) that enables the aircraft to contact local air traffic authorities.

More recently, ICAO has developed a new standard for ATN that is based on the Internet protocol suite, which is referred to as ATN/IPS, to replace the legacy OSI-based ATN. ICAO has also authored a technical manual for this new international ATM infrastructure. The manual contains the minimum communication standards and protocols that will enable implementation of ATN/IPS. The ATN/IPS has adopted the same four-layer model as defined in the Internet standard STD003. This model has four layers called the link layer, the Internet protocol (IP) layer, the transport layer, and the application layer. The manual adopts the Internet protocol version 6 (IPv6) for Internet layer interoperability [10].

1.5.2.2 VDL Systems

To expand on ACARS capability, VHF data link (VDL) schemes were developed. VDL features a bit-oriented digital transmission technology with different modes that provide various transmission capabilities.

Table 1.1 The key physical layer parameters of VDL, 25-AM, and 8.33-AM.

Technology	25-AM	8.33-AM	VDL Mode-2	VDL Mode-3	VDL Mode-4	VDL Mode-E
Modulation scheme	DSB-TC (AM)	DSB-TC (AM)	D-8PSK	D-8PSK	GFSK	D-8PSK
Pulse shaping	N/A	N/A	Raised cosine	Raised cosine	Gaussian	Raised cosine
Channel coding	N/A	N/A	(255, 249) RS Code	(72, 62) RS Code	None	(72, 62) RS Code
Data rate	N/A	N/A	31.5 kbps	31.5 kbps	19.2 kbps	31.5 kbps
Transmission mode	Voice	Voice	Voice	Voice and data	Data	Voice and data
Access method	FDMA	FDMA	CSMA	TDMA	STDMA	TDMA
Channel spacing	25 kHz	8.33 kHz	25 kHz	25 kHz/4 time slot	25 kHz/4 time slot	8.33 kHz/2 time slot
No. of radio channels	760	2280	380	3040	760	4560

Table 1.1 summarizes key physical layer parameters of different modes of VDL networks. For comparison purposes, the same information is provided for analog VHF schemes of 25-AM and 8.33-AM.

VDL Mode-2 was designed as a data-only transmission system to replace and upgrade ACARS that provides AOC, ATS (Air Traffic Services), and ATC (Air Traffic Control) communications. The VDL Mode-2 uses carrier sense multiple access (CSMA). This protocol permits statistically equal channel access to the users. Consequently, an increase in traffic translates into access delay making VDL Mode-2 unsuitable for time-critical data [11]. It should be noted that within the past three decades or so, VDL Mode-2 has emerged as the selected pre-AeroMACS digital communications standard for aeronautics.

VDL Mode-3 was developed by the FAA (Federal Aviation Administration) and industry partners, and was considered at the time as the next-generation airborne communications system (NEXCOM) for ATS. VDL Mode-3 is a digital TDMA system providing four time-slots over the existing 25 kHz voice channel, thus it quadruples the spectral capacity of the VHF system. Other key features of VDL Mode-3 are as outlined further.

- VDL Mode-3 provides the capability of simultaneous transmission of voice and data over a single RF channel.

- Since VDL-Mode 3 is spectrally compatible with the existing 25 kHz AM system, it allows a straightforward transition from the legacy technology.
- VDL Mode-3 offers improvement on safety and security relative to existing analog AM-DSB scheme.
- VDL Mode-3 features automatic channel selection, thereby reducing the pilot's workload.

VDL Mode-3 permits a smooth transition from the analog AM system in the United States. However, with the deployment of the 8.33 kHz AM system in Europe, its international implementation faced serious frequency management challenges due to its spectral incompatibility with AM-8.33 [11].

VDL Mode-4 is a data-only broadcast scheme that was originally developed for surveillance; however, it was adapted for communication use in the 1990s. VDL Mode-4 uses a self-organizing TDMA MAC layer where time slots are scheduled by a ground system to provide a nearly equal access to the channel. This scheme is based on cellular technology that requires low SNR that could increase the frequency reuse factor in the surface and terminal domains. Standards are being developed for VDL Mode-4 to be used for point-to-point communications with applications in A/A data transmission.

VDL Mode-E proposed by Rockwell Collins Inc. is a modified version of VDL Mode-3. As such it applies many protocols defined for VDL Mode-3. The most significant difference between the two technologies is that Mode-E requires a channel spacing of 8.33 kHz and each channel provides two TDMA time slots. Consequently, VDL Mode-E increases the capacity of VHF legacy system by sixfold. With six data–voice-integrated channels accommodated by a single 25-AM channel, VDL Mode-E provides the best spectral efficiency of all VDL technologies. It has been suggested that VDL Mode-E meets all of the European safety and security requirements for the next-generation integrated voice/data aeronautical communications [12]. However, spectral incompatibility of VDL Mode-E and the present 25-AM system is an issue that can create standardization problems.

LDL (L-band data link) technology is a hybrid derivative of VDL Mode-3 and Universal Access Transceiver (UAT) standards. LDL inherits its physical layer protocols from UAT, and its upper layer standards from VDL Mode 3.

1.5.3 Overlay Broadband Alternatives for Data Transmission

Following the development of VDL technologies, it was recognized that the overall capacity of VHF aeronautical communication network may be further enhanced when message and text broadcast capabilities are supported. One rather straightforward method that was considered was a transmission system that uses the entire allocated VHF spectrum while it can coexist with the legacy VHF AM network, that is, to say broadband overlay schemes. A couple of alternatives were available.

1.5.3.1 Direct-Sequence Spread Spectrum Overlay

A key property of direct-sequence spread spectrum (DSSS) signaling is its capability to overlay narrowband signals without introducing excessive interference to them, provided that a proper process gain is selected for the wideband signal. On the other hand, since DSSS signals are tolerant of interference and jamming, they will be immune from the interference effects of the narrowband AM signal to a certain extent. Several studies have indicated that under certain circumstances, it is feasible to overlay CDMA (Code Division Multiple Access) signals in various VHF aero-nautical bands while AM legacy signals occupying parts of the band [13]. The near–far problem is observed in the sense that the AM signal transmitted by an aircraft that is far away from the ground station is attenuated and becomes more vulnerable to DSSS signal interference at the ground station, which is broadcasting the spread spectrum signal. On the other hand, DSSS signal is attenuated greatly by the time it is received by an aircraft flying at some distance from the ground station and is likely to be jammed by the AM signal transmitted by the aircraft. Nonetheless, it has been shown that the overly of DSSS signal in the VHF band was a viable solution to the spectral congestion in VHF aeronautical communication.

1.5.3.2 Broadband VHF (B-VHF)

The broadband VHF (B-VHF) is an overlay scheme based on multicarrier modulation (OFDM), which was under development as a possible future aeronautical communications technology in Europe. This was a promis-ing technology that had the potential of fulfilling many functional and strategic objectives of the future long-term-integrated CNS network. Since there are always some unused 25-AM channels at any given time, as well as channels that are being used so far away that the received signal power is small enough to assume those channels are unused, spectral gaps so available in VHF band could have been used to launch B-VHF signals. In other words, B-VHF would not require a continuous part of the VHF spectrum but could operate in the VHF band spectral gaps, and therefore, would reject interference from the legacy VHF signals. On the other hand, the B-VHF overlay signal would produce a minimal amount of interference toward the legacy AM system [14].

The enabling technology in B-VHF is OFDM, which is also the technology employed in fourth-generation (4G), and perhaps will also be used in fifth-generation (5G) mobile communication. For these reasons, B-VHF was a particularly attractive technology candidate for future aviation communications. Besides, OFDM had already been successfully applied in digital audio and video broadcast systems and was considered as a candidate for future commercial mobile communication systems, thus aeronautical communications would be benefiting from hardware and software development that had already been made in OFDM technology.

1.5.4 Controller–Pilot Data Link Communications (CPDLC)

In the early 2000s, the Federal Aviation Administration launched an ICAO-compliant enroute digital communications capability, known as controller–pilot data link communications (CPDLC), into the NAS, see Section 1.7). CPDLC is a data link application supporting a number of services such as ATC communications management (ACM), clearance request and delivery (CRD), ATC microphone check (AMC), departure clearance (DCL), data link taxi (D-TAXI), oceanic clearance delivery (OCL), 4-D trajectory data link (4DTRAD), information exchange and reporting (IER), dynamic required navigation performance (DRNP), and so on. In essence, CPDLC replaces voice commands and requests with small text messages using Abstract Syntax Notation 1 (ASN-1) format.

The main objectives of CPDLC is to improve the safety and efficiency of ATM. It is well known that approach and landing are the most critical phases of a flight taking place over TMA (terminal maneuvering area). TMA is a designated area of controlled airspace surrounding an airport and characterized by high volume of air traffic, demanding higher pilot and controller workload and performance requirements. The evident advantages of using CPDLC over VHF voice communications include higher spectral efficiency over analog voice, incorporation of error control to ensure accurate reception of the messages, text display of the messages enabling the review of the messages at later time, accent-independent communications, and so on.

The CPDLC digital signals are transmitted through VDL Mode 2 at the speed of 31.5 kbps, protected by a (255, 249) Reed–Solomon code (see Table 1.1), for controller–pilot communications. In remote flight domains, the interchange may be carried out by satellite, HF, or other available suitable data link(s). General performance requirements for CPDLC links are defined by ICAO, as listed further [15].

i) The probability of nonreceipt of a message will be equal to or less than 10^{-6}.

ii) The probability that nonreceipt of a message will fail to be notified to the originator will be equal to or less than 10^{-9}.

iii) The probability that a message will be misdirected will be equal to or less than 10^{-7}.

The CPDLC messages are comprised of message elements, selected from a message set, that are used to fabricate messages that support particular operational intents. ICAO has assembled the list alongside with the definitions and functional intents of these messages in a document. The CPDLC messages are labeled with two security/priority attributes. The first is urgency attribute which defines the queuing requirements for received messages that are displayed to the end user. Urgency attributes, in the order of precedence, are classified as distress (D), urgent (U), normal (N), and low (L) types. Second is alert attribute that delineates the type of alerting required upon the receipt of the message. Alert attributes are classified, in the order of priority, as high (H), medium (M), low (L), and no alerting required (N) types. Table 1.2 provides a sample of uplink

Table 1.2 Samples of CPDLC responses, acknowledgements to messages, and attributes.

Message element	Direction	Message intent	Urgent	Alert
STANDBY	Uplink	ATC has received the message and will respond	N	L
REQUEST DEFERED	Uplink	ATC has received the request but deferred until later	N	L
REQUEST ALREADY RECEIVED	Uplink	Indicates to the pilot/crew that the request has already been received on the ground	L	N
ROGER 7500	Uplink	Notification of receipt of unlawful interference message	U	H
EXPECT DESCENT AT (time)	Uplink	Notification that an instruction should be expected for the aircraft to commence descent at the specified time	L	L
DESCEND TO REACH (level) BY (position)	Uplink	Instruction that a descent is to commence at a rate such that the specified level is reached at or before the specified position	N	M
WILCO	Downlink	The instruction is understood and will be complied with	N	M
REQUEST (level)	Downlink	Request to fly at the specified level	N	L
REQUEST DESCENT TO (level)	Downlink	Request to descend to the specified level	N	L
REQUEST VMC DESCENT	Downlink	Request that a descent be approved on a see-and-avoid basis	N	L

(pilot-to-controller) and downlink (controller-to-pilot) CPDLC messages along with their urgent and alert attributes.

The complete list of CPDLC uplink and downlink messages and responses and/or acknowledgements with their assigned urgent and alert attributes are provided in Appendix A of Ref. [16].

1.6 Selection of a Communications Technology for Aeronautics

In selecting a communications technology that would address the aeronautical spectral capacity problem and would simultaneously define transition architecture to support transformation to a long-term network-centric CNS system, a number of issues must be considered. The following presents a key, but partial list:

- *Capacity Enhancement:* A major reason for technology change was to increase spectral capacity of the aeronautical radio network.
- *Spectral Compatibility:* A given technology is efficiently implemented over a section of the RF spectrum. Clearly, this spectrum should overlap with that of VHF aeronautics.
- *System Compatibility:* The new configuration should be compatible with the old one, that is, the user should be able to utilize the old system with no difficulty.
- *Cost Efficiency:* This involves the cost of on-board and ground avionic systems.
- *Ease of Architectural Integration:* This is concerning whether the technology poses a significant technical challenge for getting integrated into a standard architecture of aeronautical communications.
- *Interoperability:* The new technology must have been interoperable with other standardized systems already in place globally.

A number of candidate terrestrial and satellite-based communications technologies, operable over the aeronautical VHF band, were identified and investigated for long-term resolution of the spectral depletion problem in the national and global aeronautical radio. These technologies that were already proven viable in other applications included narrowband VHF data link (VDL series), wideband VHF (B-VHF), various cellular communications formats, wireless LAN (the IEEE 802.11 standard family), wireless MAN (IEEE 802.16 standard-based system; WiMAX) satellite communications, public safety radio communication systems, and dedicated terrestrial or satellite-based technologies that might have been developed for aviation applications [4].

In 2004, the FAA, in close cooperation with EUROCONTROL, initiated a study that became known as future communications studies (FCS) [17]. This was essentially a technology assessment effort in which over 60 different commercial, public safety, and government communications services and standards were evaluated for applicability for communications over various aeronautical domains. Based on this technology assessment, it was recommended that, as the starting point for airport surface domain, the wireless mobile communications technology based on IEEE 802.16e (Mobile WiMAX) should be selected [18]. This led to the birth of aeronautical mobile airport communications system (AeroMACS).

Over the past few years AeroMACS has evolved from a technology concept to a deployed operating communications network over a number of major U.S. airports. It is expected that AeroMACS will be deployed globally by the year 2020. WiMAX Forum is in charge of composing profiles for WiMAX-driven technologies, including AeroMACS. The most recent version of AeroMACS System Profile that is based on WiMAX Forum Mobile System Profile Release 1 was published in 2013 [19].

1.7 The National Airspace System (NAS)

The infrastructure within which the U.S. aviation system operates is the NAS. NAS is a complex network of airports, airways, air traffic control facilities, and all the associated rules and regulation designed to supervise safety of flights for civil (commercial as well as private) and military aviation in the United States, and to manage expeditious movement of the aircraft from the point of origin to the destination in an efficient manner. As such NAS is the network (or multinetwork) of towered and nontowered airports and landing areas, communications facilities, navigation and surveillance equipment, services and applications, technical information tables and data charts, manpower, and so on, that support safety, security, and regularity of flights over the U.S. airspace. The national airspace system is supervised by the Federal Aviation Administration, and consists of three major components, area control facility (ACF) equipment, remote communications facility (RCF) equipment, and transport media. To conform to international aviation standards, the United States adopted the primary elements of the classification system developed by the ICAO.

Originally, NAS was designed for civil aviation with three hierarchical objectives in mind. First and foremost is the safety of the flight, whereas the second objective is the expeditious movement of the aircraft from the point of origin to destination. The final objective is the conduct of efficient air transportation operation. Consequently, with safety of flight being the primary concern, the use of airport facilities, the design and operation of the ATC system, the flight rules and procedures employed, and the conduct of operations are all guided by the principle that safety is the first consideration [20].

The second objective is to permit aircraft to move from origin to destination as rapidly as possible without compromising the safety of the flight. This improves the traffic-handling capacity of the system and it involves preventing conflicts between flights, avoiding delays at airports or enroute, and eliminating inefficient or roundabout flight paths. It also entails making maximum use of airport and airway capacity in order to satisfy demand, so long as safety is not compromised. If safety and capacity utilization are in conflict, the FAA operating rules require that the volume of traffic using the system be reduced to a level consistent with safety.

The third objective is about making airport and air traffic control process cost efficient, again without compromising the safety. This implies not only the optimization of monetary cost to the users, but also the minimization of penalties of delay, inconvenience, undue restriction, and this sort of items. It also entails operating the system as efficiently as possible so as to reduce transaction costs and to increase productivity, that is, to handle more aircraft or to provide better service to those aircraft with a given combination of runways, controllers, and ATC facilities.

It should be noted that although safety cannot be compromised in the interest of cost efficiency or capacity, capacity and cost efficiency may be traded off for the sake of safety. For instance, in the event of workforce reduction in the air traffic control sector, the number of aircraft allowed to use certain crowded airports and air ways at peak demand hours is lowered to a level that safety is not compromised. Clearly, this measure reduces the NAS capacity, meaning that the aggregate number of flight transactions (landing and takeoff) handled by the system's airports would be cut back.

1.7.1 Flight Control

Flight safety for civilian and military aviation over the NAS relies on the three components of CNS. In order to ensure proper functioning of CNS, an aircraft remains in continuous communication with the NAS (i.e., controllers) from the time it boards the crew and passengers at the airport of origin to the time it parks at a gate in a terminal of the destination airport. In what follows, the procedure through which an aircraft and the NAS stay in contact is explained.

Initially, the pilot is provided with preflight information from one of the flight service stations (FSS). There are currently six FSSs in operation in the NAS. The preflight information consists of data that is related to the aircraft rout of flight such as weather conditions briefing.

The U.S. airspace is divided into 22 three-dimensional "cells" or regional sectors, each sector is controlled by an air route traffic control center (ARTCC). Originally, the aircraft communicates with air traffic control tower (ATCT) of the airport from which the flight is originated. As the aircraft moves away from the airport of origin, it gets connected to its regional sector ARTCC. When the aircraft crosses the boundary of its original ARTCC into a new sector, the controller transfers the communications responsibility for the flight to the new sector's ARTCC, a process that is analogous to handover procedure in cellular networks.

When the aircraft gets to the proximity of the destination airport, the ARTCC controllers hand off the communications with the aircraft to the Terminal Radar Approach Control (TRACON) controllers. The TRACON controller is responsible for assisting the aircraft for the landing process.

Once an aircraft enters an airport area, the communications is handed off to the local ATCT controllers, who are responsible for the aircraft movement control and for supervising its final approach and landing. The ground controllers, who foresee aircraft taxing to the selected gate and the gate operation, are also a part of ATCT.

Current communications technologies supporting flight control for safety and security include classical ground-based radar systems, standard VHF and UHF radio, controller–pilot data link communications (CPDLC), ACARS, GPS, and the airport wireline cable loop system. For instance, the ground-based radar signals are interpreted, converted to digital form, and sent to computer monitors at ARTCC, TRACON, or ATCT. These technologies served the NAS adequately until the late 1990s and early 2000s. However, as time rolls by, crowded airports and runways, delays, wasted fuel, and lost revenues are evidently seen to be on the rise. FAA next-generation air transportation systems (NextGen) program promises to dramatically overhaul the current NAS by finding techniques that can combat the current challenges.

1.7.2 United States Civilian Airports

Airports are the center piece of the national airspace system. In a broad and inclusive sense, an airport is any location that is designed and equipped, or even just commonly used, for the landing and takeoff of aircraft. This all-encompassing definition covers a wide spectrum of sites, from dirt strips that are designated as airports by the FAA to Hartsfield–Jackson Atlanta International Airport, the busiest airport in the globe in the passenger traffic sense (over 101 million passengers in the year 2015). Figure 1.1 shows an aerial view of the Atlanta Airport.

Figure 1.1 An aerial view of Hartsfield–Jackson Atlanta International Airport.

Aerodromes that are exclusively utilized for helicopter landing and takeoff are called heliports. An airport for use by seaplanes and amphibious aircraft is called a seaplane base. Such a base typically includes a stretch of open water for takeoffs and landings and seaplane docks for tying-up. The minimum requirement for an area to be called an airport is the availability of a runway strip, on land or on water, that can be totally or partially used for arrival, departure, and surface movement of aircraft.

Airports may be classified in a variety of ways. In Chapter 3 it is pointed out that when characterizing airport surface radio channels, airports are identified as small, medium, or large; depending on their landmass sizes and configurations. In compliance with the Airport and Airway Development Act of 1970, the FAA maintains a master list of airport development needs for the next decade. This list, which is periodically reviewed and revised, is called the National Airport System Plan (NASP). NASP classifies nonmilitary airports, according to their aviation functions, into national, domestic air carrier, commuter, reliever, and general aviation (private aircraft). This does not imply that private aircraft only use general aviation airports. In fact, privately owned aircraft operate at all types of airports, however, general aviation airports exclusively serve private aircraft [20].

More recently, the FAA has categorized U.S. airports in accordance with their eligibility to receive AIP (Airport Improvement Plan or Program) funding. Publicly used airports that are listed in the National

Table 1.3 Primary airport classification [21].

Primary airport	Large	Medium	Small	Nonhub
Common name	Large hub	Medium hub	Small hub	Nonhub primary
Percentage of total national passenger boardings	No less than 1%	More than 0.25%, but less than 1%	At least 0.05%, but less than 0.25%	More than 10,000 boardings but less than 0.05%

Plan of Integrated Airport Systems (NPIAS) are eligible to receive this funding. From the legal point of view "An airport is defined in the law as any area of land or water used or intended for landing or takeoff of aircraft including appurtenant area used or intended for airport buildings, facilities, as well as rights of way together with the buildings and facilities" [21]. The law categorizes airports by type of activities, including commercial service, primary, cargo service, reliever[3], and general aviation airports [21]. *Commercial service airports* (CSA) are defined as publicly owned airports with at least 2500 annual passenger boardings, including passengers who continue on an aircraft in international flight that stops at an airport in any of the 50 states for a nontraffic purpose, such as refueling or aircraft maintenance. Passenger boardings at airports that receive scheduled passenger service are referred to as *enplanements*. Commercial service airports are further classified into *primary* and *nonprimary* airports. Nonprimary airports are small airports with no more than 10,000 annual passenger boardings. Primary airports, on the other hand, are further classified in accordance with their level of air-traffic handling, that is, the percentage of the total national passenger boardings. This classification is presented in Table 1.3.

U.S. airports may also be classified as towered and nontowered airports. A large number of general aviation airports are nontowered, while all commercial airports are towered. It is estimated that there are about 12,000 nontowered airports in the United States. Currently, the FAA plans to install AeroMACS networks only on towered airports. As far as AeroMACS is concerned, the total number of towered airports and how they are distributed across the contiguous part of the United States land is an important factor for computation or estimation of the level of interference that AeroMACS imposes onto coallocated applications. In particular, AeroMACS interference to the feeder links of non-geostationary satellite

---

3 Reliever airports are special airports, designated by the FAA, to relieve congestion at CSAs and to provide improved general aviation access to the overall community [21].

systems in the mobile satellite service (MSS) is a critical AeroMACS design issue that limits the output power level, as well as the orientation, of AeroMACS antennas on the surface of airports. The Globalstar Satellite Constellation is an example of an existing operational MSS system that operates in the same band that is also allocated for AeroMACS [22]. Chapter 8 provides an extensive coverage on the subject of AeroMACS interference to coallocated applications. As of 2014, there were 497 towered airports in the United States.

The United States airports may also be categorized as domestic and international. While international airports carry domestic flights as well, domestic airports are exclusively used for internal flights. A significant percentage of air traffic volume in international airports is dedicated to air carrier flights operating between the United States and foreign countries. International airports are among largest, busiest, and best equipped airports in terms of runways, air traffic control facilities, and landing-aid equipment. There are 153 international airports in the United States as of December 2016. Table 1.3 provides a summary of some traffic-related data, as well as landmass sizes, for the top ten busiest U.S. international airports. Airport ranking is based on the total number of aircraft operations (landing and takeoff). The main sources for numerical data provided in Table 1.3 are FAA reports and statistical information [5], as well as individual airport traffic reports and information [23].

These airports with their landmass expanse and the volume of air traffic, as presented in Table 1.4, will require extensive AeroMACS network infrastructure containing multiple AeroMACS (cellular) base stations. The figures designated as "air carrier operations" show the number of aircraft operations related to airlines passenger flights and exclude operations related to air taxi, general aviation, and military flights. The table also provides IATA[4] airport codes (location identifier code, or station code) for these airports.

1.8 The Next Generation Air Transportation Systems (NextGen)

Early on in the twenty-first century, both in the United States and the European Union, long-term initiatives were taken for implementation of advanced air traffic management to support enhanced safety, increased capacity, and efficiencies of the air transportation system. In the United States, the initiative is termed the Next Generation Air Transportation

---

4 IATA (International Air Transport Association) code is a three-letter code used for identification of many airports around the world.

Table 1.4 Air traffic information for top ten busiest U.S. international airports.

Airport	Location	Total no. of passengers served	Total air carrier operations	Total aircraft operations	Approximate land size (km²)
Hartsfield–Jackson (ATL)	Atlanta, GA	101,491,106 [22]	780,326	882,497	19.02
O'Hare (ORD)	Chicago, IL	76,949,336	597,750	875,136	29.14
Dallas Forth-Worth (DFW)	Arlington, TX	65,712,163	506,095	681,261	69.67
Los Angeles (LAX)	Los Angeles, CA	74,937,004	570,445	654,501	14.64
Denver (DEN)	Denver, CO	54,472,514	424,930	547,648	139.99
Charlotte Douglas (CLT)	Charlotte, NC	44,876,627	363,667	543,944	24.28
McCarran (LAS)	Las Vegas, NV	37,687,870	349,606	524,878	11.34
John F. Kennedy (JFK)	New York, NY	56,827,154	407,460	446,644	19.95
Sky Harbor (PHX)	Phoenix, AZ	44,006,205	360,675	440,411	12.14
San Francisco (SFO)	San Francisco, CA	50,067,094	354,576	430,518	19.82

System (NextGen) and in Europe it is called Single European Sky ATM Research (SESAR). Both initiatives are quite similar in what is planned to be achieved. We discuss NextGen concept exclusively, with the understanding that SESAR project shares, more or less, the same goals and plans to modernize the air transportation system. In fact, AeroMACS can be considered as a component of both NextGen and SESAR visions.

In 2003, the U.S. Congress passed the "Vision 100 - Century of Aviation Reauthorization Act," which authorized the creation of Joint Planning and Development Office (JPDO) within the Federal Aviation Administration to manage work related to the creation of a next generation air transportation system. *"JPDO has responsibility for coordinating the research efforts of its government partner agencies which include the Departments of Transportation (DOT), Commerce (DOC), Defense (DOD), and Homeland Security (DHS); the FAA; the National Aeronautics and Space Administration (NASA), and the White House Office of Science and Technology Policy, to coordinate funding with the Office of Management and Budget. Additionally, JPDO has responsibility to consult with the public; to coordinate federal goals, priorities, and programs with those of aviation and aeronautical firms; and to ensure the participation of stakeholders from the private sector, including commercial and general aviation, labor, aviation research and development entities, and manufacturers. JPDO is jointly funded through FAA and NASA"* [24].

1.8.1 The Nextgen Vision

In January 2004, the DOT announced the plan for the Next Generation Transportation System as a multiagency, multiyear modernization of the air traffic system. More specifically, NextGen is a U.S. government-sponsored program aimed at modernization of NAS by means of a multistage implementation plan across the United States between the years 2012 and 2025 to meet the challenges and the goals of national and global aviation in the twenty-first century, to enhance safety and security of flights and airports, and to reduce flight delays and the negative environmental effects of aviation.

In a similar fashion, the European Union has put in place a program known as the Single European Sky ATM Research in anticipation of growth in air traffic volumes that will far outstrip the capacity of existing ATM systems. The crux of the NextGen and SESAR idea is the transformation of ATC system from the traditional radar-based system to a satellite/GPS-based radio communication network. GPS technology will be exploited for shortening the aviation routs, improving the traffic-handling capacity of the NAS, increasing the safety margin for air traffic controllers by monitoring and managing the aircraft movement, reducing flight delays and fuel consumption, and so on.

In the meanwhile, the United States and the European Union are closely cooperating to ensure interoperability and harmonization between NextGen and SESAR. These efforts are, in part, in support of ICAO GANP with Aviation System Block Upgrade (ASBU) program [25]. NextGen and SESAR projects have recognized the necessity of integration of air and ground components of the traffic management system, and the need for harmonious sharing of accurate information. In this manner, the United States–European Union joint harmonization work facilitates global modernization and advancements in air transportation systems and supports cooperation, clear communication, unified operations, and optimally safe practices [26].

1.8.2 Nextgen Key Components and Functionalities

NextGen (as well as SESAR) is not a single program, but consists of a series of initiatives. The NextGen (SESAR) project has several recognizable components. In what follows, some major components of Nextgen are briefly described.

1) A key GPS-related component of Nextgen is *automatic dependent surveillance–broadcast* (ADS–B). As the name implies, this constitutes the surveillance component of the Nextgen CNS. ADS-B broadcasts information on precise aircraft location to network of ground stations (air traffic controllers) and pilots using GPS–satellite technology. In the legacy NAS this type of information is provided by radar-based systems, which cannot match the accuracy afforded by GPS by any stretch of imagination. Thus, ADS-B presents a paradigm shift that essentially transforms the NAS into a more efficient one by replacing ground-based radar systems with satellite/GPS technology. The ADS-B on-board system operates by receiving satellite signal from the aircraft GPS receiving device and combining it with additional data furnished by other aircraft avionics, thereby providing a very accurate data on aircraft's location, altitude, ground speed, and many other quantities. This data is then broadcast to ground stations, aircrafts, and any other entity in the area that has proper receiving system in place. ADS-B also provides traffic and weather information directly to the cockpits of aircrafts that are equipped properly. This clearly raises the situational awareness for the pilot, in particular, and for the NAS, in general. ADS-B functionalities are divided into "ADS-B In" and "ADS-B Out." ADS-B Out provides capabilities related to broadcasting of critical flight data such as aircraft location, ground speed, and altitude. On the other hand, ADS-B In functionalities will provide the aircraft cockpit display with real-time information on traffic and weather conditions. While ADS-B In is considered to be an optional capability, the FAA requires that aircraft operating in the U.S. NAS must be equipped with ADS-B Out on-board system by January 1, 2020. According to FAA, as of October 2016, more than 24,000 general aviation aircrafts and 720 commercial aircrafts have been equipped with ADS-B Out avionics [27]. ADS-B consists of four independent components.

- ADS-B is a satellite-based technology, the GNSS[5] Satellite Constellation is exploited for aircraft onboard GPS device to continuously receive data. This data is interpreted and sent to ADS-B ground stations.
- Ground stations form the second component of ADS-B concept. FAA plans to install at least 700 ground stations in the United States

---

5 GNSS (global navigation satellite system) refers to a satellite-based navigation system with global reach. As of December 2016, the U.S NAVSTAR Global Positioning System (GPS), the Russian GLONASS, and the E.U. Galileo are the only operational GNSSs in the world.

that receive satellite data and transmit the data to air traffic control stations.

- Instrument flight rules (IFR) Certified Wide Area Augmentation System WAAS is required in the onboard aircraft avionics for ADS-B to function.
- For aircraft flying above 18,000 feet a 1090 MHz, extended squitter link with a Mode-S transponder is required. For aircraft flying below 18,000 feet (mostly general aviation), a 978 MHz UAT (Universal Access Transceiver) system is needed. In both cases, the additional devices are for use with an existing transponder.

2) *Next Generation Data Communication,* often referred to as *Data Comm,* is another major component of NextGen that corresponds to the communications module of the NextGen CNS. Data Comm defines a new method for pilots and controllers to access and communicate data, that is, via digital communications techniques, while currently push-to-talk voice communications is the predominant mode of information exchange. Specifically, Data Comm enables the transfer of clearances, approach procedures and instructions, all other routine pilot-to-controllers and controller-to-pilot exchanges, as well as advisories in digital communications form, predominantly in the form of text exchange. For instance, using Data Comm air traffic controllers and pilots can communicate through computer text messages instead of voice communications, which enables rapid and unambiguous (accent-free) communication of critical information such as clearance information, on the one hand; and more efficient use of available spectrum, on the other hand [28]. The schedule for Data Comm rollout is still in planning stages. The FAA is in collaborative talk with airports and operators to work out the details. However, it is believed that by 2020 the infrastructure for Data Comm will be in place and aircraft will be equipped with required onboard devices by then, as well.

3) The digital data distribution backbone of NextGen is *System Wide Information Management* (SWIM). SWIM, in effect, is an information management infrastructure that combines existing and new information systems and applications that interact through SWIM services. SWIM can be viewed as a means of providing user's access to NAS database, through either subscription or publication. The primary objective of SWIM is to provide the right information to various constituencies when needed. SWIM connects producers and users of data in a near real-time fashion with a common language and single point of contact to access data such as aeronautical information, flight information, and weather condition information [29].

4) *Performance-based navigation* (PBN) uses satellite-based navigation system, as well as onboard devices for selection of optimum and safe

routs. This clearly serves as the NextGen's CNS navigation module. PBN enables the aircraft to utilize navigation procedures that are more accurate than ground-based standard navigation aids, for more direct and shorter flight routes from departure to arrival. Consequently, PBN saves time and fuel, and therefore reduces the negative environment effects of aviation and provides more overall time efficiency. PBN also allows the realization of closely spaced parallel routs [30].

5) It is well known that substantial percentage of flight delays are caused by inclement weather and turbulent skies. The Next Generation Network Enabled Weather (NNEW) will be employed in the NextGen to significantly reduce weather-related delays. NNEW vision aims at combining weather forecasting models, climate observations, and data from airborne, land-based, and marine sources into a single national (or global) weather information system that becomes available to all NAS constituencies [31].

It should be mentioned at this point that AeroMACS, as a new broadband data link that has the ability to support the ever-expanding range of ATM communications requirements, is emerging under the modernization initiatives of NextGen and SESAR, and therefore, is considered to be an integral and enabling part of both NextGen and SESAR visions.

For a rather thorough list of NextGen components and modules, the FAA web page for NextGen [32] should be consulted. The FAA also posts progress reports and updates on implementation of various parts of NextGen on its website periodically; Ref. [27] is the most recent (as of December 2016) report.

1.9 Auxiliary Wireless Communications Systems Available for the Airport Surface

In addition to VHF AM radio that is available for establishing communications links between various nodes across the airport surface, there are a number of other data/voice transmission systems in place that may be used for information exchange at an airport. Some of these systems are wireless and even mobile, while the others are fixed and wired. The airport surface fiber-optic and copper cable loops are examples of wired transmission systems that support flight security and safety. These systems are parts of airport information exchange infrastructure. In Section 1.10.1, fiber-optics cable loop system is explored. In this section, some wireless communication systems that are available for use on the airport surface are briefly reviewed.

1.9.1 Public Safety Mobile Radio for Airport Incidents

Public safety (PS) mobile radio systems play a critical role in providing effective response to emergency situations and in the events of catastrophic man-made and natural disasters. The key challenge of disaster management is the minimization of the impact to individuals, assets, and the environment. In order to coordinate the relief efforts and to develop situational awareness, it is essential that the first responders be able to exchange voice and data in a timely manner. The PS communications system may be used at airports in case there is an unforeseen incident, or a catastrophic natural and/or man-made event, at the airport surface. The key advantage of this medium is the possibility of interoperable communications with first responders such as law-enforcement agencies, fire departments, paramedics, ambulances, and so on. In what follows, the PS communications systems, which is currently undergoing rapid expansion and evolution toward a network-centric broadband system in the United States, is briefly explored.

1.9.1.1 Public Safety Communications (PSC) Systems Architecture and Technologies

The architecture for PSC has been conventionally similar to that of "precellular" mobile communications (see Chapter 2) that may be viewed as a "single-cell" system where mobile users connect to a single high-power base station that provides radio coverage to a large zone. However, new systems for the first responder are being developed and deployed that commensurate to various environments, identified by the Department of Homeland Security.

In recent years, PSC community has witnessed increased attention. This is primarily due to better awareness of the need for reliable communications for the first responders in emergency situations, particularly in the aftermath of September 11 terrorist attack on New York City and hurricane Katrina's assault on New Orleans. Hence, wideband communications technologies capable of providing services such as image transfer, video streaming, geolocation, and so on have been promoted for PS applications.

In the United States, Project-25, also known as P-25 and APCO-25 (Association of Public-Safety Communications Official International), is the dominant narrowband standard for digital wireless communication that is used for PS applications. APCO-25 is essentially a suite of communications standards used by federal, state, and local PS agencies in the United States and Canada. APCO-25 has been developed, in collaboration with TIA (Telecommunications Industry Association), with four objectives in mind. The primary objective was to improve spectrum efficiency in comparison with the legacy analog FM land mobile radio (LMR) networks.

Secondly, it was to provide enhanced equipment functionalities; third, to offer open system architecture to promote competition between various vendors; and finally, to allow effective, efficient, and reliable intra-agency and interagency communications [33].

APCO-25 continues to be a dominant PSC technology in the United States, as well as several other parts of the world, in part because of its adaptability to the changing user's need. Phase I of P-25 features radios with 12.5 kHz bandwidth capable of operating in analog, digital, or mixed modes. Phase II of Project-25 features radios operating with 6.25 kHz bandwidth, which was developed in anticipation of the FCC's narrowbanding mandate [34]. It should be noted that APCO-25 is a noncellular narrowband trunked communications network that requires a fixed infrastructure. APCO-25 provides voice and limited data communications at rates up to 9.6 Kbits/s. APCO 25 offers a rich set of services, including messaging, group calls, broadcast call, and others, through a simple and direct device-to-device "walkie-talkie"-type radio, over short range of about 5 miles [33].

Similarly, in most parts of Europe, TETRA (Terrestrial Trunked Radio) is the leading PS communications technology. TETRA, which was deployed for the first time in 1997, is a telecommunications standard for private mobile digital radio (PMDR) systems developed and commercialized by European Telecommunications Standards Institute (ETSI) to meet the needs of PS communications users in Europe [33]. Later version of TETRA known as "*enhanced packet and data service TETRA Release 2,*" referred to as "TEDS," was published by ETSI that provides enhanced packet and data service with data rate up to 473 Kbits/s [35]. TETRA is also a narrowband trunk telecommunications network.

1.9.1.2 Public Safety Allocated Radio Spectrum

Several bands of frequencies are allocated for PS communications. In fact, the available spectrum for PSC is rather vast and is scattered across the RF spectrum and includes VHF, UHF, and C-band spectra. Traditionally, the UHF "800 MHz band" has been home to three applications: commercial cellular phone, private mobile radio such as SMR (Specialized Mobile Radio) and ESMR (Enhanced SMR), and PSC. Segments of the 800 MHz band have been the primary spectral bands for narrowband PSC as the main means of effective communications between dispatchers and their corresponding first responders, or between the first responders themselves. One issue of concern for PSC systems operating in 800 MHz band has been the increasing levels of interference from commercial cellular radio and private/professional mobile radio (PMR) systems operating in the same band. The interference problem in the 800 MHz band is caused by adjacent channel interference of fundamentally incompatible communication technologies. On the one hand, commercial mobile wireless systems conforming to cellular architecture, that is, multiple cells with low-power base station antennas, and on the other hand, noncellular PSC systems using a single base station with high-power antennas within a desired coverage area are mixed with spectral proximity. To combat the effects of this harmful interference, the FCC has ordered a reconfiguration, "rebanding", of the 800 MHz band, moving PS licensees to lower segments of the band and commercial cellular networks to higher segments, separated by an expansion band and a guard band [36]. Table 1.5 presents various spectral bands allocated for PSC, along with some characterizing notes for each band and their FCC designated allocation and application.

In 2002, the FCC allocated 50 MHz of spectrum, known as 4.9 GHz band, for fixed and mobile services excluding aeronautical applications. 4.9 GHz band may be used to support PSC. The stipulation is that nontraditional PS entities, such as utilities and the Federal Government may enter into sharing arrangements with eligible traditional PS agencies to use the band in support of their missions regarding homeland security.[6] One issue with this relatively new PSC band is the short wavelength of the signal that bears rapid attenuation. Consequently, this band may be suitable for PS services that require extensive bandwidth, but not practical for wide area coverage.

1.9.1.3 700 MHz Band and the First Responder Network Authority (FirstNet)

Certain segments of the 700 MHz band are allocated for PSC and the remaining spectrum is assigned to commercial wireless communication systems. Signals over this band, relative to higher frequency bands, can penetrate buildings and walls rather easily, and for less obstructed terrains, they can provide coverage to large geographical areas.

In July 2007, the FCC issued an order that would allow PSST (Public Safety Spectrum Trust) Corporation to enter into leases of spectrum usage rights with commercial licensees/operators of the so-called 700 MHz "D Block" (758–763 MHz/788–793 MHz). A Second R&O (Report and Order) included rules for the D Block auction winner(s) to build a nationwide PS-shared wireless broadband network that would be paid for by the networks and not by the PS community or the taxpayers. The FCC rules are intended to ensure that PS will have priority access in emergencies and that the network would be continually refreshed with the latest technical improvements.

---

6 http://www2.fcc.gov/pshs/public-safety-spectrum/4-9GHz-Public-Safety-Band.html.

Table 1.5 Allocated spectral bands for PS communications.

Spectrum	Notes/PS bandwidth	Band designation	FCC main allocation/ application
25–50 MHz	Nontrunked, susceptible to "skip interference"/6.3 MHz	VHF low band	Private land mobile state Highway patrol
138–144 MHz 148–174 MHz	Narrowband PS spectrum, less affected by skip interference and noise/ 3.6 MHz	VHF high band	Private mobile radio
450–460 MHz	Narrowband PS spectrum, virtually immune from skip and environmental noise interferences/3.7 MHz	UHF band	Land mobile radio and PS
470–512 MHz	Currently used in 11 US metro areas to support critical PSC/42 MHz	T-band	The FCC is to auction off this band during 2021–2023.
758–763 MHz/ 788–793 MHz "D-Block"	Allocated for a nationwide broadband PS network (FirstNet)/10 MHz	700 MHz band	PS/commercial wireless
763–768 MHz 793–798 MHz	Broadband PS systems: FirstNet, guard bands are provided/10 MHz	700 MHz band	Land mobile PS/ commercial wireless
768–769 MHz 798–799 MHz	Guard band/2 MHz	700 MHz band	PS/commercial wireless
769–775 MHz 799–805 MHz	Narrowband PS systems/ 12 MHz	700 MHz band	Land mobile PS/ commercial wireless
806–809 MHz 851–854 MHz	NPSPAC band, local, state and regional PS use/6 MHz	800 MHz band	Land mobile PS/ PMR/cellular phone
809–815 MHz 854–860 MHz	National PS band/10 MHz	800 MHz band	Land mobile PS/ PMR/cellular phone
815–816 MHz 860–861 MHz	Expansion band may be used PS/2 MHz	800 MHz band	PS/PMR/cellular phone
816–817 MHz 861–862 MHz	Guard band 2 MHz of bandwidth	800 MHz band	PS/PMR/cellular phone
4.94–4.99 GHz	Supports broadband applications for homeland security missions/50 MHz	4.9 GHz C-band	WLAN, mobile data VoIP; and hoc Nets

1.9.2 Wireless Fidelity (Wifi) Systems Applications for Airport Surface

Local area networks (LAN) are computer networks that provide device connectivity within a limited area, such as inside a single building or a group of buildings. In order to extend the outreach of a LAN, it can be connected to other Lans over any distance to form a larger telecommunications network. Several different transport protocols can be adopted for a LAN infrastructure, which includes Ethernet, token ring, ATM, frame relay, and so on. The adopted networking technology determines which data transmission methods can be implemented and defines the maximum possible transmission speed (maximum data rate and throughput). Therefore, the selection of the networking technology is a critical design matter that influences the capabilities and capacity of the LAN.

The Wireless Fidelity (WiFi) technology is an IEEE 802.11 Standard-based wireless alternative to wireline LAN. Originally, this wireless local area network (WLAN) was developed as an extension of Ethernet over license-free ISM (industrial scientific medical) bands, primarily, for private applications. However, WiFi has gradually become commonplace in public locations for creation of so-called "hot spots," offering the user easy, and often free-of-charge, broadband access to the Internet [37]. In order for WiFi technology to become as commonplace as it is today, the technology had to address four major technical challenges. Simplicity of operation and use has been always paramount for any publicly used electronic system. The second issue is the security and interference over ISM bands. Resource planning and bandwidth allocation cannot be guaranteed over ISM bands, and essentially there are no protections against interference, aside from the fact that FCC limits the output power of the systems that might operate over these bands. For instance, the domestic and commercial (restaurant) microwave ovens operate near 2.45 GHz, this is within the most popular ISM band of 2.4–2.5 GHz.

The current state of the art of WiFi technology allows pleasure and business travelers to continually stay connected through WiFi networks in their residences, restaurants, airport waiting halls, and on aircraft. It is also possible to acquire a cost-effective and convenient global WiFi access across smartphones, tablets, and laptops. By the year 2018, it is estimated that on average, there will be one WiFi hot spot for every 20 inhabitants of the planet. In 2014, 52% of the global IP traffic was carried through wireline networks, 44% via WiFi devices, and cellular phones' share only accounted for 4% of this traffic. It is expected that the WiFi proportion of traffic for Internet access will increase in the future [38]. The WiMAX networks may be used to provide backhaul[7] support for mobile and stationary WiFi hot spots. Originally, WiFi hot spots were connected to the Internet via a wireline backhaul networks such as digital subscriber line (DSL). However, by using a wireless backhaul network such as WiMAX, the costly wireline infrastructure can be avoided [39].

WiFi technologies have several categories of applications in air transportation systems, in general, and in the airport area, in particular. One is a commercial application that provides broadband Internet access to passengers, crew, and airport labor forces, within the airport waiting halls, gate areas, and even inside the aircraft while it is parked in front of a terminal gate. This service is provided to the users free of charge in large percentage of United States and international airports. WiFi services are also commonplace aboard aircraft, providing broadband IP access to passengers while the aircraft is in flight.

The WiFi hot spots, distributed on the surface of airports, a "WiFi system", may be viewed as an ancillary wireless network that might be applied to support, or provide backup support, for some aviation functionalities. For instance, Ref. [40] describes an interesting application of WiFi networks in positioning system, considered as an alternative to that of GPS-based system. The article introduces an approach for automatically calibrating the WiFi positioning system to improve accuracy and to achieve performance levels that are comparable to those of GPS system. A prototype of this system has been implemented and tested in Munich airport in Germany. It is anticipated that WLAN applications, in various aspects of air transportation systems, will expand in the future.

1.10 Airport Wired Communications Systems

The current wired airport surface communications system primarily consists of buried copper or fiber-optic cables. The principal purpose of the airport cable system is to provide the physical media that enables interconnectivity of all communications systems and "communications rooms" across the airport. In this section, we briefly explore the dominant wireline signal exchange media on airport surface, that is, the fiber-optic cable loop communications systems (CLCS).

---

7 Backhaul networks are intermediate networks that provide connection between the core (backbone) networks and the small subnetworks that need to access the core network. In this case, the backbone network is the IP network and subnets are WiFi hot spots, and WiMAX as a backhaul network connects the WiFi hot spots to the IP core network for Internet access.

In late 1980s, the FAA initiated a modernization plan by introducing fiber-optic technology to airport surface communications, a technology that is still the dominant mode of information exchange on the airport surface. The fiber-optic transmission system is complemented by VHF radio for point-to-point transmission, as well as by airport auxiliary communications networks; hence, in fact, a hybrid communications approach is in place. The fiber-optic cables interconnect the air traffic control system to communications, navigation, and surveillance facilities, as well as to other airport constituencies. The goals for this modernization were stated to be increasing the capacity of the NAS, improving the safety of airspace operations, increasing the productivity of FAA facilities, and enhancing the overall cost effectiveness of the NAS operations [41].

Employing fiber-optic cable loop enables simultaneous bidirectional signal transmission using multiple fiber strands in the cable. This provides redundancy and enhances reliability. Additionally, a number of benefits are realized when the backbone of airport surface communications infrastructure is fiber optics, simply because the information is transported via optical signals. A partial list of these benefits is provided further.

- *Extremely High Bandwidth:* Single-mode fiber-optic cables can carry signals at speeds in excess of 1 Tbps.
- *Immunity to Electromagnetic Interference:* Since the signal is carried by light, fiber-optic cables are immune to electromagnetic interference.
- *Low Signal Loss:* These cables induce very low loss as optical signal travels through. For instance, a cable specified by FAA for use in airport circuits, FAA-E-2761a, has signal loss as low as 1 dB/km [39]. However, in modern single-mode fiber-optic cables, the typical loss is less than 0.3 dB/km. For this reason, single-mode fiber-optic cables can sustain transmission over distances of up to 140 km before regeneration is needed [41].
- *Electrical Isolation of Transmitter from the Receiver:* This simplifies link design as there are no ground loops.
- *Small Physical Size and Weight:* Cables containing six fibers in protective jackets can be as small as one half of inch in diameter [41]. In today's state-of-the-art fiber-optics technology, a fiber cable may contain up to a thousand fiber strands.
- *Security and Cross Talk:* There is no signal leak out of a fiber optic, consequently no cross talk between the fibers or fiber-optic cables exists [42].
- *High Electrical Resistance:* Fiber cables are not affected by close by high-voltage equipment. In fact, fiber-optic cables for airport surface communications are buried with power cables in the same trenches [41].
- *Wavelength Division Multiplexing* (WDM): For duplex communications, fibers are normally used in pairs, with one fiber carrying the signal

in one direction and the other in the opposite direction. However, simultaneous transmission of several signals over a single strand is possible by using different wavelengths and appropriate coupling/splitting devices. This is the essence of wavelength division multiplexing that enables a single-fiber strand to bear an aggregate data rates measured in terabits per second.

Some of problems and disadvantages of utilizing fiber optics for data transmission, in general, and for airport surface communications, in particular, are discussed further.

- *High Installation Cost:* The initial investment for installation of fiber-optic infrastructure is still high, although these costs are dropping significantly every year. It should be noted that despite significant initial expenses, over the lifetime of fiber-optics communications system, they prove to be more economical than similar wired electrical transmission systems, that is, they are of lower cost in the long run.
- *Optical Transmitters and Receivers:* The optical transmitter/receiver systems are more expensive than those of electrical systems that are used for the likes of coaxial cables.
- *Susceptibility to Physical Damage:* Fibers are compact and tiny transmission media (on the order of few to around 100 μm in diameter[8]) and are highly susceptible to getting cut, broken, or damaged during installation or construction activities.
- For the case of airport surface communications, an additional challenge crops up when the cable network needs to be repaired, upgraded, or expanded. Since cables are installed underneath the airport surface, it may be required that a section of the airport surface be blocked and be placed out of service temporarily. This could cause the shutdown of some airport runways and taxiways, which translates into lower capacity, airport congestion, and loss of revenue. This is one of the reasons that a network-centric wireless standard, such as AeroMACS, is preferred for airport surface over CLCS.

Fibers are made out of either glass or transparent plastic; however, for long-distance communications, glass fibers are used exclusively owing to lower optical absorption in glass fibers.

---

8 Normally, the diametric size of an optical fiber is given by three numbers that represent the outer diameter of the core (central part of the fiber through which light is transmitted), diameter of core and cladding (an outer layer glass that facilitates the transmission of light down the fiber), and diameter of core and cladding and coating (multilayers of plastic applied to provide protection against shocks and from abrasion) put together, respectively. For instance, by fiber size of 100/150/250, it is meant a fiber with core diameter of 100 μm, cladding diameter of 150 μm, and coating diameter of 250 μm.

1.10.1 Airport Fiber-Optic Cable Loop System

Fiber-optic cable loop system provides interconnectivity between the "communication rooms" through cabling routed between each of the "rooms" that are distributed throughout the airport's area, and from the communications rooms to the users' workstations. The communications rooms serve as the distribution points for the end users of various airport systems [43]. The phrase "loop system" means a closed-loop transmission system that provides an inherent redundancy in the event of any single link being detached from the loop. Fiber-optic loop systems have an additional advantage of simplicity, in that they can be implemented using as few as only a pair of fibers, whereas traditional copper cable loops require multiple pairs of facilities [41].

Two basic configurations are defined for loop systems. The simplest one is when a loop is shared by only two facilities, and the second is when the number of facilities connected to the loop is at least three. A loop that is shared by three or more facilities must have a protocols suite designed to manage the operation of the loop, for determining which node is allowed to receive or transmit at any given time, while preserving the integrity of the transmitted data. For airport fiber-optic loop systems, two protocol suites have been recommended, time division multiplexing (TDM), where all facilities have equal and orderly access to the loop, and fixed master–slave protocol (MSP) with multiple slaves, where a slave is addressed and responds according to a predetermined procedure [41].

The airport facilities, such as ATCT, ASR (airport surveillance radar), security devices, various remote transmitters and receivers, and so on, are distributed across the airport area according to the initial installation and the airport ensued evolutionary development path. In MSP, each node receives all the transmitted data on the loop, but accepts and reacts only to the information addressed to it [41]. The operation of a MSP using counter-rotating rings[9] proceeds as follows:

- The ATCT, which plays the role of master, transmits interrogation and command signals into the loop. The signals are received by a facility, that is, a slave, regenerated and transmitted to the next facility in the loop.
- The signals typically consist of a facility address code followed by the message for that facility. A facility's response to an ATCT interrogation takes the form of an immediate message into the loop. When the ATCT has received a response, it repeats the sequence until all facilities have been interrogated and have responded.

---

9 In counter-rotating ring configurations, there are two rings in use. One ring transmits in a clockwise direction, while the other transmits in a counterclockwise direction. This provides a fully redundant system that is unaffected by a single failure in either loops.

- If a facility fails to respond within a prescribed time and after a predetermined number of polls, diagnostic operations are initiated at the ATCT to determine the failure mode. When a message becomes distorted during transmission, a request to retransmit will be generated.
- The ATCT receiving device is configured to prevent retransmission of signals.
- Both the clockwise and counterclockwise transceivers receive signals from their respective directions, and in turn, they retransmit the signals. Although the two signals are identical, they are not synchronous at any given facility because the lengths of the transmission paths are different. Therefore, some accommodation is necessary, using one of two methods. One is to correct for the difference in arrival times. The other method is to preferentially accept either of the two signals based on a criterion such as signal quality or some other predetermined preference, or a combination of the two methods.

As far as optical network architecture is concerned, the loop configuration is based on synchronous optical network (SONET) ring architecture. SONET defines optical signals and synchronous frame structure for multiplexed digital traffic [42].

1.10.2 Applications of CLCS in Airport Surface Communications and Navigation

Currently, the fiber-optic CLCS supports a number of applications on the airport surface. As was mentioned earlier, the fiber-optic cable loop, primarily, forms the backbone of communications and information exchange for various airport functionalities supporting voice/data/and video services. The loop system also distributes voice and video signals across the airport terminal field.

The second category of services supported by fiber-optic CLCS is in airline and airside[10] operations, and in aircraft and passenger's operations through the airport's airside. These services include FAA air traffic control and Navigation Aids (NAVAID) systems, aircraft refueling, visual docking guidance, runway monitor, gate management, parking information, baggage handling, and so on.

Airport landside operations comprise all essential services needed to support operations for travelers, luggage, and cargo to and from the

---

10 An airport is divided into two areas of landside and airside. Airport spaces related to physical access to the airport, such as access roads, parking decks, and train stations are called landside areas. Airside areas are airport locations that are accessible to aircraft, such as runways and taxiways. Airside operations consist of ramp control, aircraft gate control, and access control.

ground transportation. These services include but are not limited to parking systems, taxi dispatch, and surface vehicle monitoring. Fiber-optic CLCS supports these services as well.

Another airport operation supported by fiber-optic CLCS is in airport safety and security operations. Security systems include fire alarms, explosive detections, perimeter sensors, close circuit television (CCTV), access control, and so on [43–44].

1.11 Summary

In this chapter, the evolution of airport wireless communications systems from the legacy analog VHF AM system to the making of the AeroMACS concept, is sketched. The background material that explains the reasons behind the emergence of this innovative aviation technology is described. The technical details of AeroMACS networks are covered in the following chapters. AeroMACS will operate within the larger context of NAS, as well as in the International Airspace System. A concise overview of the current NAS is presented. The Federal Aviation Administration's NextGen vision, planned to transform and modernize the NAS, is discussed in some details. NextGen promises to vastly improve the precision of aircraft navigation, enhance the quality of aircraft position data that is made available to controllers, and reduce the burdens of air-to-air and air-to-ground communications. The emergence and evolution of pre-Aero-MACS digital communications techniques in aviation are discussed. ACARS and VDL technologies are briefly reviewed. The major technologies that currently play complementary and auxiliary roles to VHF radio and fiber-optic loop in the airport communications infrastructure are discussed. The airport fiber-optic cable loop system is briefly reviewed. It is pointed out that although communications networks based on fiber optic cable buried underneath the airport area are available in the majority of the U.S. airports, they are found to be expensive to install and maintain. In addition, repairing, upgrading, and updating fiber-optic cable system is a challenging task that may require blocking a section of the airport surface and putting it out of service temporarily. The Aeronautical Mobile Airport Communications System overcomes some of the challenges associated with the fiber-optic cable loop system.

References

1 P. Camus and J. Rascol, "Data Link System Between an Aircraft and the Ground and Procedure for Recovering from a Failure," US Patent, 6173230 B1, January 2001.

2 D. W. Matolak and J. I. Rodenbaugh, "Optimum Detection of Differentially-Encoded 8-Ary Phase Shift Keying in a Dispersive Aeronautical Channel: New Results and Approximation," *Proceedings of 33rd Southeastern Symposium on System Theory*, March 2001.

3 E. Haas, "Aeronautical Channel Modeling," *IEEE Transactions On Vehicular Technology*, 51(2), 254–264, 2002.

4 B. Kamali, "An Overview of Civil Aviation Radio Networks and the Resolution of Spectrum Depletion," *Proceedings of IEEE ICNS 2010*, Herndon VA, May 2010.

5 Federal Aviation Administration, *"Air Traffic Activity System (ATADS),"* 2016. Available at http://aspm.faa.gov/opsnet/sys/Main.asp.

6 R. J. Kerczewski, "CNS Architectures and System Research and Development for the National Airspace System," *Proceedings of the IEEE Aerospace Conference*, pp. 1636–1643, March 2004.

7 ITT Industries, "Technology Assessment for the Future Aeronautical Communications System," Report to NASA, May 2005.

8 Raytheon Network Centric System, Description and Requirements Assessment for Terminal Class B and Class C Airspace Communications to NASA GRC for ACAST Project, Mid Term Review-Phase-I Draft, April 25, 2005.

9 T. L. Singore and M. Girard, "The Aeronautical Telecommunications Network (ATN)," *Proceedings of IEEE Military Communications (MILCOM)*, October 1998.

10 ICAO, "Manual on the Aeronautical Telecommunication Network (ATN) using Internet Protocol Suite (IPS) Standards and Protocol", 2nd Edition, January 1, 2015.

11 M. S. Pereira, "Commercial Aviation VDL Choices," *IEEE AESS System Magazine*, June 2003.

12 F. Studenberg, "A Digital TDMA System for VHF Aeronautical Communications on 8.33 kHz Channels," *Proceedings of IEEE Digital Avionics Conference*, 2004.

13 O. Drugge et al., "DS-CDMA Overlay for a VHF-AM Communication System," Luleal University of Technology, May 2000.

14 D. W. Matolak and J. T. Neville, "Direct-Sequence Spread Spectrum Overlay in the Microwave Landing System Band," *Proceedings of IEEE Aerospace Conference*, 2004.

15 ICAO, DOC 9694, "Manual of Air Traffic Services Data Link Applications," 1999.

16 ICAO, Document 4444, "Procedures for Air Navigation Services/Air Traffic Management (PANS ATM)" 2001.

17 EUROCONTROL/FAA, "Future Communications Study Final Conclusions and Recommendations," Report Version 1.1, November 2007.

18 J. M. Budinger and E. Hall, "Aeronautical Mobile Airport Communications System (AeroMACS)," NASA Technical Memorandum; TM-2011-217236, October 2011.

19 RTCA, "Aeronautical Mobile Airport Communications System Profile", March 2, 2010. (Revised on June 28, 2013.)

20 Chapter 3, "The National Airspace System," in *Airport and Air Traffic Control System*, Princeton University Press, available online.

21 Federal Aviation Administration, *"Airports Categories."* Available at https://www.faa.gov/airports/planning_capacity/passenger_allcargo_stats/categories/ (accessed March 3, 2016).

22 B. Kamali, J. D. Wilson, and R. J. Kerczewski, "Application of Multihop Relay for Performance Enhancement of AeroMACS Networks," *Proceeding of IEEE ICNS 2012*, Herndon VA, April 2012.

23 Hartsfield–Jackson Atlanta International Airport, "Monthly Airport Traffic Report," December 2015. Available online.

24 U.S. Government Accountability Office, Testimony Before the Subcommittee on Space and Aeronautics, Committee on Science, House of Representatives, "NEXT GENERATION AIR TRANSPORTION SYSTEM, Preliminary Analysis of the Joint Planning and Development Office's Planning, Progress, and Challenges" March, 2006.

25 ICAO, "2013–2028 Global Air Navigation Plan" Montreal, Canada, 2013. Available online.

26 U.S.-E.U. MOC Annex 1 – Coordination Committee, "NextGen – SESAR State of Harmonization Document," December 2014.

27 Federal Aviation Administration, "NextGen Update: 2016." Available at https://www.faa.gov/nextgen/update/progress_and_plans/adsb/.

28 Federal Aviation Administration, Data Communications (Data Comm.). Available at https://www.faa.gov/nextgen/programs/datacomm/.

29 J. S. Meserole and J. W. Moore "What is System Wide Information Management (SWIM)?" *IEEE Aerospace and Electronic Systems Magazine*, 22(5), 2007.

30 Federal Aviation Administration, "Performance Based Navigation," 2016. Available at: https://www.faa.gov/nextgen/update/progress_and_plans/pbn/.

31 Federal Aviation Administration, "NextGen Network Enabled Weather (NNEW)," November 2009. Available at https://www.faa.gov/nextgen/programs/swim/documentation/media/demo_tim_4/nnew.

32 Federal Aviation Administration, "NextGen." Available at https://www.faa.gov/nextgen/ (accessed May 29, 2018).

33 G. Baldini, S. Karanasios, and D. Allen, "Survey of Wireless Communication Technologies for Public Safety" *IEEE Communications Surveys and Tutorials*, 16(2), 2014.

34 A. Paulson, "A Review of Public Safety Communications: From LMR to Voice over LTE (VoLTE)," *Proceedings of IEEE International Symposium on Personal, Indoor, and Mobile Radio Communications*, 2013.

35 ETSI TR 102 580 V1.1.1, "Terrestrial Trunk Radio (TETRA); Release 2; Designer's Guide; TETRA High-Speed Data (HSD); TETRA Enhanced Data Service (TEDS)", October 2007.

36 FCC Document, "Rebanding of 800 MHz," FCC-04-168A1, 2004.

37 P. S. Henry and H. Luo, "Wifi: What's Next?" *IEEE Communications Magazine*, 40(12), 66–72, 2002.

38 H. A. Omar, K. Abboud, N. Cheng, K. R. Malekshan, A. T. Gamage, and W. Zhuang, "A Survey on High Efficiency Wireless Local Area Networks: Next Generation WiFi," *IEEE Communications Survey and Tutorials*, 18(4), 2315–2344, 2016.

39 D. Niyato and E. Hossain, "Integration of WiMAX and WiFi: Optimal Pricing for Bandwidth Sharing," *IEEE Communications Magazine*, 45(5), 140–146, 2007.

40 F. Gschwandtner and P. Ruppel, "Automatic Calibration of Large-Scale WiFi Positioning System," *Proceedings of the IEEE Globecom Conference*, 2008. Available at http://globecom2008.ieee-globecom.org/downloads/ DD/DD16W2%20Localization/DD16W2.pdf.

41 Federal Aviation Administration, "Airport Fiber Optics Design Guidelines," 1989.

42 J. C. Palais, *Fiber Optic Communications*, 5th edn, Prentice Hall, 2004.

43 Airport Consultant Council (ACC), "Airport Information Technology & Systems (IT&S)," July 2008.

44 Airport Consultant Council, http://www.acconline.org January 2012.

2

Cellular Networking and Mobile Radio Channel Characterization

2.1 Introduction

Over the past few decades, a new technological paradigm has been defined by wireless communications that has truly permeated all aspects of life on this planet. Form ubiquitous voice communication, to Internet access, to facilitating wireless connections between electronic devices that are in proximity of each other, and to applications in such diverse areas as universal and unlimited voice/text/data exchange, remote control, tele-medicine, smart homes and appliances, and soon to emerge smart cities, all are parts of this paradigm shift. The decades-old dream of "information superhighway" in which all three major networks created in the twentieth century, that is, data networks, telephone networks, and cable television networks emerging in a single pervasive digital network, has come to realization in the second decade of the twenty-first century to a large extent. AeroMACS (Aeronautical Mobile Airport Communication System) is the reflection of this paradigm shift in management of the next generation of air transportation systems that facilitates wireless communications between control towers, aircraft, airport terminals, airlines offices, and all other units involved in the airport service operation; for better administration, increased safety and security, fewer delays, lower operational cost, nondestructive extension of the communication network in the event of airport expansion, and efficient resolution of airport incidences.

Historically, or at least in the first few decades of its emergence, the growth witnessed in wireless mobile communications was rather slow, until the enabling technologies became available. The two main technical breakthroughs that brought about sudden and exponential growth in mobile radio were the development of cellular concept in the 1970s and the advent of highly reliable miniature solid-state radio frequency (RF) digital hardware in the 1970s and 1980s.

AeroMACS: An IEEE 802.16 Standard-Based Technology for the Next Generation of Air Transportation Systems, First Edition. Behnam Kamali.
© 2019 the Institute of Electrical and Electronics Engineers, Inc. Published 2019 by John Wiley & Sons, Inc.

This heuristic review chapter provides the background information necessary for understanding of WiMAX (Worldwide Interoperability for Microwave Access) and AeroMACS technologies. The chapter begins with a brief review of the cellular concept and Erlang traffic theory, leading to the design of classical cellular network for a given grade of service. The wireless radio channel is reviewed next. Three categories of signal attenuation/fading are highlighted. First is *long-term signal attenuation called path loss*, which is directly related to distance between the mobile station (MS) and the base station (BS), as well as the spectrum over which the wireless network operates. This reflects the fact that as the MS moves away from the BS the mean value of the received signal power is reduced. Several analytical and empirical-statistical models for calculation and estimation of path loss are discussed. Second is *large-scale fading* caused by shadowing and foliage effects, which can also be viewed as medium-scale attenuation since it effects the signal for some limited duration of time and it will disappear once the MS is "out of the shadow." Empirical data has shown that this attenuation has log-normal distribution. Consequently, the combined long-term and large-scale fading (medium-scale attenuation) have a log-normal distribution whose mean value is equal to the distance-dependent path loss. Third is short term fading or *small-scale fading*, which subjects the radio signal to rapid envelop and phase fluctuations. The short-term fading is caused by the multipath nature of the received signal as well as motion in the mobile channel. The multipath fading channels are then classified based on how fast the signal is transmitted relative to various measures of fading channel bandwidth, and how quickly the wireless channel characteristics change as a result of motion existed in the channel. The last part of the chapter is dedicated to the elementary description of the challenges faced when broadband signals are transmitted through the airport surface channel. The statistical characterization of the airport surface channel over C-band (5 GHz band) is presented in some detail in Chapter 3.

2.2 The Crux of the Cellular Concept

The cellular technology presents a major breakthrough that, for the most part, has addressed the challenges of spectral scarcity and spectral congestion in wireless networks. In this manner, the cellular architecture is the backbone of large number of contemporary wireless networks. A central issue in design, implementation, and deployment of wireless networks is radio resource management (RRM). It is interesting to observe that the preliminary steps in defining and understanding RRM in present-day "telephone" networks are embedded in the cellular concept and cellular layout of the wireless network.

In order to appreciate the salient features of cellular systems and how and why the cellular concept has revolutionized mobile communications, we briefly explore the architecture of the mobile communication systems that were in place before the development of cellular communications. It should be mentioned that the architectures very similar to that of "precellular" systems are still widely applied in public safety communication systems and in support of mission-critical voice communications. Terrestrial Trunked Radio (TETRA (European)) and Project-25 or APCO-25 (North American) are narrowband public safety communication standards of this sort.

2.2.1 The "Precellular" Wireless Mobile Communications Systems

A number of mobile radio technologies were in use prior to the advent of cellular mobile telephony. Since they were the predecessors of the cellular telephones they may all be called precellular systems, or as they are sometimes referred to "zeroth generation" of mobile telephony; 0G. Some of these early networks were part of the public switched telephone (telecommunications) networks (PSTN), and a few, although not a part, but were connected to PSTN. However, there were (and there are) wireless mobile networks that were not (and are not) connected to the PSTN at all. Examples of these types of wireless networks are taxi dispatch systems and public safety communication networks. What is considered here as precellular system, though, is the core part of all these networks regardless of their mode of connection to the PSTN. The key design objective of these networks is to provide coverage for the largest possible area with a single high-power/high tower transceiver antenna.

In order to develop a wireless mobile network, whether with cellular architecture or not, the allocated band of frequencies are divided into radio channels; to be exact, into pairs of radio channels with maximum possible spectral separation for "uplink," that is, MS to the zone station, and "downlink," that is, zone station to MS, connections. In this fashion, a set of duplex radio channels are created. The steps needed to be taken for the development and then deployment of the "precellular" mobile network are outlined further.

1) Each given locale requiring coverage is divided into several "zones."
2) Each zone possesses a *high-power transceiver* station with an omnidirectional antenna located ideally in the center of the zone, which provides coverage to the entire zone, as shown in Figure 2.1.

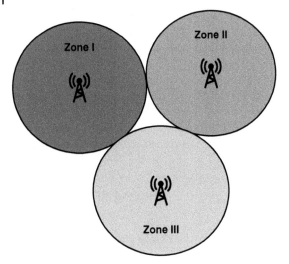

Figure 2.1 A precellular mobile communications system: the service area is divided into zones with high-power/high-tower transceivers to to provide coverage for the zones.

3) A set of duplex radio channels is permanently assigned to each zone. Each set is carefully selected such that they do not include neighboring channels, in an attempt to minimize RF interference.

As the MS moves away from the zone's station antenna, the received signal weakens. Around the edges of the zone signal quality degrades to the extent that eventually the call has to be dropped and reinitiated for connection to the new zone's antenna.

At least, three deficiencies can be identified in the precellular radio network. First and foremost is the spectrum inefficiency that manifests itself in the form of low-network capacity. As we will see later, wireless network capacity may be defined in a variety of ways. When the capacity is defined as the maximum possible number of users who may be simultaneously accommodated by the network, the capacity of the precellular network is equal to the number of available duplex radio channels. The cellular network overcomes this limitation by introducing the concept of *frequency reuse*. Second, this network is unable to provide continuous communication when the mobile station crosses the zone boundary. Cellular networks address this shortcoming by invoking a procedure called *handoff* or *handover*. Last, the precellular network does not provide a uniform quality of communication (QoC). The mobile users who are closer to the zone station antenna receive stronger signals than the ones that are farther away. The cellular networks equalize the QoC by *power control* and handoff protocols. A number of other advantages of cellular communication systems over precellular networks will be discussed in the next sections.

2.2.2 The Core of the Cellular Notion

Consider the "precellular" system just reviewed. The crux of the cellular idea is the replacement of each of the large zones with several smaller hexagonal-shaped areas, called cells, and the replacement of high-power zone transceivers with as many low-power cell transceivers as needed, in fact one for each cell, each providing coverage to the corresponding cell area and not beyond. A set of duplex radio channels are then carefully divided into subsets of radio channels, and each cell is provided with one of these subsets. Cells that are "far away" from each other may use the same radio channel subsets. In this manner, the *spectrum is reused* and spectral efficiency of the cellular network is dramatically improved relative to that of the precellular system.

Before presenting the details of cellular concept, a brief description of cell classifications is provided. Cells may be categorized according to the size of their coverage area. A *macrocell* covers a relatively large area with an antenna that is mounted on a high tower. A macrocell serves a large cell in the ubiquitous mobile phone network with a cell radius of greater than 1 km and a transceiver antenna output power ranging from 40 to 160 W. A *microcell* is a smaller cell in a mobile phone network, which typically provides coverage for small sections of urban areas such as shopping malls and airport terminals. The microcell radius is between 250 and 1000 m with the output power of 2–20 W. A *picocell* provides coverage for a smaller area than that of a microcell and is normally used for indoor applications. The radius of a picocell is about 100–300 m with an output power that ranges from 250 mW to 2 Watts. Finally, the *femtocell* that is used for in-building applications. The radius of femtocells is in the range of 10–50 m with antenna power output of 10–200 mW [1,2]. Cells in AeroMACS network are of either macrocell or microcell type.

According to the classical cellular network design principals, for a given geographical area over which cellular service is contemplated, a number of steps shall be taken and careful planning is deliberated before the network is laid out. In what follows, the cellular architecture is briefly described. The treatment is brief and in outline form, as the intent is a quick review. For more extensive discussion on cellular concept, the reader is referred to Ref. [3–5].

1) Each service area is divided into a number of hexagonal-shaped areas called *cell*. Each cell has a transceiver system called *base station (BS)*, *base transceiver station (BTS)* or *cell site*.

2) The available spectrum is divided into a set of duplex radio channels. The set is further divided into a number of nonadjacent radio channel subsets. Different subsets of radio channels are assigned to neighboring cells' base stations. This is to reduce *adjacent channel interference*, which is the interference between spectrally adjacent radio channels in the cellular network.

3) The cell coverage area conceptually has a hexagonal shape. The actual cell coverage area, which is generally different from the conceptual one, is called *cell footprint* as shown in Figure 2.2. Cell footprints are determined by field measurements and signal propagation models. Figure 2.2 also illustrates that in actual cellular networks, there might be areas in which the received signal is so weak that coverage is not available, these are the so-called *blind spots*.

4) Cells with a single BS located at their centers are called *center-excited* cells. Cells with transceiver stations on their vertices are called *edge-excited* cells. Each station provides coverage for a section of the cell.

5) The power output of each BS is adjusted for the cell coverage and not beyond. This provision is carried out by *power control* protocols.

6) Directional antennas may be selected for base stations. The underlying cell is divided into, normally, three or six sectors. Each sector is served by a single-directional antenna. This configuration reduces the interference, thus increases the network capacity. This process is called *sectorization*.

7) As the MS moves away from the BS, the signal loses its strength; the cellular network may automatically interface the MS to a new BS with a new assigned radio channel. This process, called *handover* or *handoff*, sustains the quality of communications at a predefined level. Handoff procedures constitute an important subset of cellular network mobility protocols. Depending on the architecture of the network, handoff may occur between adjacent cells, adjacent sectors, smaller and larger cell, and so on. The entire handoff process is transparent to the mobile user.

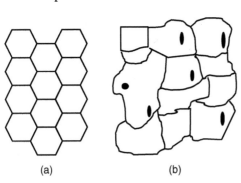

Figure 2.2 (a) Conceptual cellular plan with hexagonal-shaped cells. (b) Actual cellular layout with cell footprints and blind spots shown by darkened areas.

(a)   (b)

8) When the traffic increases in a cell, the cell is divided into four or more smaller cells, each new cell will be able to handle almost as much traffic as the original cell. This is known as *cell splitting*, which is another key feature of cellular communications. Cell splitting may be implemented in static (permanent) or dynamic fashion.

9) The same set of radio channels are used in more than one cell. This is one of the key distinguishing features of cellular communications compared to conventional mobile communications, and is called *frequency reuse*. Cells with identical allocated radio channels are called *co-channel cells*. Reusing an identical radio channel in different cells in the cellular network creates *co-channel interference* (CCI). CCI and adjacent channel interference (ACI) eventually limit the cellular network capacity for a given allocated spectrum.

10) In order to assemble an operational network that may also be connected to other telecommunications networks, several BSs are connected to a Mobile Switching Center (MSC) or Mobile Telephone Switching Office (MTSO). MSC-BS connection is formed through stationary links such as coaxial cables, terrestrial microwave, or fiber optic lines. MSCs are essentially software machines that act like the nerve centers of the network. MSCs are in charge of channel assignment to BSs and MSs, handoff processes, and a host of other protocols. MSCs are connected to other MSCs and to public telecommunication networks such as PSTN and ISDN (integrated services digital network) as shown in Figure 2.3.

11) Signaling data and control information in cellular networks are transmitted through a small set of dedicated radio channels that are separate from traffic channels. This technique is known as *common channel signaling*. In other words, radio channels are divided into *traffic channels* and *control channels*.

12) Smart antenna systems with multiple-input multiple-output (MIMO) configuration are commonly incorporated into modern cellular networks as a standard feature. MIMO techniques improve system throughput and performance at the expense of computational intensity and hardware/software complexity.

It is evident that the cellular concept, supported by frequency reuse notion, represents a major breakthrough in resolving the spectral scarcity and consequently has substantially increased the system capacity with a limited spectrum allocation. In summary, frequency reuse, mobility management, trunking, and the MIMO concept are fundamental ideas that form the foundation of modern wireless networks. We now further investigate key aspects of cellular network, which is the architecture adopted by AeroMACS as well.

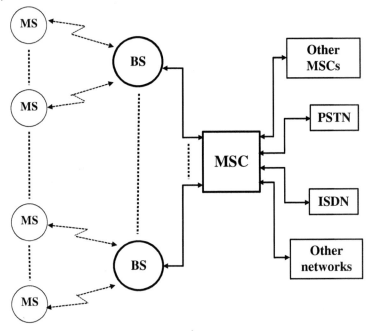

Figure 2.3 Cellular network organization. The solid connections represent fixed links and the broken lines signify wireless mobile links.

2.2.3 Frequency Reuse and Radio Channel Multiplicity

It was argued earlier that the efficient use of electromagnetic spectrum through *frequency reuse* concept is a key feature of cellular networks. At the outset, the coverage area is divided into nonoverlapping hexagonal-shaped cells. The hexagonal cells form a grid in the coverage area with no gaps or overlaps. In some analyses, the cell shape is assumed to be circular for convenience of computation. Circular shapes represent the constant signal-level contour from an omnidirectional antenna located at the center of the cell in a homogeneous propagation environment. The actual cell coverage area, following the deployment of the cellular network, as depicted in Figure 2.2, is the cell footprint. A basic building block of a cellular network is a collection of N cells in a defined pattern that is called a *cluster*. All available duplex radio channels are carefully divided between the N cells of a cluster as either a traffic channel (voice and data) or a control and signaling channel. Clusters are duplicated as many times as needed until the service area is covered. Every time a cluster is added to the network the same radio channels are reused in a different geographical part of the network coverage area, this is the disposition of frequency reuse that allows the same spectrum be used over and over in the cellular network. Figure 2.4 illustrates a service area that is covered by a cellular network of cluster size 3. The cluster is duplicated nine times for complete coverage of the area, which amounts to spectral multiplicity of 9.

2.2.3.1   Co-Channel Reuse Ratio (CCRR), Cluster Size, and Reuse Factor

In this section, we elaborate on how frequency reuse enables dramatic increase in the number of radio channels in the cellular system. Suppose the total number of available duplex channels before reuse is given by M and the cluster size is N, then the number of radio channels allocated for each cell is given by

$$q = \frac{M}{N} \tag{2.1}$$

When a cluster is replicated $\mathfrak{R}$ times to cover a given geographical area, the total number of available radio channels in the network Γ, which may be viewed as a definition for network capacity, is determined by

$$\Gamma = \mathfrak{R} \cdot M = \mathfrak{R} \cdot q \cdot N \tag{2.2}$$

Equation 2.2 expresses a direct relationship between the network capacity and cluster size in classical terrestrial cellular networks. If the cluster size is reduced while cell size is kept constant, more clusters are needed to cover the same area, this increases the network capacity. However, the co-channel cells are placed closer to each other, resulting in higher level of co-channel interference. Clearly, larger cluster size is more desirable in the sense that it reduces the co-channel interference (CCI). Thus, selection of cluster size N involves a trade-off between network capacity and the intensity of CCI.

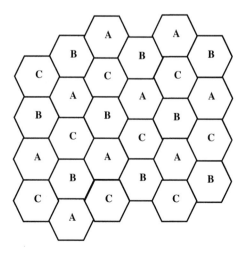

Figure 2.4 A cellular network of cluster size 3. The cluster is duplicated nine times, that is, each radio channel is used nine times in this network, a spectral multiplicity of 9.

The inverse of cluster size *1/N* called *frequency reuse factor* (FRF) or just *reuse factor* is a parameter that determines what fraction of all available radio channels are assigned to each cell. As we will see later, in WiMAX and AeroMACS networks, this parameter may be made adaptive in conjunction with a spectrum reuse technique called *fractional frequency reuse* (FFR).

With the assumption of uniform co-channel interference for all hexagonal-shaped cells in the network, and by invoking Euclidean geometry, the following two properties can be shown to hold true [3].

1) The cluster size N in cellular layout is given by the following equality, where i and j are small positive integers, sometimes referred to as shift parameters.

$$N = i^2 + j^2 + ij \tag{2.3}$$

This implies that only cluster sizes that fit into this equation are legitimate.

2) The distance between the center of two neighboring co-channel cells in a cellular layout D and the radius of a single cell R (center-to-vertex distance) for hexagonal cells are related by

$$D = \sqrt{3N} \cdot R \tag{2.4}$$

The ratio of D to R is an important design parameter and is called *co-channel reuse ratio* (CCRR) sometimes referred to as *reuse distance*.

$$\ell \equiv \frac{D}{R} = \sqrt{3N} \tag{2.5}$$

The four leftmost columns of Table 2.1 present values of ℓ for practical cluster sizes, along with corresponding shift parameters.

2.2.3.2 Signal to Co-Channel Interference Ratio (SIR)

The objective of this section is finding an approximate expression for signal to co-channel interference ratio (SIR) as a function of co-channel reuse ratio ℓ. The approach follows that of Rappaport [4]. CCI is the primary disturbance that places limit on the capacity of cellular networks, therefore, understanding and tracking the quantities of SIR within a conceptual cell boundary is crucial at the initial stages of cellular network design.

It is well known that the average measured received signal strength in a radio channel decays with certain "power law" of transmitter receiver (TR) distance, as long as the receiving antenna is in the *far-field region*[1] of

---

1 Far-field or Fraunhofer region of an antenna is defined by the distance range of $2D^2/\lambda \leq d \leq \infty$ from the transmitting antenna. D is the largest linear dimension of the antenna and λ is the wavelength of the radio wave.

Table 2.1 SIR versus cluster size and CCRR, for $\nu = 4$.

i	j	N	$\ell = D/R$	SIR	SIR (dB)
1	0	1	1.73	0.13	−8.86
1	1	3	3.00	4.79	6.80
2	0	4	3.46	12.32	10.9
2	1	7	4.58	49.44	17
3	0	9	5.20	87.97	19.44
2	2	12	6.00	165.52	22
3	1	13	6.24	197.26	22.95
4	0	16	6.93	310.85	24.92
3	2	19	7.55	449.56	26.53
4	1	21	7.94	557.49	27.46
3	3	27	9.00	947.83	29.77

the transmitting antenna [6]. For instance, signal average power in a free-space propagation environment attenuates proportional to the inverse of square of TR distance. The exponent in this power law is called propagation *path loss exponent* and is shown by ν. For free-space propagation $\nu = 2$, for outdoor land mobile environment path loss exponent is greater than 2. Since the propagation terrain is normally nonhomogeneous, the path loss exponent is not a constant throughout a cell footprint.

To facilitate the approximate computation of SIR, the following assumptions are made.

1) Cells have hexagonal shape.
2) The output power of transmit antenna of all base stations is equal.
3) Only interference from the first-tier co-channel cells is significant, that is, the CCI from second tiers and beyond are assumed to be negligible.
4) The propagation environment is homogeneous, that is, path loss exponent ν has the same value throughout the coverage area.

Recognizing that in a hexagonal-shaped cellular layout there are exactly 6 first-tier co-channel interfering cells, the SIR in the forward channel is given by

$$\frac{S}{I} = \frac{S}{\sum\limits_{k=1}^{6} I_k} \tag{2.6}$$

Here S is the effective received power and I_k is the interfering power from the kth co-channel cell in the first tier. For the worst-case scenario, that is, when mobile moving on the cell boundary, the SIR can be approximated by

$$\frac{S}{I} = \frac{P/\alpha R^{\nu}}{\sum\limits_{k=1}^{6} P/\alpha D_k^{\nu}} = \frac{R^{-\nu}}{\sum\limits_{k=1}^{6} D_k^{-\nu}} = \frac{1}{\sum\limits_{k=1}^{6} \left(D_k/R\right)^{-\nu}} \qquad (2.7)$$

Here P is the output power of the BS antenna, D_k is the distance between the MS and the BS of the kth co-channel cell, and α is the constant of proportionality.

As an example of worst-case approximate computation of SIR, consider Figure 2.4, which illustrates a part of a cellular layout for cluster size $N = 7$. Equation 2.8 calculates an approximate measure for SIR in this scenario [4].

$$\frac{S}{I} = \frac{R^{-\nu}}{2(D-R)^{-\nu} + (D-R/2)^{-\nu} + (D+R/2)^{-\nu} + (D+R)^{-\nu} + D^{-\nu}} \qquad (2.8)$$

As Equation 2.8 indicates, the distances between the MS and various centers of co-channel cells, shown by double-arrowed lines in Figure 2.5, are approximated in terms of D (the distance between neighboring co-channel cell centers) and R (the radius of each cell) as follows. Beginning from the thick solid line and moving to dashed lines counterclockwise, respectively,

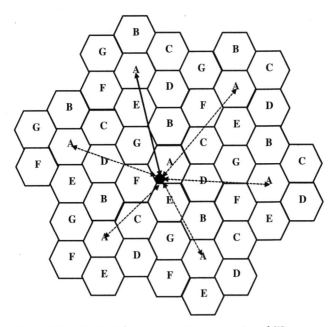

Figure 2.5 A visual aid for approximate computation of SIR.

approximate distances are

$$D + \frac{R}{2}, \quad D, \quad D - R, \quad D - R, \quad D - \frac{R}{2}, \quad D + R$$

The latter equation can be expressed in terms of co-channel reuse ratio ℓ as given by Equation 2.9.

$$\frac{S}{I} = \frac{1}{2(\ell - 1)^{-\nu} + (\ell - 0.5)^{-\nu} + (\ell + 0.5)^{-\nu} + (\ell + 1)^{-\nu} + \ell^{-\nu}} \tag{2.9}$$

Table 2.1 presents approximate computed values of SIR, according to Equation 2.9, in direct and logarithmic scale in its two rightmost columns. These computations are carried out for some practical values of CCRR, with the assumption that path loss exponent $\nu = 4$.

Numerical Example 1
Cellular service is to be provided for a metropolitan district of area of 15,000 km^2. The resources and key cellular network parameters are given in the following list.

1) Total available bandwidth: 27 MHz
2) Required bandwidth for traffic and control channels in each direction: 30 kHz
3) Bandwidth dedicated to control and signaling channels: 3 MHz
4) Cell radius 12 Km
5) Path loss exponent $\nu = 4$

For four cluster sizes of 4, 7, 12, and 19, determine the following:

1) Number of traffic channels per cell
2) Number of control channels per cell
3) The cellular network capacity
4) Worst-case SIR

Preliminary Computations
$a_c = 3\sqrt{3}R_c^2/2 = 3\sqrt{3}(12)^2/2 \cong 374.12$ km^2 Cell area
$n_c = 15000/374.12 \cong 40$ Total number of required cells
$(BW)_{T.C.} = 27 - 3 = 24$ MHz Bandwidth available for traffic channels
$M_T = 24{,}000/2 \times 30 = 400$ Total number of available duplex traffic channels
$M_C = 3000/2 \times 30 = 50$ Total number of available duplex control channels

Solution for N = 4
$q_T = M_T/N = 400/4 = 100$ (Equation 2.1) Total number of available duplex voice channels per cell
$\Re = 40/4 = 10$ Number of clusters
$\Gamma_T = \Re \cdot q_T \cdot N = 10 \times 100 \times 4 = 4000$ (Equation 2.2) Network voice capacity

$\ell \equiv \sqrt{3N} = \sqrt{12} \cong 3.46$ (Equation 2.5) Co-channel reuse ratio

$\text{SIR} \cong 12.32 \cong 10.9$ dB (Equation 2.9 or Table 2.2) worst-case signal to co-channel interference ratio

$n_{cc} = 50/4 \cong 13$ Number of control channels per cell

Solution for N = 7

$q_T = M_T/N = 400/7 \cong 57$ (Equation 2.1) Total number of available duplex traffic channels per cell

$\mathfrak{R} = 40/7 \cong 6$ Number of clusters

$\Gamma_T = \mathfrak{R} \cdot q_T \cdot N \cong 6 \times 57 \times 7 \cong 2394$ (Equation 2.2) Network traffic capacity

$\ell \equiv \sqrt{3N} = \sqrt{21} \cong 4.58$ (Equation 2.5) Co-channel reuse ratio

$\text{SIR} \cong 49.56 \cong 17$ dB (Equation 2.9 or Table 2.2) worst-case signal to co-channel interference ratio

$n_{cc} = 50/7 \cong 7$ Number of control channels per cell

Solution for N = 12

$q_T = M_T/N = 400/12 \cong 34$ (Equation 2.1) Total number of available duplex traffic channels per cell

$\mathfrak{R} = 40/12 \cong 3$ Number of clusters

$\Gamma_T = \mathfrak{R} \cdot q_T \cdot N \cong 3 \times 34 \times 12 \cong 1224$ (Equation 2.2) Network traffic capacity

$\ell \equiv \sqrt{3N} = \sqrt{36} = 6$ (Equation 2.5) Co-channel reuse ratio

$\text{SIR} \cong 165.57 \cong 22$ dB (Equation 2.9 or Table 2.2) worst-case signal to co-channel interference ratio

$n_{cc} = 50/12 \cong 4$ Number of control channels per cell

Solution for N = 19

$q_T = M_T/N = 400/19 \cong 21$ (Equation 2.1) Total number of available duplex traffic channels per cell

$\mathfrak{R} = 40/19 \cong 2$ Number of clusters

$\Gamma_T = \mathfrak{R} \cdot q_T \cdot N \cong 2 \times 21 \times 19 \cong 798$ (Equation 2.2) Network traffic capacity

Table 2.2 The key results of Example 1.

N	Γ_T	SIR(dB)
4	4000	10.9
7	2394	17
12	1224	22
19	798	26.53

$\ell \equiv \sqrt{3N} = \sqrt{57} = 7.55$ (Equation 2.5) Co-channel reuse ratio
$\text{SIR} \cong 449.56 \cong 26.53$ dB (Equation 2.9 or Table 2.2) worst-case signal to
co-channel interference ratio
$n_{cc} = 50/19 \cong 3$ Number of control channels per cell

Results and Conclusions

Table 2.2 summarizes the key results of this example. As the table indicates, if cell size remains fixed, with an increase in cluster size, capacity decreases whereas SIR increases. This example serves to provide a general understanding on relationship between three key parameters of cellular networks, namely, cluster size N, capacity Γ_T (defined as the total number of available traffic channels in the network for a fixed cell size), and SIR.

However, a number of important practical issues remain to be addressed, some of which are listed below.

1) How many users can a network, such as the one in example 1, accommodate (this is also known as network capacity)?

2) What is the overall quality of service (QoS) in this network? We have one indication about the QoS expressed by SIR values, but what is the overall QoS and grade of service (GoS) in this network?

3) Should the cell size be assumed to be fixed at the outset of the cellular network design, or should it be considered a parameter at the disposal of the network designer?

4) How do we determine the actual area in which a given BS provides coverage, that is, how does one determine the cell footprint? The cell footprint depends not only on the transceiver antenna height and output power, but also on the nature of surrounding terrain, that is, how cluttered the area is, whether it is an urban area or suburban area or rural area, whether it is hilly or flat or the combination of the two, and so on.

5) What are the levels of power that are needed for each BS and what is the relationship between the BS output power and cell footprints?

6) Given a cell and its footprint, what percentage of the area inside the cell footprint provides acceptable QoS, and what percentage of the cell footprint is in blackout conditions? In other words, where are the blind spots in the cell footprint?

Some of these questions will be addressed in Section 2.2.4.

2.2.3.3 Channel Allocation

Earlier, it was described that the allocated radio spectrum for a cellular network is initially divided into disjoint radio channels. These radio channels are further distributed between various cells within a cluster.

Radio channels, generically identified as a key transmission resource, may be derived in a variety of ways. One method is direct division of the spectrum into nonoverlapping frequency bands separated by some guard bands, which leads to frequency division multiplexing (FDM) for which the corresponding access technology is referred to as frequency division multiple access (FDMA). Orthogonal frequency division multiplexing (OFDM) and multiple access (OFDMA) are resulted when the selected radio channels are perpendicular in the Fourier domain. Time division multiple access (TDMA) schemes split the system time resource into dedicated "time slots" that are exploited for radio channel formation. Yet another technique of channel creation is joint frequency and time division. When spread-spectrum techniques are employed, where the entire spectrum is used by the MS, code division multiple access (CDMA) is the method that defines radio channels and the access technology.

Channel allocation in cellular networks involves the assignment of radio channels to various cells, with the main objective of optimizing the system spectral efficiency while mitigating the effects of CCI and ACI. Channel allocation schemes may be classified in different ways. We briefly discuss channel allocation algorithms that are directly related to co-channel separation. In this respect, there are three major categories for assigning radio channels to the BSs of various cells: Fixed Channel Allocation (FCA), Dynamic Channel Allocation (DCA), and Hybrid Channel Allocation (HCA).

In FCA, a set of specific radio channels are assigned to each BS for their exclusive use within the corresponding cell area, therefore, there is a one-to-one relationship between the channel sets and the cells. Each set is normally assembled on the basis of minimization of ACI, that is, the maximization of spectral separation between each pair of radio channels. The simplest form of FCA format assumes uniform traffic throughout the network and hence assigns the same number of radio channels to each cell. However, traffic distribution in the network may become temporally or spatially nonuniform. Under these circumstances, the uniform channel allocation may block some users from accessing the network, while there are idle radio channels in the neighboring cells. This clearly results in an inefficient channel utilization. In nonuniform FCA schemes, the number of channels assigned to each cell depends upon their estimated traffic loads. *Channel borrowing schemes* form a popular suite of nonuniform FCA algorithms. The idea is that a cell that has used up its nominal channels may borrow idle channels from the neighboring cells, provided that the borrowed channels do not pose interference to the ongoing communications [7].

In DCA schemes, unlike FCA formats, there is no defined relationship between channels and cells. Instead all radio channels are placed in a common pool of resources. As calls are initiated within various cells, the channels are dynamically assigned to the corresponding BSs in a manner that frequency reuse requirements are not violated and ACI constraints are satisfied. Once a call is completed, the assigned channel returns to the common pool of resources. Channel selection strategy from an available pool of resources sits at the heart; and is the main feature of DCA algorithms. Normally, the "cost" of using each candidate channel is evaluated and the channel that minimizes the cost is selected, provided that CCI and ACI constraints are met. The selection of the cost function is what differentiates DCA algorithms [7,8]. These algorithms are computationally intense and may require large computing resources for real-time operation. This is one of the drawbacks of DCA techniques. DCA algorithms may be classified into two groups. In *call-by-call* DCA strategies, the channel assignment decisions are made solely based on current channel usage situation in the service area. In contrast to that, *adaptive* DCA techniques select channels based on previous as well as present channel usage conditions [9]. Depending upon the type of control protocol they employ, DCA algorithms may also be classified as *centralized* or *distributed* [7].

HCA techniques form the third category of channel allocation protocols and are a mixture of FCA and DCA techniques. Many channel allocation methods have been proposed and implemented that fall within this category. The general idea in HCA is to divide the entire available radio channels into two sets: the fixed channel set and the dynamic channel set. The fixed set contains nominal channels that are assigned to each cell in exactly the same fashion as in FCA schemes. The set of dynamic channels is placed in a common pool to be shared by all users in the network. When a call is initiated in a cell, a search is conducted primarily for one of the cell's nominal channels. In case all associated channels are busy, a channel from dynamic set is assigned to the caller based on any of the standard DCA strategies. An important performance parameter in HCA schemes is the fixed to dynamic channel ratio. This ratio is normally a function of traffic load and varies with time [7].

2.2.4 Erlang Traffic Theory and Cellular Network Design

There are many subscribers in a cellular telephone network, much greater than the number of available duplex radio channels. The same is the case inside any particular cell, that is, in general, there are a large number of MSs moving inside the boundary of a cell footprint compared to the relatively small number of duplex radio channels that are offered by the BS in conjunction with the corresponding cell. Also, in a given area in which cellular services are provided, the Mobile Switching Centers are connected either through small set of wireline circuits or small number of fixed microwave channels to serve a relatively large number of users associated with these MSCs. However, every cell phone needs be connected to other phones only few times a day and for relatively short period of time. This enables the accommodation of large number of users with relatively small number of duplex radio channels.

2.2.4.1 Trunking, Erlang, and Traffic

Like a single trunk of a tree that supports a huge number of branches, *trunking* means organizing a pool of limited resources to accommodating a large number of users. For telecommunications networks, trunking means pulling together a limited number of communication links to support a large number of users. A link that may be temporarily borrowed by any user from a common pool is called a *trunk*. The concept of trunking has been applied in many fields, in particular in transportation and telecommunication. *Trunking theory, or traffic engineering*, is the analytical tool, based on queuing theory, that serves to determine the quantity of the required resources for a given GoS.

In the context of cellular communications, trunking involves placing of a limited number of traffic channels in a common pool to support a large number of subscribers by taking advantage of their statistical behavior. In this manner, a traffic channel that has served a user for short duration of time returns to the common pool to support other subscribers. The management of this process is one of the major tasks of the dedicated control channels in the network. In short, trunking allows a large number of users to share a relatively small pool of available radio channels inside a cell, or between the MSCs, to gain access to the network. Each radio channel in the pool is referred to as a trunk. A key design issue in cellular networks is the determination of the required number of trunks, that is, radio channels, in various cells (as well as the number of links connecting various MSCs) for a desired GoS.

A call that is denied access to the network due to unavailability of resources is said to be *blocked*. A critical design issue is to provide as many trunks as needed in each cell in order to ensure that the probability that a call is blocked, or the probability that a call is delayed in a queue beyond certain time duration before it receives service, is below a certain level at peak traffic time. The traffic engineering, which is based on the Danish mathematician Erlang's statistical developments, provides reasonable solutions related to the quantity of required channels in each cell for a required QoS. We start with defining traffic measurement unit and traffic intensity.

Erlang: Erlang (Er) is the unit for measurement of traffic intensity and is a dimensionless quantity. If a telephone line is totally occupied all the time, that is, the line is busy 60 min in the hour, it is said that the line carries 1 Erlang of traffic. Erlang is used to measure traffic in both wireline and wireless telephone systems. For instance, if a radio channel is occupied, on average, 48 min in the hour, it carries 0.8 Erlang of traffic. Thus, *the traffic through a telecommunication line or channel in Erlang is equal to the average of the fraction of an hour that the line/ channel is occupied by the user.* For instance, if the traffic over a link is 0.3 Erlang, that means on the average the line is occupied by the user 18 min per hour. In other words, Erlang is a measure that corresponds to the average percentage of time occupancy of a link.

Traffic Intensity: Traffic intensity is the expected value of channel time occupancy measured in Erlang. Traffic intensity may be used to measure the average time utilization of a single channel or that of multiple channels. Traffic intensity, or just traffic, in a network may be described as either *offered traffic* or *carried traffic*. Offered traffic, simply, is the traffic requested by users from the cellular network. However, due to resource limitations all requests would not be honored. The portion of the offered traffic that passes through the network is the carried traffic or carried load. Carried traffic may be measured by determining the average number of calls simultaneously in progress during a particular period of time. Traffic intensity is denoted by A.

The statistical behavior of phone service users can be characterized by the following two parameters.

1) *Call Holding Time:* The expected value of time duration of a typical telephone call is branded as call holding time and is denoted by H, which may be expressed in hours, minutes, or seconds.
2) *Call Request Rate, Calling Rate, or Average Call Arrival Rate:* Call request rate is the average number of call requests per unit time, normally an hour, and is denoted by μ. From the perspective of queuing theory, this is the rate at which call request is placed by a single user.

The traffic intensity offered, or requested, by each user is then calculated from the following formula, provided that matched units are used for H and μ.

$$A_{user} = \mu H \quad Er \tag{2.10}$$

Often time μ is expressed in call request rate per hour, and H is given in seconds. Under these conditions, the offered traffic per user is calculated from Equation 2.11.

$$A_{user} = \mu H / 3600 \, Er \tag{2.11}$$

With Ω users in the system the total offered traffic intensity in Erlang is given by Equation 2.12.

$$A = \Omega \cdot A_{user} \tag{2.12}$$

2.2.4.2 The Grade of Service

The GoS is a probabilistic measure of the ability of a user to access the radio channels available in the trunking system during peak traffic hours. The GoS, therefore, is related to the level of congestion in the cellular network. As it was alluded to earlier, the offered (requested) traffic may exceed the maximum carried traffic that can be afforded by the trunking system. Under these circumstances some calls are bound to be blocked, that is, either dropped or placed on a queue. There is a trade-off between the number of available telephone channels in the trunking system and the likelihood of a user's call being blocked at the peak traffic time. The GoS specifying the desired performance of the trunk system is defined by the probability of a call being blocked, or a call being delayed beyond a specified amount of time, at peak traffic time.

Blocking Probability is defined as the expected value of the ratio of the number of blocked calls to total number of calls. In certain call handling strategies, grade of service is defined by blocking probability.

2.2.4.3 Blocked Calls Handling Strategies

There are two general strategies concerning how to handle the blocked calls in cellular telephone networks. In order to facilitate statistical analysis and formulation of call handling strategies, the following assumptions are made.

1) Call arrival time has Poisson distribution

   This implies that the probability that n calls are initiated within time duration of τ is given by

   $$Pr(\# \text{ of calls in } t \text{ unit time} = n) = \frac{e^{-\kappa t}}{n!} (\kappa t)^n \text{ for } n = 0, 1, 2, \ldots \tag{2.13}$$

   Here κ is the dimensionless constant in the Poisson random variable.
2) Calls are made statistically independently.
3) The probability of channel occupancy duration by a user has an exponential distribution, that is, longer calls are less probable than shorter calls by an exponential proportion.
4) There are a finite number of channels available.
5) There is infinite number of users in the system.

Conjectures 1 and 3 suggest realistic probabilistic models for call arrival time and duration. Assumption 2 is reasonable, at least approximately, and the last assumption is necessary if one desires to apply the results of queuing and Erlang theories for various traffic-related computations.

Blocked Calls Cleared Strategy

In this strategy, every user that requests service is given an immediate access to the network. If no radio channel is available, the call is dropped and the user may reinitiate the call later. As such this strategy offers no queuing for call requests. The grade of service in this strategy is defined by probability of blocking. A key issue is the required number of radio channels in a given cell for the network to guarantee a desired level of GoS, which is defined as probability of blocking for this strategy. Based on the assumptions made earlier, Erlang traffic theory is invoked to establish an equation, often translated into a family of curves for quick reference, that expresses traffic versus probability of blocking (GoS) in Blocked Calls Cleared strategy [10].

The offered traffic intensity by each user is given by Equations 2.10 and 2.11. With Ω users in the system, the total offered traffic load, or traffic capacity, is calculated from Equation 2.12. If the number of available traffic channels is designated by C, and the traffic capacity is denoted by A, the probability of blocking, that is, GoS for Blocked Calls Cleared strategy, is given by the *Erlang B formula* as stated in Equation 2.14.

$$Pr(B) = \frac{A^C/C!}{\sum\limits_{k=0}^{C} A^k/k!} \quad Erlang\ B\ formula \qquad (2.14)$$

Here C is the number of available traffic channels, A is the traffic capacity and $Pr(B)$ is the blocking probability which represents GoS for this strategy. For instance, the AMPS (Advanced Mobile Phone Services, analog 1G network in North America) system, which uses "Block Calls Cleared" strategy, is designed for a GoS of 2% blocking.

Blocked Calls Delayed Strategy

In this strategy if a channel is not available to provide service, the blocked call is delayed and placed on a queue until a channel becomes available. Therefore, in contrast with the first strategy, this trunking system offers a queuing arrangement for holding the calls that are blocked. The GoS in this approach is measured by the probability that a call does not receive service after waiting a specific length of time in the queue. In others words, the probability that a call will have to wait for more than, let us say, t seconds, before a radio channel can be made available for it, defines the GoS for this strategy.

It is desired to find a formula that expresses traffic versus probability of blocking for Blocked Calls Delayed strategy, as well. The probability that a call does not have immediate access to the network is given by *Erlang C formula*, as expressed in Equation 2.15.

$$Pr[delay > 0] = \frac{A^C}{A^C + C!\left(1 - \dfrac{A}{C}\right)\sum_{k=0}^{C-1}\dfrac{A^k}{k!}} \quad Erlang\ C\ formula$$

(2.15)

If no channel is available, the probability that the call is forced to wait more than t seconds, that is, the GoS, is given by Equation 2.16

$$\begin{aligned} \text{GoS} &\equiv Pr[delay > t] = Pr[delay > 0]Pr[delay > t|delay > 0] \\ &= Pr[delay > 0]exp[-(C - A)t/H] \end{aligned}$$

(2.16)

One can calculate the average delay δ for all calls as given in Equation 2.17.

$$\delta = Pr[delay > 0]\frac{H}{C - A}$$

(2.17)

The average delay for those calls that are queued is given in Equation 2.18

$$\delta_Q = \frac{H}{C - A}$$

(2.18)

The Erlang B and Erlang C formulas are normally plotted in graphical form, which are used for determination of GoS for either of the strategies. Since the formulas involve three variables, a three-dimensional space is required. However, as it is customary, this three-dimensional function can be represented by a class of two-dimensional curves in which one of the variables is kept constant, that is, as a parameter. In these cases, the number of radio channels is taken as the parameter.

2.2.4.4 Trunking Efficiency

As it was described earlier, trunking is a technique used in cellular systems for the purpose of providing network access for many users by sharing relatively small number of radio channels. Trunking efficiency is a measure related to how many users can be accommodated, at a given GoS, with various trunking configurations. A numerical example at this point serves to better understand the concept of trunking efficiency.

Suppose the Blocked Calls Cleared strategy is used in a single cell of a cellular network. The objective is to compute traffic capacity of the cell for four different GoS levels, assuming ten possible sets of available traffic channels. The computation is based on Erlang B formula, Equation 2.14, and the results are illustrated in Table 2.3.

Table 2.3 Traffic intensity in Erlang offered by Blocked Calls Cleared Strategy.

C	GoS = 0.01	GoS = 0.005	GoS = 0.002	GoS = 0.001
2	0.153	0.105	0.065	0.046
4	0.869	0.701	0.535	0.439
5	1.360	1.130	0.900	0.762
10	4.460	3.960	3.430	3.090
20	12.000	11.100	10.100	9.410
40	29.000	27.300	25.700	24.500
60	46.900	44.750	42.350	40.750
80	65.350	62.650	59.700	57.800
100	84.100	80.900	77.400	75.200

As Equation 2.14 (Erlang B formula) indicates, the relationship between the three quantities C, A, and GoS is highly nonlinear. For instance, if we explore the Table 2.3 column 2, related to GoS = 0.01, by changing the number of channels in trunk from 2 to 4, the traffic capacity changes from 0.153 to 0.869, by a factor of almost 6. Furthermore, if we compare the following two cases

$$\text{GoS} = 0.01 \quad and \quad C = 2 \rightarrow A = 0.531\,Er$$
$$\text{GoS} = 0.01 \quad and \quad C = 100 \rightarrow A = 84.1\,Er$$

Here $0.153 \times 50 = 7.65\,Er$. In other words, 50 trunks each with 2 channels in the trunk carries 7.65 Er of traffic intensity whereas a single trunk with 100 channels in it is able to support 84.1 Er of traffic, this accounts for more than an order of magnitude. The important conclusion is that the *trunking efficiency* of trunks with larger number of channels in them is much higher than that of the trunking arrangements with the same number of channels divided into trunks with small number of channels in them. Therefore, how radio channels are allocated into various trunking configurations is a critical factor in determining network capacity and the number of subscribers that may be supported by the cellular systems.

There are several approaches for quantification of trunking efficiency; we adopt the method given in Ref. [5]. In Blocked Calls Cleared strategy, GoS is the probability of blocking, that is, $[1 - Pr(B)] = [1 - \text{GoS}]$ represents the fraction of the users that receive immediate service from the network. Consequently, the carried traffic load for various trunking configurations with the same number of radio channels in the configuration is calculated from Equation 2.19.

$$A_{carried} = A_T(1 - \text{GoS}) = A_T[1 - Pr(B)] \tag{2.19}$$

Here A_T denotes total offered traffic by the users. The trunking efficiency is defined as the ratio of total carried traffic to the number of traffic channels, as expressed in Equation 2.20.

$$\xi = \frac{A_T}{C}[1 - Pr(B)] = \frac{A_T}{C}[1 - \text{GoS}] \qquad (2.20)$$

Numerical Example 2

Suppose 100 traffic channels are organized in 6 different trunked arrangements for Blocked Calls Cleared strategy, as shown in Table 2.4. The GoS is assumed to be 0.01 for all cases. The table also shows the number of subtrunks, traffic A for each subtrunk, total traffic A_T, and trunking efficiency ξ. The last column of Table 2.4 lists the number of subscribers that can be supported by each trunking configuration for a nominal value of the requested traffic per user of $A_{user} = 0.2$.

This example demonstrates how substantially trunking efficiency alters the network capacity; expressed here as the number of subscribers supported by the network. A trunked system with 100 traffic channels in a single bunch supports significantly more users, in fact by more than one order of magnitude, than a trunked system consisting of 50 subtrunks with two channels in each.

A common scenario in which the capacity is partially lost, on account of trunking efficiency decrease, is when cell sectorization is implemented.

2.2.4.5 Capacity Enhancement through Cell Splitting

A key issue in cellular systems is network capacity, while the most important mechanism through which a large number of radio channels are established in a cellular setting is frequency reuse and channel multiplicity, there are other methods for capacity enhancement. Among these methods are the use of more efficient modulation, adaptive

Table 2.4 Trunking efficiency in Example 2.

Channels/ subtrunk	# of sub trunks	Traffic/Sub trunk (Er)	Total offered traffic: A_T (Er)	Trunking efficiency ξ	# Users for $A_{user} = 0.2$ (Er)
2	50	0.153	7.65	0.075735	38
4	25	0.869	21.725	0.215077	109
5	20	1.36	27.2	0.26928	136
10	10	4.46	44.6	0.44154	223
20	5	12.00	60.00	0.594	300
100	1	84.10	84.10	0.8316.	421

modulation and coding, exploitation of more effective source and speech coding, reduction of cell radius, and so on [11]. In this section and the next we briefly discuss two elementary methods of capacity expansion for cellular networks.

Initially, a cellular network is designed to provide a reliable service to uniformly distributed users in a target service area. As demand for cellular service grows, often times randomly and asymmetrically, the number of users may reach saturation in certain sections of the coverage area. Consequently, in certain cells capacity expansion becomes necessary. One classical remedy for this sort of traffic congestion is to subdivide the congested cell area into two or more zones to form new cells. Traditionally, a start-up cell is first split into four cells, and the process may continue until other network consideration prohibits further divisions. This process that is called *cell splitting*, is one of the standard methods of enhancing cellular network capacity. Each new cell has its own base station with adjusted (lowered) transmitting antenna output power, and often reduced antenna height. The reduction in the output power is necessary in order to maintain an acceptable level of interference to co-channel and adjacent cells [3,12]. An example of cell splitting is demonstrated in Figure 2.6. The Figure shows a cellular layout with 7 cell clusters. In this example, a congested cell, cell A is in the center of the layout and parts of its surrounding cells are divided into seven new smaller cells.

Figure 2.6 illustrates that the cellular network capacity has increased as a result of the fact that the higher channel multiplicity is ensued. Hence, cell splitting, the process of subdividing a traffic-congested cell into smaller cells, leads to capacity enhancement for the overloaded cell and its proximity. The implementation of cell splitting comes with its own "price tag" and disadvantages. Primarily, the network expenditures are increased since new base stations must be installed and additional equipment becomes required. In this regard, it is imperative that a cost–benefit analysis be conducted to compare overall cost of cell splitting versus other available alternatives for handling the increased traffic in the cell. Second, decreasing cell radii implies that cell boundaries will be crossed more often. This results in more handoffs per call and a higher processing per subscriber. It can be shown that a reduction in cell radius by a factor of four will increase the handoff rate per subscriber by about tenfold [13]. Moreover, since after cell splitting the network will support higher traffic, the call processing load will increase as well. It is well known that call processing load increases geometrically with number of subscriber [14].

There are two methods of cell splitting: *static cell splitting* and *dynamic cell splitting*. Our focus is on static cell splitting that is the process of

(a)

(b)

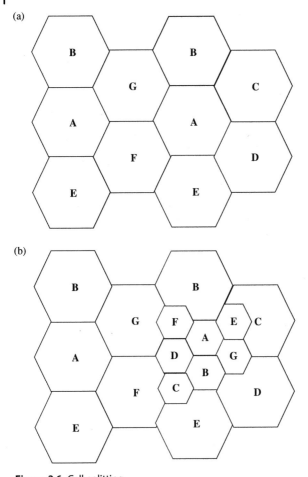

Figure 2.6 Cell splitting.

permanently dividing an existing cell into smaller cells, such that the new created cells can meet the growing capacity demands in the original parent cell. Since the radius and the coverage area of the created cells are reduced, it is conceivable that the original macrocells may be converted to microcells. Under these circumstances, techniques such as *antenna down tilting* may be used to limit the radio coverage to the microcell area. Antenna down tilting essentially focuses the radiated electromagnetic energy of the BS transmit antenna toward the ground rather than toward the horizon [4]. This technique has also been suggested as a method of combating AeroMACS interference into mobile satellite system (Globalstar) that shares the same operating spectrum with AeroMACS (see Chapter 8).

From the network design perspective, cell splitting provides a design trade-off between cellular network capacity, on the one hand, and MSC data processing intensity and network cost, on the other hand.

2.2.4.6 Capacity Enhancement via Sectorization

An alternative method for capacity improvement (or equivalently reduction of CCI) is sectorization. Sectorization is essentially a method of enhancing SIR without changing the cluster size. On the other hand, sectorization may be applied to enhance network capacity, while keeping SIR fixed. In this technique, instead of using omnidirectional antenna at the center of the cell, directional antennas are used for the base stations. Each antenna covers a sector of the cell. Consequently, the CCI is reduced as the effective number of first-tier co-channel interferer is reduced, as will be discussed later. In classical cellular networks, two sectorization configurations are commonly used, as described further.

- *Three-Sector Format:* It is an arrangement in which a cell is divided into three sectors and the set of allocated radio channels for the cell is divided between these sectors. Each sector has a directional antenna that radiates over a space angle of $120°$. Figure 2.7 shows part of a three-sector cellular network of cluster of size seven. Sectorization is shown only for cell A in the center of the layout and its first-tier co-channel cells. Consider sector 1 of cell A in the center of Figure 2.7. The directional antenna that feeds this sector radiates electromagnetic wave over the $120°$ space angle shown for this sector. Same is the case for all other sector 1 in the network. As a consequence, and as Figure 2.7 demonstrates, sector 1 of cell A in the center of the figure is exposed to CCI only from two sectors located in cells A_1 and A_2 on the left of cell A. This is clearly the case for other sectors as well. Therefore, the three-sector configuration reduces the number of CCI sources from six to two, leading to higher value for SIR. An approximate calculation shows that a gain of 7 dB in SIR can be achieved through conversion to three-sector configuration relative to the nonsectored network [15]. This gain can be translated into capacity improvement, further reduction in the cell size and/or cluster size, while maintaining SIR fixed.
- *Six-Sector Configuration:* In this format radio channels allocated to a cell are divided in six groups, one for each sector. Each sector has a directional antenna that forms a radiation beam over a space angle of $60°$. A similar argument may be made for the case of six-sector format and demonstrate that the number of CCI sources is further reduced to only one. An approximate calculation may be conducted to show that an additional 5 dB gain in SIR is achieved relative to the three-sector

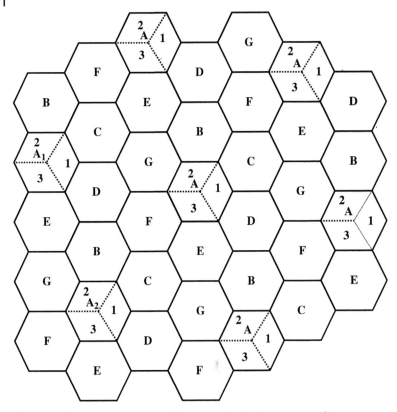

Figure 2.7 A seven-cell-cluster cellular network with three-sector split.

format [15]. Once again, this gain may be translated into various network enhancements, among them capacity increase.

Sectorization is classified as a standard elementary technique for capacity enhancement in the cellular network. However, several concerns about sectorization process need to be pointed out at this juncture. Primarily, since the allocated radio channels for a cell are divided into a group of three or six channel subsets, the available number of radio channels for each sector is dramatically reduced, leading to a large decline in trunking efficiency that has an adversarial effect on capacity increase in the network. Second, we have been implicitly advocating the use of ideal directional antennas. Practical antennas cannot form focused beams that radiates electromagnetic waves into a perfect 120° or 60° sectors, there are always sidelobes associated with the main lobe. The sidelobes introduce additional interference into sectors. More importantly, the breaking up of cells into sectors means that mobile stations, in addition to crossing cell borders, may have to move from one sector to another sector of the same cell or the neighboring cell, while a call is in progress. This requires more hand off procedures be initiated in the network, and therefore more data processing load is imposed on the MSCs and BSs. To alleviate the computation and data processing intensity in a sectored network, the concept of "microcell zone" has been proposed [4].

The traditional sectorization idea discussed so far has been applied in classical as well as modern cellular networks, such as WiMAX, Aero-MACS, and LTE-based cellular networks. However, more recently, higher order sectorization (beyond three-sector and six-sector configurations) has been explored as an alternative to MIMO technology for enhancing system spectral efficiency. It has been demonstrated that sectoring a cellular base station antenna radially, into multiple (up to 12) sectors, assuming the ideal case that there is no intersector interference, the mean throughput of the BS increases proportional to the number of sectors [16].

2.3 Cellular Radio Channel Characterization

Radio channel characterization is, perhaps, the prime challenge in analysis and design of wireless communication systems. At the outset, it should be pointed out that throughout this text our focus will be on the outdoor wireless channel that is more relevant to the case of AeroMACS networks. Unlike wireline channels that are assumed, by and large, to be linear, time invariant, and stationary, wireless cellular channels are time varying, dynamic, unpredictable, and are plagued with a number of interferences and other limiting factors. In other words, the mobile communications link in the framework of the cellular architecture suffers from a number of channel impairments and interferences, as described in the following sections. For AeroMACS networks, additional challenges are brought about by airport surface channel peculiarities and C-band signaling.

2.3.1 Cellular Link Impairments

A signal traveling across a cellular mobile communications link, experiences a number of channel impairments:

1) **Interference**
   - *Self-Interference:* The receiver receives several copies of the same signal with different magnitudes, phases, and latencies; this is known as *multipath effect*, which is the main culprit in causing "short term fading."

- *Co-Channel Interference:* As it was discussed earlier, this is inter-ference from the same radio channels that are being used in other cells in the network. Co-channel interference defines the ultimate limitation on the classical cellular network capacity.
- *Adjacent-Channel Interference:* Spectrally neighboring radio chan-nels leak energy into each other's passband. Proper spectrum planning can minimize the effect of adjacent channel interference.

2) **Doppler Frequency Shift and Frequency Spread**

Motion causes frequency shift in the received signal. Different echoes of the signal may be subject to different Doppler frequency shift. This creates Doppler frequency spread that may affect the signal in the form of an FM noise.

3) **Long-Term Attenuation: Path Loss**

As the MS moves away from the BS, the mean value of the signal power is reduced. This is clearly due to distance between the mobile user and the base station. The long-term signal decay is not only a function of TR distance but also is directly proportional to the signal frequency. This suggests that path loss is much more intense in the C-band, over which AeroMACS operates, than it is in the VHF band within which the legacy aeronautical communications systems func-tion. Path loss is also known as *slow fading.*

4) **Large-Scale Fading (Medium-Term Attenuation): Shadowing and Foliage**

Unlike the long-term attenuation, which is essentially a function of MS–BS distance and frequency, large-scale fading is influenced by terrain profile, vegetation density and nature, and environmental disposition (urban, suburban, rural), through which the radio signal travels. Consequently, medium-term attenuations are associated with phenomena such as *shadowing* and *foliage.* The characterization of this type of fading leads to some important information about cell coverage area, including the estimation of the percentage of covered area and the percentage of areas with inadequate coverage (blind spots) within a cell footprint.

5) **Small-Scale Fading (Short-Term Attenuation)**

Small-scale fading is the rapid fluctuation of the envelope of a signal over a short period of time or short travel distance. The short period of time or short travel distance is defined such that the path loss due to propagation and medium-scale attenuation owing to shadowing can be neglected over the considered time period and/or distance. Major factors influencing short-term fading are multipath interference, speed of the mobile station, speed of the surrounding objects, and the transmission bandwidth of the signal. Small-scale fading is also known as *short term fading* but most commonly just *fading* [4].

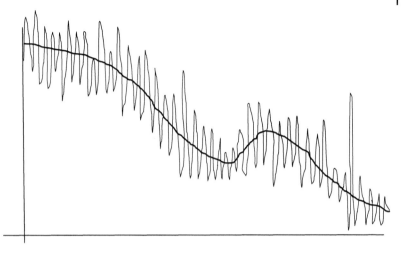

Figure 2.8 An exaggerated plot for radio signal envelope that illustrates the three types of signal fluctuations in a wireless mobile channel. The middle line shows the combine effect of long-term and medium-term attenuations around which short-term fast variation takes place.

Figure 2.8 is an attempt to portray the three types of signal envelope attenuation observed in wireless mobile channels.

6) **Noise**

In addition to thermal (Johnson) noise, signals in radio channels may be afflicted by other natural and man-made noise. In broadband systems, such as AeroMACS networks, noise can be a major problem in the sense that the system noise floor is raised as a result of wideband signal reception.

Figure 2.9 illustrates various impairments associated with the cellular mobile channel, along with characterization of the resulting harmful effects for most of these impairments.

2.3.2 Path Loss Computation and Estimation

In order to properly plan, design, and deploy a wireless mobile communications network, one needs to have an educated estimation about the signal-to-noise ratio (SNR) as a function of TR distance in the area on which the network is to be laid out. Propagation and path loss models are used extensively to provide some estimation of the range of SNR over various parts of the service area. More specifically, path loss models are used when conducting feasibility studies, network planning, interference prediction, network coverage range, and finally network deployment.

The path loss determining models may be classified into categories of deterministic, empirical, stochastic, and the combination of thereof. The

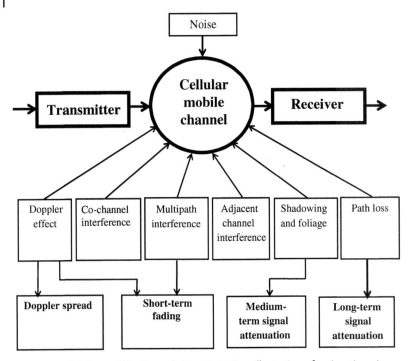

Figure 2.9 Cellular mobile channel characterization: illustration of various impairments and their effects on the radio signal.

deterministic models make use of the laws of physics and electromagnetic theory to calculate the radio signal power at a given locale. Free-space propagation model and ray tracing techniques are examples of this category. The double-ray ground reflection model provides an approximate deterministic path loss computation technique, which has a ray tracing flavor to it as well. Empirical models are exclusively based on measurement and data collection. However, in most practical cases one has to rely on computer simulation or measurement-based empirical models for estimation of path loss. The motivation behind developing various path loss models is to enable the network designer to have access to proper tools for designing cost-effective networks that meet or exceed the quality of service that the client requires.

Okumura–Hata model is the most widely applied empirical path loss estimator. Stochastic models involve casting the propagation environment into a series of random variables. These models are the least accurate, however, they require minimum amount of information about the environment to generate path loss predictions [17].

The propagation and attenuation of radio waves in various environments are strongly influenced by area characteristics such as density, height, location and size of buildings, density and type of vegetation, terrain variations, proximity to large bodies of water, and so on. However, both deterministic and measurement-based empirical models have indicated that the signal power in a radio channel decreases logarithmically with distance. Path loss may be defined, in a broad sense, as the ratio of the transmitted signal power to the received signal power expressed in decibels, that is, $P\ell = 10\log_{10}(P_t/P_r)$. Owing to the fact that the propagation environment is seldom homogeneous, path loss is a quantity that is highly direction sensitive. In other words, path loss for a fixed distance from the BS antenna, but in different directions, is a random variable. For this reason, empirical models for path loss computation provide either median or mean value for path loss. The expected value of this random variable is expressed by Equation 2.21 [4].

$$E\left[P\ell(d)_{\text{dB}}\right] = E\left[P\ell(d_0)_{\text{dB}}\right] + 10\nu\log_{10}(d/d_0) \qquad (2.21)$$

Here $E[P\ell(d)]$ is the expected value of the path loss at TR distance d, ν is *path loss exponent*, and $E[P\ell(d_0)]$ is the expected value of path loss at distance d_0 from the BS antenna. d_0 is the so-called *close-in reference distance*, which is determined from measurements close to the transmitter (1–10 m for indoor and 10–100 m for outdoor cases) but in the far-field region of the transmit antenna. The value of the path loss exponent ν is environment dependent. For free-space LOS propagation, ν is 2 and when obstructions are present, ν can assume values in the range of 2–6 [4].

2.3.2.1 Free-Space Propagation and Friis Formula

The simplest path loss model is corresponding to the case in which the signal travels along an interference-free unobstructed line-of-sight (LOS) path from the transmit antenna to the receive antenna, that is, the free-space propagation scenario. In this case the precise value of the path loss may be calculated from a simple deterministic equation, known in physics as Friis Formula, expressed in Equation 2.22.

$$P_r(d) = \frac{g_r g_t \lambda^2}{(4\pi)^2 d^2} P_t \qquad (2.22)$$

Here d is the TR distance in meters, g_r and g_t are directional gain of receiver and transmitter antennas, respectively, λ is the wavelength of the carrier signal in meters, P_r and P_t are received and transmitted power, respectively, in Watts. Hardware and filter losses have been ignored in this equation and if existed should be accounted for separately. Furthermore, it should be noted that this is a valid equation only if distance d falls in the "far-field" region of the transmitting antenna. Cases in which free-space model may be applied includes satellite repeater-to-earth station link and line-of-sight terrestrial microwave transmission systems.

Path loss for this deterministic model is computed from Equation 2.22, as expressed in Equation 2.23.

$$P\ell = 10 \log_{10}^{P_t} - 10 \log_{10}^{P_r} = 20 \log_{10}^{4\pi} + 20 \log_{10}^{d} - 20 \log_{10}^{\lambda} - 10 \log_{10}^{g_r}$$
$$-10 \log_{10}^{g_t} \cong 22 + 20 \log_{10}^{d} - 10 \left(\log_{10}^{g_r} + \log_{10}^{g_t} \right)$$

$$(2.23)$$

If the Friis equation is expressed in terms of frequency, that is, $P_r(d) = g_r g_t c^2 / (4\pi d f)^2$, the free-space path loss may be expressed in the form given in Equation 2.24 [12].

$$P\ell = 20 \log_{10} f + 20 \log_{10} d - 10 \log_{10} g_t - 10 \log_{10} g_r - 147.6 \, \text{dB}$$

$$(2.24)$$

Note that Equation 2.23 ensures that path loss is expressed as a positive quantity by defining it as $10 \log_{10}^{P_t} - 10 \log_{10}^{P_r}$ and not as $10 \log_{10}^{P_r} - 10 \log_{10}^{P_t}$.

2.3.2.2 The Key Mechanisms Affecting Radio Wave Propagation

In the majority of land mobile radio networks, including AeroMACS, LOS component may not even reach the receiver antenna, and even if it does, it may not be the only ray arriving at the receiver. In the absence of LOS ray, the propagation mechanisms, in particular, reflection, diffraction, and scattering, provide means for radio communications over the wireless channel.

Reflection occurs when a traveling electromagnetic wave collides with an object whose dimensions are much larger than the wavelength of the radio wave. In land mobile radio networks, reflections off the surface of the earth, buildings, and stationary and moving objects produce additional echoes that may reach the receiving antenna and create the "multipath effect" that may have constructive or destructive effect on the received signal. In general, when an electromagnetic wave impinges upon another medium with different dielectric constant, part of the wave's energy transmits through and part of it is reflected back. Proportion of transmitted and reflected energies depend upon medium's electromagnetic properties. For instance, if a plain wave in air collides with a perfect conductor perpendicularly, the entire energy of the radio signal is reflected back and no energy will be absorbed by the conductor.

Diffraction happens when radio waves strike a large obstructing object that has sharp edges. Diffraction, in effect, is the bending of the electromagnetic waves around sharp edges of the colliding object. As a result, secondary waves are generated that radiate throughout the space and even behind the obstruction, producing diffracted echoes that infiltrate into the shadowed areas created by the obstruction. Often times the diffracted copies of the radio wave provides a strong enough signal for MS in shadowed areas. Theoretically, the phenomenon of diffraction can be explained by the Huygens–Fresnel diffraction principle. Huygens–Fresnel principle states that all points on a wave front can be considered as point sources for the production of secondary wavelets, and these wavelets combine to produce a new wave front in the direction of propagation. Estimation of signal attenuation caused by diffraction is mathematically very difficult, if not impossible. The general method for the estimation of diffraction loss is to make theoretical approximation modified by necessary empirical corrections. In simple cases in which a single obstruction (such as a building or a hilltop) is involved, the attenuation caused by diffraction can be estimated by treating the obstruction as a diffracting knife-edge. For more complicated scenarios, multiple knife-edge diffraction models may be used. Approximate equations are developed based on knife-edge diffraction model that leads to estimation of attenuation caused by diffraction [4].

Scattering takes place when electromagnetic waves collide with objects that are smaller than the wavelength of the radio wave. Consequently, the reflected electromagnetic energy is spread out in all directions. Common objects that cause scattering in wireless channels are trees, lampposts, traffic signs, traffic lights, and so on. Since scattering spreads the signal energy in multiple directions, it can provide additional radio energy at a receiver, and in some occasions, it may be the only mechanism that brings the signal to the MS receiving antenna. Flat surfaces that have much larger dimensions than a wavelength may be modeled as reflective surfaces; however, the roughness of such surfaces often induces propagation effects that are different from that of just reflection. This can be explained by observing that the surface not only has a reflective effect but it also has a scattering impact [4].

Most radio propagation models that are used in practice to predict large-scale path loss are based on the three mechanisms of reflection, diffraction, and scattering. Although empirical models, to be discussed next, are based on measured data, curve fitting, analytical expressions, or the combination thereof, they indirectly account for reflections, diffractions, and scatterings that take place in the wireless channel.

In summary, the MS antenna, in general, receives several copies of the transmitted waveform that may or may not include the LOS component. This is the multipath propagation, which is caused by a large number of reflection, diffraction, and scattering processes in the wireless channel. Clearly, it becomes too complicated to account for the effect of each individual occurrence on the received signal. The most important sought-after parameter is the received signal power that determines the range of BS antenna for an acceptable QoS. Probabilistic description of the range of values assumed by the received signal power and empirical methods that estimate the statistical averages of this parameter are all that the wireless network planners and designers have at their disposal.

2.3.2.3 The Ray Tracing Technique

Ray tracing method, a deterministic path loss model, predicts radio signal strength based on site-specific propagation information. Ray tracing adopts a geometric optics approach in the sense that it considers various paths that the radio signal possibly can take from the transmitter to the receiver as if they were rays of light. As such, ray tracing technique incorporates all signal components that reach the receiver directly, or through reflections, diffraction, and scattering, for calculation of the received signal envelope magnitude and power. Clearly, this model requires detailed information about the physical wireless channel of the site and the predicted results may not be applicable to other locations [18].

The majority of computer simulation packages, used for wireless channel modeling for indoor environments, make use of ray tracing method for path loss prediction. A number of models may be created from the ray tracing concept. The well-known double-ray ground reflection model in which two paths exist between the transmitter and receiver, that is, a direct LOS path and a ground reflected path, presents an example of application of ray tracing technique, which is discussed in some detail in the next section.

2.3.2.4 Ground Reflection and Double-Ray Model

The free-space propagation along a line-of-sight path, although acceptable for few cases, it is not realistic for mobile channels of any aeronautical domain, perhaps, with the exception of oceanic domain in which satellite links are used for communications. A double-ray ground reflection model provides a more accurate and closer to reality description of propagation path loss over some aeronautical domains.

This model, which represents an example of geometric optics-based ray tracing technique, assumes two paths for signal propagation across the wireless channel. The first is a line-of-sight path that directly connects transmit and receive antennas. The second ray corresponds to signal reflection bouncing off the ground surface and reaching the receiver antenna. This model has been found reasonably accurate for several scenarios, including air-to-ground domain of aeronautical channel, that is, the pilot to control tower link when the aircraft is away from the airport, although more often than not, Rayleigh fading model has been used for characterization of this link.

Let us denote the distance between the bases of transmit and receive antennas by d_0. Assume that d_0 is short enough to presume that the earth surface remains flat. Figure 2.10 illustrates this model with various distances and heights. The height of transmit and receive antennas are designated as h_t and h_r, the length of the LOS path is d, and d_1 and d_2 are the distances between ground reflection point and transmitter and receiver antennas, respectively.

Since the principals of Cartesian Optics are applied, a number of statements may be made regarding this propagation model. For instance, Snell's law is applicable to maintain that incident angle θ_i is equal to reflection angle θ_r. The objective is to find an equation similar to Friis equation for double-ray ground reflection model. In order to facilitate this computation, the following reasonable assumptions are made that lead to a simplified approximate mathematical expression for path loss attenuation for this ray tracing model.

1) The distance d is much greater than the height of transmitter and receiver antennas, that is, $d \gg h_t$, $d \gg h_r$.
2) The incident angle θ_i and the reflection angles θ_r are very small.
3) $d_1 + d_2 \cong d$
4) Generally speaking, the earth is not a "perfect conductor" but a good conductor. The implication is that most of the energy of the incident wave is reflected and a very small fraction of it is transmitted through the earth [19]. However, it is assumed that the earth is a perfect conductor and the entire energy of the incident wave is reflected off the surface of the earth. The assumption of the perfect reflecting conductor also implies that the phase of the reflected signal is changed by 180°. It is noted that if the reflecting surface is a body of water, such

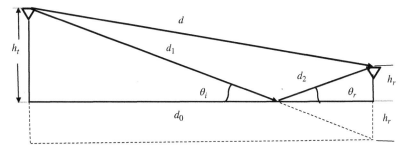

Figure 2.10 The ground reflection double-ray propagation model. This model is based on ray tracing technique that follows the principle of geometric optics. The additional geometry shown in the figure is for understanding of the analysis that ensues.

as lakes and oceans, significant portion of the signal energy is lost to the reflecting surface and the conjecture of perfect conductor is not valid.

With these approximations in mind, we first establish the following relationships based on geometry shown in the Figure 2.10.

$$d \cong \left[d_0^2 + (h_t - h_r)^2\right]^{1/2} \tag{2.25}$$

$$d_1 + d_2 \cong \left[d_0^2 + (h_t + h_r)^2\right]^{1/2} \tag{2.26}$$

In order to obtain an approximate value for the path length difference of the LOS ray and the ground reflection ray, we expand Equations 2.25 and 2.26 into their Taylor series and approximate each series with the first two terms by virtue of the assumption 1 listed already.

$$d \cong d_0 \left[1 + \frac{(h_t - h_r)^2}{d_0^2}\right]^{1/2} \cong d_0 \left[1 + \frac{(h_t - h_r)^2}{2d_0^2}\right] \tag{2.27}$$

$$d_1 + d_2 \cong d_0 \left[1 + \frac{(h_t + h_r)^2}{d_0^2}\right]^{1/2} \cong d_0 \left[1 + \frac{(h_t + h_r)^2}{2d_0^2}\right] \tag{2.28}$$

$$\Delta d \triangleq (d_1 + d_2) - (d) \cong \frac{2h_t h_r}{d_0} \tag{2.29}$$

For the simplicity of the analysis we assume that a single-tone signal, in the form of $s(t) = A \cos(2\pi f_c t)$, is transmitted. The received double-path signal components may, therefore, be expressed as given in Equations 2.30 and 2.31.

$$r_{\text{LOS}}(t) = A_{\text{LOS}} \cos\left[2\pi f_c(t - d/c)\right] \tag{2.30}$$

$$r_{\text{GR}}(t) = A_{\text{GR}} \cos\left[2\pi f_c\{t - (d_1 + d_2)/c\} - 180°\right] \tag{2.31}$$

In these equations c is the speed of electromagnetic waves in free space. The double-path signal received at the front end of the MS/SS antenna (the sum of the two signals) is given by Equation 2.32.

$$r(t) = r_{\text{LOS}}(t) + r_{\text{GR}}(t) = A_{\text{LOS}} \cos\left[2\pi f_c(t - d/c)\right] \\ + A_{\text{GR}} \cos\left[2\pi f_c\{t - (d_1 + d_2)/c\} - 180°\right] \equiv A_r \cos(2\pi f_c t + \phi) \tag{2.32}$$

The Friis equation indicates that in free-space propagation the path loss exponent is equal to 2, therefore, the signal power decays proportional to the square of the TR distance. In the case of sinusoidal signals that translates to a signal amplitude drop-off that is inversely related to TR distance. In other words, the amplitudes of the two components at the receive antenna may be determined by Equation 2.33.

$$A_{\text{LOS}} = \frac{A}{d}, \quad A_{\text{GR}} = \frac{A}{d_1 + d_2} \tag{2.33}$$

Recall that under the assumptions that were made, the ground reflection ray incurs a phase change of $-180°$ but otherwise preserves its entire strength at the reflection point. This ray essentially propagates through the free space and traverses a path that is longer than that of the LOS ray, therefore, it is justified to assume that the amplitude of this ray at the receiving antenna be given in accordance with Equation 2.33. We now take advantage of the second assumption, that is, $d \cong d_1 + d_2$, to further simplify Equation 2.32:

$$r(t) \cong \frac{A}{d} \left\{ \cos\left[2\pi f_c\left(t - \frac{d}{c}\right)\right] - \cos\left[2\pi f_c\left(t - \frac{d_1 + d_2}{c}\right)\right]\right\} \tag{2.34}$$

Here we have also used the trig identity: $\cos(\alpha - 180°) = -\cos(\alpha)$. Applying the trig identity $\cos(\alpha) - \cos(\beta) = 2\sin\left(\frac{\alpha+\beta}{2}\right)\sin\left(\frac{\beta-\alpha}{2}\right)$ in Equation 2.34

$$r(t) = \frac{2A}{d}\sin\left[\pi f_c\left(\frac{d_1 + d_2}{c} - \frac{d}{c}\right)\right]\sin\left[2\pi f_c t - 2\pi f_c \frac{d}{c} + \pi f_c \frac{d_1 + d_2}{c}\right] \tag{2.35}$$

Substituting Equation 2.29 for $(d_1 + d_2) - d$ into Equation 2.35 and using the trig identity $\sin(\alpha) = \cos(\alpha - 90°)$, Equation 2.35 can be put into the following form:

$$r(t) = \frac{2A}{d}\sin\left(\frac{2\pi f_c h_t h_r}{cd_0}\right)\cos\left(2\pi f_c t - \pi f_c \frac{d}{c} + \pi f_c \frac{d_1 + d_2}{2} - 90°\right) \tag{2.36}$$

Comparing Equation 2.36 with the general form of the received signal $r(t) = A_r \cos(2\pi f_c + \phi)$, it is concluded that

$$A_r = \frac{2A}{d}\sin\left(\frac{2\pi f_c h_t h_r}{cd_0}\right) = \frac{2A}{d}\sin\left(\frac{2\pi h_t h_r}{\lambda d_0}\right) \tag{2.37}$$

The received power can, therefore, be approximated by

$$P_r = \frac{A_r^2}{2} = \frac{2A^2}{d^2}\sin^2\left(\frac{2\pi h_t h_r}{\lambda d_0}\right) \tag{2.38}$$

We next find an equation that expresses A, the amplitude of the transmitted single-tone signal, in terms of other parameters. Noting that if we consider just the LOS component, the power of the received signal would be calculated from the Friis equation, that is,

$$P_{\text{LOS}} = \frac{P_t g_t g_r \lambda^2}{(4\pi d)^2} \quad (L = 1) \tag{2.39}$$

On the other hand, we know that $P_{LOS} = 1/2(A/d)^2$, equating this expression with Equation 2.39, A can be calculated from Equation 2.40.

$$A^2 = \frac{2P_t g_t g_r \lambda^2}{(4\pi)^2} \quad (2.40)$$

Replacing Equation 2.40 into Equation 2.38 we obtain a general approximate equation for received power for the double-ray ground reflection model.

$$P_r = \frac{4P_t g_t g_r \lambda^2}{(4\pi d)^2} \sin^2 \left(\frac{2\pi h_t h_r}{\lambda d_0} \right) \quad (2.41)$$

To appreciate the implications of the latter equation, we put the equation into a more familiar form by making two additional reasonable approximations.

$$d_0 \cong d \gg \frac{2\pi h_t h_r}{\lambda} \quad \Rightarrow \quad \sin \left(\frac{2\pi h_t h_r}{\lambda d_0} \right) \cong \sin \left(\frac{2\pi h_t h_r}{\lambda d} \right) \cong \frac{2\pi h_t h_r}{\lambda d} \quad (2.42)$$

Replacing the latter equation into Equation 2.41, we conclude

$$P_r \cong \frac{P_t g_t g_r h_t^2 h_r^2}{d^4} \quad (2.43)$$

Sometimes Equation 2.41 is referred to as "exact equation" (even though it is an approximation) for calculation of the received power for the double-ray ground reflection model, and Equation 2.43 is dubbed as approximate equation [4].

Comparing Equations 2.42 and 2.22, the following important observations, regarding the path loss for the double-ray model, are made.

1) The path loss exponent has risen to four for the double-ray model. In other words, the received power (path loss) decays (increases) inversely proportional to distance raised to power four.

2) The received power (path loss) is independent from frequency (wavelength) of the electromagnetic wave. Equation 2.42 is an approximation, nevertheless, it implies no dependency or very weak dependency of path loss on wavelength of the electromagnetic wave.

3) The height of receiver and transmitter antennas play direct roles in the intensity of the received power, whereas in Friis equation for free-space propagation indicates that the received power is independent from these parameters.

Therefore, for long distance, that is, $d \gg h_t + h_r$ or $d \gg \sqrt{h_t h_r}$, the received power decays in accordance with the fourth power law. Referring to Equation 2.42, the term $d^4/h_r^2 h_t^2$ represents the actual path loss for the double-ray ground reflection model. Equation 2.44 shows this path loss in decibel.

$$(P\ell)_{DR} = 10 \log_{10}\left(\frac{d^4}{h_r^2 h_t^2}\right) = 40 \log_{10}d - 20 \log_{10}h_r - 20 \log_{10}h_t \quad (2.44)$$

2.3.2.5 Empirical Techniques for Path Loss (Large-Scale Attenuation) Estimation

We have explored two possible techniques for computation/estimation of long-term signal attenuation over the wireless channel. The first model corresponds to single-path communication between a receiver and a transmitter in the LOS mode. The second is a double-path model consisting of a LOS path and a "ground reflection" path. In outdoor land mobile communication systems, the signal can arrive at the receiving antenna through several propagation paths that may or may not include the LOS path. Moreover, signal energy is absorbed by propagation environment that includes large number of man-made and natural objects. The aggregate of multipath phenomena, terrain profile, and other physical effects imposes a significant attenuation on the traveling electromagnetic wave that cannot be predicted by the simple models that we have discussed so far. Standard empirical statistical methods are available that may be invoked to provide some estimation for path loss over a given propagation environment.

It is important to realize that there is not a single model that is universally applicable to all situations. In other words, a particular model might be appropriate for application in a certain environment and not suitable for others. In general, the accuracy of a particular model in a given service area depends on the fit between the parameters required by the model and those of the given case. Furthermore, all of these path loss estimation techniques have frequency restriction, that is, they are valid only over a defined range of frequencies.

Various path loss predicting models have limited range of applicability not only with respect to frequency, but also with respect to distance, terrain profile, environmental nature, and so on, since they only provide approximate values. Nevertheless, they are important tools that could determine initial network layout and provide insight into how the proposed system would roughly perform in a given service area, and how much equipment is required for the dimensioning and planning of the radio network.[2]

---

2 Dimensioning is the process of estimating the required equipment and number of base stations for providing coverage and adequate capacity for the target service area.

2.3.2.6 Okumura–Hata Model for Outdoor Median Path Loss Estimation

Perhaps, the most widely applied methods for path loss (long-term attenuation) estimation of radio signals in the urban area are the technique proposed by Okumura and the schemes that were later developed based on his initiative [20]. Okumura model has been constructed upon an extensive collection of measured data in the metropolitan Toyo over the VHF and UHF bands, extended from 150 to 1920 MHz. This model that estimates the median path loss in large cells was later extrapolated for up to 3000 MHz. Okumura model calculates the free-space loss between any two points, assuming BS antenna height of 200 m and MS antenna height of 3 m, and subsequently adds extra losses to account for urban area attenuation that is looked up from a family of attenuation graphs that are functions of distance and frequency. Several other correction factors are added to account for discrepancies and deviation from standard Okumura model. Most of these corrections are furnished by graphs that are functions of frequency, distance, and terrain profile.

Despite the early success and wide acceptance of Okumura's method for estimation of urban area propagation loss, this model suffers from a couple of shortcomings. Primarily, correction factors must be incorporated for every possible deviation from the standard scenario. Second, owing to its empirical nature, the incorporation of Okumura model into software design tools is very difficult.

To overcome these limitations, Hata proposed an analytical formulation for median propagation path loss based on data provided in the Okumura model [21]. Hata model is valid over VHF and UHF bands from 150 to 1500 MHz. The second restriction on Hata model is the requirement that TR separation must be at least 1 km. Several extensions of Hata model have been developed over the ensuing years, a process that is still unfolding. Some of these extensions will be discussed later. Since Hata model is expressed by equations; providing approximation for Okumura graphic data, it is well suited to be incorporated into software design tools.

Hata technique provides a standard formula for estimation of median path loss in urban environments. The correction equations are developed to adapt the model for application in suburban and rural areas. The key parameters in Hata standard equation and correction formulae are distance, carrier frequency of the waveform, and the antenna heights of BS and MS units. The standard Hata formula is given in Equation 2.45.

$$(P\ell_{md})_{ur} = 69.55 + 26.16 \log_{10} f_c + \left(44.9 - 6.55 \log_{10} h_t\right)$$
$$\times \log_{10} d - 13.82 \log_{10} h_t - \alpha(h_r) \qquad (2.45)$$

Here $(P\ell_{md})_{ur}$ is the median path loss in dB, f_c is the carrier frequency of the radio signal in MHz, d is the TR distance in km, h_t and h_r are the heights of BS and MS antennae in meters, and $\alpha(h_r)$ is the correction factor for the MS antenna height. The index md signifies median value of path loss. The $\alpha(h_r)$ formulae for large metropolitan area and medium to small cities are given in Equations 2.46 and 2.47, respectively.

$$\alpha(h_r) = 3.2\left[\log_{10}(11.75h_r)\right]^2 - 4.97\,(\text{dB}) \quad f_c \geq 400\,\text{MHz} \quad (2.46)$$

$$\alpha(h_r) = \left[1.1\log_{10}f_c - 0.7\right]h_r - 1.56\log_{10}f_c + 0.8\,(\text{dB}) \quad (2.47)$$

Hata model provides formulae for estimation of suburban and rural area median propagation path loss, expressed in Equations 2.48 and 2.49, respectively:

$$(P\ell_{md})_{sub} = (P\ell_{md})_{ur} - \left[\log_{10}\left(\frac{f_c}{28}\right)\right]^2 - 5.6\,(\text{dB}) \quad (2.48)$$

$$(P\ell_{md})_{rur} = (P\ell_{md})_{ur} - 4.78\left[\log_{10}f_c\right]^2 + 18.33\log_{10}f_c - 40.94\,(\text{dB}) \quad (2.49)$$

It should be noted that $(P\ell_{md})_{ur}$, in both Equations 2.48 and 2.49, corresponds to urban area path loss in medium to small city environment.

The prediction of Hata model closely approximates the estimation provided by the original Okumura technique. What is more, the formulae provided by Hata model are of significant practical values so long as the TR distance exceeds 1 km. For applications in which TR distance is shorter than 1 km, or the operating frequency is outside of the model's spectral range, modified or extended Hata models may be developed.

Numerical Example 3
The TR distance in a large metropolitan area is 10 km. Assume that the BS and MS antenna heights are 30 and 3 m, respectively. Furthermore, assume that the signal carrier frequency is 1000 MHz. Calculate large-scale propagation loss using the following models.

1) Hata model
2) Double-ray ground reflection approximate model

Solution
1) Using Equation 2.46

$$\alpha(h_{MS}) = 3.2\left[\log_{10}(11.75h_r)\right]^2 - 4.97$$

$$= 3.2\left[\log_{10}(11.75 \times 3)\right]^2 - 4.97 \cong 2.69\,\text{dB}$$

Replacing this value into Equation 2.45

$$(P\ell_{md})_{ur} = 69.55 + 26.16\log_{10}f_c + \left(44.9 - 6.55\log_{10}h_t\right)\log_{10}d$$

$$-13.82\log_{10}h_t - \alpha(h_r) = 69.55 + 26.16\log_{10}(1000)$$

$$-13.82\log_{10}(30) - 2.69 + \left[44.9 - 6.55\log_{10}(30)\right]\log_{10}(10) = 158.16\,\text{dB}$$

This is the estimation of the median propagation loss according to Okumura–Hata model.

2) Assuming the double-ray ground reflection model for this case, one can apply Equation 2.44

$$(P\ell)_{DR} = 40\log_{10}d - 20\log_{10}h_r - 20\log_{10}h_t = 40\log_{10}(10000)$$
$$- 20\log_{10}(3) - 20\log_{10}(30)$$
$$\cong 160 - 9.54 - 29.54 = 120.92 \text{ dB}$$

Note that the Hata equation estimates the path loss median, whereas double-ray model approximates the path loss itself. Having said that, we recognize that there is 38 dB of difference between these two computations, which represents a significant disagreement between two closely related quantities. This is explained by observing that the double-ray model accounts for only a single reflection and the role of other propagation mechanisms is disregarded all together. This implies that double-ray is an unrealistic model for estimation of path loss for land mobile radio channels in the urban environment. Okumura–Hata empirical estimation model is based on measured data in an urban area, thus it implicitly accounts for the possibility of signal being affected by various propagation mechanism several times, and therefore provides a more accurate estimation than that predicted by the double-ray model.

2.3.2.7 COST 231-Hata Model

The original Okumura–Hata path loss model is applicable over the frequency range of 150 MHz to 1.5 GHz. In the early 1990s, the European Cooperation for Scientific and Technical (COST) Research developed an extension to the Hata model. The proposed model covers the frequency range of 1.5–2 GHz (for personal communications systems PCS); a range for which Hata model was known to underestimate the path loss. This scheme is referred to as COST 231-Hata model [22]. The model provides good path loss estimates for microcells and macrocells. A wide selection of parameters such BS height in the range of 30–200 m, MS antenna height of 1–10 m, distances from 1–20 km, and various propagation environments such as urban, dense urban, suburban, and rural areas are supported by this model.

The formula for estimation of median path loss for COST 231-Hata model is given by Equation 2.50:

$$(P\ell_{md})_{ur} = 46.3 + 33.9\log_{10}f_c - 13.82\log_{10}h_t - \alpha(h_r)$$
$$+ (44.9 - 6.55\log_{10}h_t)\log_{10}d + \Psi$$

(2.50)

Ψ is an adjustment parameter that is equal to 0 dB for medium- to small-sized cities and suburban areas, and is equal to 3 dB for large metropolitan centers. All other variables and parameters, in the latter equation, as well as their measurement units, are the same as those in Hata Equation 2.45. The mobile antenna correction factor $\alpha(h_r)$ is given by Equations 2.46 and 2.47.

At least one study suggests that Cost 231-Hata path loss model provides the closest prediction to some collected data for urban areas over 2–6 GHz band [23]. This frequency spectrum slice includes the C-band of 5.091–5.150 GHz over which AeroMACS operates.

2.3.2.8 Stanford University Interim (SUI) Model: Erceg Model

This model was jointly developed by the 802.16 IEEE working group and Stanford University with the objective of creating a channel model for WiMAX applications in suburban environments for frequencies above 1900 MHz [24]. In order to compute path loss, SUI model classifies the propagation environment into three categories. Category A, corresponding to the highest level of path loss, is a hilly terrain with moderate to heavy tree densities. Category B relates to either a hilly environment but skimpy vegetation, or high vegetation but flat terrain. Category C is associated with flat terrain with light tree densities. The basic path loss formula is given by Equation 2.51.

$$Pl_{md} = 10\nu \log_{10}(d/d_0) + B + \alpha(f) + \alpha(h_r) + X \qquad (2.51)$$

Here ν is path loss exponent whose value depends on terrain categories defined in SUI model, d is the distance between the transmit and receive antennae, d_0 is the close-in reference distance, B is the free space path loss from transmit antenna to the close-in reference point, $\alpha(f)$ is the correction factor for frequencies above 2 GHz, $\alpha(h_r)$ is the receiving antenna height correction factor, and X is a log-normally distributed correction factor that accounts for shadowing, foliage, and other clutter losses. The values for X are in the range of 8.2–10.6 dB [25].

Formula for calculation of B and ν are given by Equations 2.52 and 2.53, respectively.

$$B = 20 \log_{10}(4\pi d_0/\lambda) \qquad (2.52)$$

$$\nu = a - bh_t + \frac{c}{h_t} \qquad (2.53)$$

The values of the constants in the latter equation are dependent upon the category of propagation terrain and are given in Table 2.5. Also, h_t is the height of BS transmitting antenna above the ground level, with values ranging from 10 to 80 m [25].

Frequency correction factor is computed from Equation 2.54, where f is expressed in MHz.

$$\alpha(f) = 6 \log_{10}\left(\frac{f}{2000}\right) \qquad (2.54)$$

Table 2.5 Constants for path loss exponent in SUI model.

Constant	Terrain category A	Terrain category B	Terrain category C
a	4.6	4.0	3.6
b (m^{-1})	0.0075	0.0065	0.005
c (m)	12.6	17.1	20.0

Receiving antenna height correction factor is calculated from Equation 2.55 for terrain categories A and B, and Equation 2.56 for terrain category C.

$$\alpha(h_r) = -10.8 \log_{10}\left(\frac{h_r}{2000}\right) \tag{2.55}$$

$$\alpha(h_r) = -20 \log_{10}\left(\frac{h_r}{2000}\right) \tag{2.56}$$

It is interesting to note that SUI model, in addition to estimating the long-term path loss, accounts for the medium-term signal attenuation as well, that is, the effect of shadowing and foliage is incorporated in the path loss formula. The SUI model is a suitable propagation model for WIMAX, AeroMACS, and BFWA (Broadband Fixed Wireless Applications) applications.

2.3.2.9 ECC-33 Model

Okumura–Hata model is, perhaps, the most widely used path loss estimator for UHF bands. However, its upper limit frequency is 1500 MHz. The COST-231 model has extended this frequency limit up to 2 GHz. Furthermore, the Okumura model is based on measurements in and around the city of Tokyo and the model subdivides the urban areas into "large cities" and "medium to small cities." The model also provides correction factors for "suburban" and "open" areas. In that context, most European and, perhaps, North American cities may be classified as "medium cities" [24]. Upon a recommendation by International Telecommunications Union (ITU) (ITU-R Recommendation P.529 [26]), regarding the extension of Okumura–Hata model for up to 3.5 GHz, the Electronic Communication Committee developed a path loss model that has become known as ECC-33 model [27]. The median path loss formula for ECC-33 path loss model is given by Equation 2.57:

$$P\ell_{md} = P\ell_{fs} + P\ell_{bmd} - \alpha(h_t) - \alpha(h_r) \tag{2.57}$$

Here $P\ell_{md}$ is the median path loss predicted by the model, $P\ell_{fs}$ is the free-space propagation loss, $P\ell_{bmd}$ is the "basic median loss," $\alpha(h_t)$ is the transmitter (BS) antenna height correction factor, and $\alpha(h_r)$ is the receiver (MS) antenna height correction factor. Equations 2.58–2.60 provide formulae to compute the latter three quantities, respectively.

$$P\ell_{fs} = 92.4 + 20\log_{10}d + 20\log_{10}f \tag{2.58}$$

$$P\ell_{bmd} = 20.41 + 9.83\log_{10}d + 7.894\log_{10}f + 9.56\left(\log_{10}f\right)^2 \tag{2.59}$$

$$\alpha(h_r) = \left(42.57 + 13.7\log_{10}f\right)\left(\log_{10}h_r - 0.585\right) \tag{2.60}$$

In all of these equations, f is frequency expressed in GHz, d is the distance between BS and MS antennas in kilometers, h_t is the BS antenna height in meters, and h_r is the MS antenna height in meters.

In this section, we have briefly reviewed commonly applied path loss models. Reference [28] provides a list, and some descriptions of most path loss models that have been considered for WiMAX networks up to the date of the article's publication.

2.3.3 Large-Scale Fading: Shadowing and Foliage

In the previous section, several large-scale signal attenuation models were explored. The common thread in all of these path loss estimators is the fact that they are all functions of TR distance, they exhibit strong dependency on frequency, and they predict the median of the signal power loss rather than the absolute value of the path loss. The main reason that the absolute value of path loss for a given TR distance cannot be determined, as was alluded to earlier, is that for a given TR separation, path loss might be greatly different from one location to another. For instance, for the same TR distance, in one direction the signal might encounter a large number of obstructions, while in another direction there might be no clutter and obstruction and free-space propagation between the transmitter and the receiver takes place. Therefore, two locations having exactly the same TR distance could have vastly different path losses. In other words, the path loss for the same TR distance at different location, that is, in different directions, is a random variable. The expected value of this random variable may be assumed to be given by the path loss predicted by large-scale models discussed in the previous section, or be given by Equation 2.21. The connection between path loss models and Equation 2.21 is that path loss predictions by the models can be used for estimation of path loss exponent, ν, in Equation 2.21.

The signal attenuation associated with environmental clutter is essentially due to two phenomena of foliage and shadowing. The existence of randomly distributed leaves, branches, and trees, acting like scatterers that can also absorb the energy of the radiated electromagnetic wave, causes attenuation in the signal. This is called foliage. Normally, foliage loss is evaluated in the context of one of the three scenarios: a tree, a line or multiple lines of trees, and a forest [29]. Signal attenuation due to natural or man-made objects, such as mountains and tall buildings, obstructing the propagation path between the transmitter and the receiver, is called shadowing. Notwithstanding the obstructions, a weaker version of the signal reaches the MS antenna due to diffraction and the existence of other signal echoes in the shadowed area. The shadowing may vary as the mobile station turns a corner, moves behind a building or a hilltop, or even moves into a building. Signal attenuation caused by foliage and shadowing is different from that of path loss, in the sense that it may not exist ubiquitously all over the propagation environment, and its effect might even disappear. For instance, the MS could move out of the shadow of a tall building, as a consequence the receiving antenna may even receive a LOS component from the BS, resulting in a stronger signal, whereas path loss gradually builds up as TR distance increases. It has been shown that path loss changes due to shadowing is rather gradual [5], in contrast to fast fluctuation that is caused by multipath effect, a subject to be discussed later in this chapter.

2.3.3.1 Log-Normal Shadowing

The path loss models provide estimation for median signal attenuation. However, providing an acceptable level of QoS in a wireless network cannot be achieved merely by the knowledge of estimation of median path loss. One needs to have access to the estimation of the range of SNRs as a function of TR distance. In other words, path loss model predictions must be modified in order to enable the calculation of range of actual signal loss that is encountered at any given distance from the transmitter. This implies that in addition to large-scale path loss, the effect of foliage and shadowing must also be taken into account.

Empirical data and measurements have shown that for a given value of TR distance d, the combination of path loss and shadowing at a given location is a random variable. When the overall signal attenuation (path loss plus shadowing effect) is expressed in decibel the random variable has a normal distribution, that is, the actual path loss added to shadowing attenuation has a log-normal distribution[3] [30]. The expected value of this log-normal random variable is reasonably estimated by large-scale path loss models or by Equation 2.21. A probabilistic model for the total path loss may, therefore, be expressed as in Equation 2.61.

$$(P\ell)_{dB} = (P\ell_m)_{dB} + (X)_{dB} \qquad (2.61)$$

---

3 A random variable X is said to have log-normal distribution if random variable $\log_{10} X$ has a Gaussian (normal) distribution.

Here, $(P\ell)_{dB}$ is a non-zero mean Gaussian random variable representing the overall path loss, $(P\ell_m)_{dB}$ is the path loss mean value in dB given by large-scale path loss models, $(X)_{dB}$ is a zero-mean Gaussian random variable characterizing the shadowing effect. It is the random variable X that when expressed in straight values has a log-normal distribution. The assumption of log-normal distribution for random variable X is consistently supported by measured data [4,5,30].

Since $(X)_{dB}$ is a zero-mean random variable, its probability density function (PDF) is given by Equation 2.62:

$$f_{X_{dB}}(x) = \frac{1}{\sqrt{2\pi}\sigma_X} e^{-\frac{x^2}{2\sigma_X^2}} \tag{2.62}$$

σ_X is the standard deviation of the Gaussian random variable $(X)_{dB}$. When the PDF of a random variable is available probability calculations become a straight forward matter. In this case, for complete characterization of the PDF, a value for σ_X is required. In practice, the value of σ_X is determined from measured path loss data over a wide range of locations and TR separations, and with the application of linear regression that minimizes the mean square error between the measured data and estimated model. Application of this technique on data measured and collected from four cities in Germany has resulted in a value for σ_X that is equal to 11.8 dB [31]. However, Ref. [5] claims that empirical studies and measured data have indicated that $\sigma_X = 8$ dB for a large number of urban and suburban areas, and the value of this standard deviation increases by $1-2$ dB for dense urban areas, and is lowered by $1-2$ dB in open rural environments. Having a value for σ_x makes it possible to compose a formula for the computation of the received signal power at distance d, as expressed in Equation 2.63.

$$[P_r(d)]_{dB} = (P_t)_{dB} - [P\ell(d)]_{dB} = (P_t)_{dB} - [P\ell_m(d)]_{dB} - (X)_{dB} \tag{2.63}$$

In this equation, all other losses and gains that might exist in the system, such as antenna directivity gains, have been disregarded, or can be assumed to have been included in the term $(P_t)_{dB}$. The expected value of the received power at distance d can be expressed by the first two terms on the right side of Equation 2.63.

$$E\left[P_r(d)_{dB}\right] = (P_t)_{dB} - [P\ell_m(d)]_{dB} \triangleq \mu_{dB}(d) \tag{2.64}$$

Therefore, we conclude that

$$[P_r(d)]_{dB} = \mu_{dB}(d) - (X)_{dB} \tag{2.65}$$

Hence, the received power at distance d, when expressed in dB, has a non-zero Gaussian distribution with mean value $\mu_{dB}(d)$, and a PDF that is given by Equation 2.66.

$$f_{P_r}(d) = \frac{1}{\sqrt{2\pi}\sigma_X} e^{-\frac{[P_r(d) - \mu_{dB}(d)]^2}{2\sigma_X^2}}$$ (2.66)

With this PDF at hand, the probability that the received power at distance d from the BS antenna is smaller than or equal to a required value of p_{dB} can be calculated as follows.

$$Pr\left[P_r(d)_{dB} \leq p_{dB}\right] = \int_{-\infty}^{p_{dB}} f_{P_r}(d)dP_r(d) = \int_{-\infty}^{p_{dB}} \frac{1}{\sqrt{2\pi}\sigma_X} e^{-\frac{[P_r(d) - \mu_{dB}(d)]^2}{2\sigma_X^2}} dP_r(d)$$

(2.67)

Similarly, the probability that the received signal power at distance d is greater than a desired threshold value p_{dB} is given by Equation 2.68.

$$Pr\left[P_r(d)_{dB} > p_{dB}\right] = \int_{p_{dB}}^{\infty} f_{P_r}(d)dP_r(d) = \int_{p_{dB}}^{\infty} \frac{1}{\sqrt{2\pi}\sigma_X} e^{-\frac{[P_r(d) - \mu_{dB}(d)]^2}{2\sigma_X^2}} dP_r(d)$$

(2.68)

Since there is no closed-form equation for the integral of the Gaussian function, the values of these integrals are normally given either by the Q-function[4] or the complementary error function expressions. With a simple change of variable, the Gaussian functions inside the integral signs in Equations 2.67 and 2.68 can be converted to standard Gaussian PDF, and consequently the integrals can be shown to have the values given by Equations 2.69 and 2.70.

$$Pr\left[P_r(d)_{dB} \leq p_{dB}\right] = Q\left(\frac{\mu_{dB}(d) - p_{dB}}{\sigma_X}\right)$$ (2.69)

$$Pr\left[P_r(d)_{dB} > p_{dB}\right] = 1 - Q\left(\frac{\mu_{dB}(d) - p_{dB}}{\sigma_X}\right) = Q\left(\frac{p_{dB} - \mu_{dB}(d)}{\sigma_X}\right)$$ (2.70)

The second part of Equation 2.70 is obtained when the Q-function property of $Q(-x) = 1 - Q(x)$ is applied. Often times, to make proper

---

4 The Q-function is used for computation of probabilities for standard Gaussian random variable, it is defined by $Q(x) \triangleq \frac{1}{\sqrt{2\pi}} \int_x^{\infty} e^{-\frac{y^2}{2}} dy$. As such the Q-function calculates the area under the upper tail of the standard Gaussian PDF. The values of $Q(x)$ is normally looked up from a table of the Q-function. For $x > 3$ approximate equation of $Q(x) \cong \frac{e^{-\frac{x^2}{2}}}{\sqrt{2\pi}x}$ might be used.

interpretation of these probabilities they are translated into percentages. For instance, $Pr\left[P_r(d)_{\text{dB}} > -100\,\text{dBm}\right] = 0.8$ means if signal power is measured at large number of points on a circle of radius d centered on BS transmitting antenna, about 80% of data would show a power level greater than -100 dBm. If d is the approximate radius of the underlying cell, and if -100 dBm is the receiver sensitivity,[5] then this probability value indicates that 80% of the cell boundary receives adequate signal power for the delivery of a predefined level of signal-to-noise ratio.

2.3.3.2 Estimation of Useful Coverage Area (UCA) within a Cell Footprint

The log-normal shadowing, discussed in the previous section, provides an "average solution" to the problem of unpredictable presence of obstructions to signal propagation in wireless channels. However, the random effect of shadowing makes it impossible for the cellular network designer to guarantee that the received signal is adequate everywhere throughout a cell footprint. In practice, the cellular systems are designed to guarantee coverage over a predefined percentage of the service area. As Figure 2.2b shows there are always "blind spots" within the service area. In this section, we develop an equation that estimates the percentage of UCA within a cell footprint, as a function of fraction of cell boundary coverage. In other words, the boundary coverage determines the percentage of area coverage within the footprint.

Let us assume that the receiver sensitivity in decibel is given by γ dB. The probability that the received signal power at a distance d from the BS antenna exceeds the threshold value of γ dB can be calculated from Equation 2.70, as given in Equation 2.71.

$$Pr\left[P_r(d)_{\text{dB}} > \gamma_{\text{dB}}\right] = Q\left(\frac{\gamma_{\text{dB}} - \mu_{\text{dB}}(d)}{\sigma_X}\right) \tag{2.71}$$

We next define the boundary coverage (BC) as the probability that the received power at the cell boundary is greater than γ dB. Assuming that the cell has an approximate circular shape with radius R, BC is calculated form the following equation.

$$BC \triangleq Pr\left[P_r(R)_{\text{dB}} > \gamma_{\text{dB}}\right] = Q\left(\frac{\gamma_{\text{dB}} - \mu_{\text{dB}}(R)}{\sigma_X}\right) \tag{2.72}$$

In fact, BC represents the percentage of cell boundary over which the received power exceeds the threshold value of γ dB. The probability that the received signal power within the cell footprint is greater than γ dB can

---

5 Receiver sensitivity is the minimum required received signal power for the receiver to perform at an acceptable level, or to generate a predefined signal-to-noise ratio.

be found from Equation 2.71. Plugging Equation 2.64 into Equation 2.71, the following equation is resulted:

$$Pr\left[P_r(d)_{\mathrm{dB}} > \gamma_{\mathrm{dB}}\right] = Q\left(\frac{\gamma_{\mathrm{dB}} - (P_t)_{\mathrm{dB}} + [P\ell_m(d)]_{\mathrm{dB}}}{\sigma_X}\right) \tag{2.73}$$

Following some alteration, this equation will be used to determine percentage of the area within the cell footprint that receives sufficient signal power for acceptable performance, that is, the *useful coverage area* (UCA). In fact, it will be demonstrated that percentage of UCA can be calculated from the knowledge of BC and the path loss exponent ν. It should be noted that in actual course of design of a cellular network, the process takes place in the reverse order, that is, the designer starts from an acceptable level for percentage of UCA and measurement-based estimated value for ν, she then proceeds to determine the BC, which would lead to computation of average required power over the cell boundary [5]. The calculated BC is a key factor in determination of the required output power of the BS antenna.

As the MS moves away from the cell boundary and toward the center of the cell, it is anticipated that the expected value of the received power should increase. The path loss gain relative to cell boundary is given by

$$\frac{P_r(d)}{P_r(R)} = \frac{R^\nu}{d^\nu} = \left(\frac{R}{d}\right)^\nu \rightarrow 10\log_{10}\left[\frac{P_r(d)}{P_r(R)}\right] = 10\nu\log_{10}(R/d) \tag{2.74}$$

Therefore, the path loss inside the cell boundary at distance d from the BS antenna is smaller than the path loss on the cell boundary by a factor of $10\nu\log_{10}(R/d)$dB, it then follows that

$$[P\ell(d)]_{\mathrm{dB}} = [P\ell(R)]_{\mathrm{dB}} - 10\nu\log_{10}\left(\frac{R}{d}\right) \tag{2.75}$$

Replacing Equation 2.75 into Equation 2.73, Equation 2.76 is resulted.

$$Pr\left[P_r(d)_{\mathrm{dB}} > \gamma\right] = Q\left(\frac{\gamma - (P_t)_{\mathrm{dB}} + [P\ell(R)] - 10\nu\log_{10}(R/d)}{\sigma_X}\right)$$
$$\tag{2.76}$$

Recalling that

$$(P_t)_{\mathrm{dB}} - [P\ell(R)]_{\mathrm{dB}} = E[P_r(R)] \tag{2.77}$$

Replacing the latter equation into the former, Equation 2.78 is obtained.

$$Pr\left[P_r(d)_{\mathrm{dB}} > \gamma_{\mathrm{dB}}\right] = Q\left(\frac{\gamma_{\mathrm{dB}} - E[P_r(R)]_{\mathrm{dB}} - 10\nu\log_{10}(R/d)}{\sigma_X}\right)$$
$$\tag{2.78}$$

This equation computes the probability that a point inside the cell footprint at distance d from the BS antenna receives power that exceeds the receiver sensitivity. Furthermore, the same equation is used to calculate an important cellular network parameter. This parameter is the percentage of UCA. Assuming a receiver sensitivity of γ dB, the UCA percentage is denoted by $UA(\gamma)$.

Computation of $UA(\gamma)$ is carried out by first determining the area within the cell boundary for which the received power exceeds γ dB, that is, the UCA, and then dividing the result by the cell area. Recall that for this computation the cell footprint is assumed to have an approximate circular shape of radius R, and therefore, the cell area is equal to πR^2. For calculation of UCA, we first define an infinitesimal area within the cell footprint such that $Pr[P_r(d)_{\text{dB}} > \gamma_{\text{dB}}]$ remains constant within this area. Applying polar coordinate system, the infinitesimal area can be expressed by

$$dA = x\,dx\,d\theta \tag{2.79}$$

Here x is the radial distance and θ is the angular variable. With this in mind, we can calculate UCA as follows.

$$UCA = \int Pr[P_r(x)_{\text{dB}} > \gamma_{\text{dB}}]\,dA = \int_0^{2\pi} \int_0^R Pr[P_r(x)_{\text{dB}} > \gamma_{\text{dB}}]\,x\,dx\,d\theta$$

$$\tag{2.80}$$

To determine an expression for the percentage of useful area, $UA(\gamma)$, we simply divide UCA by the cell area.

$$UA(\gamma) = \frac{1}{\pi R^2} \int_0^{2\pi} \int_0^R Pr[P_r(x)_{\text{dB}} > \gamma_{\text{dB}}]\,x\,dx\,d\theta$$

$$\tag{2.81}$$

$$= \frac{1}{\pi R^2} \int_0^{2\pi} \int_0^R Q\left(\frac{\gamma_{\text{dB}} - E[P_r(R)]_{\text{dB}} - 10\nu \log_{10}(R/x)}{\sigma_X}\right) x\,dx\,d\theta$$

Equation 2.81 can be simplified and plotted to provide a family of curves presenting percentage of useful coverage area against σ_X/ν while BC is kept as a parameter. These curves are shown in Figure 2.11. This set of curves originally appeared in Ref. [32].

For instance, if 80% of the cell boundary is covered, that is, the received power is above the threshold value over 80% of the cell boundary, while $\sigma/n = 5$, then according to the graphs of Figure 2.11, $UA(\gamma)$ is 0.88. This means that the received power over 88% of the cell footprint is above the threshold value of γ.

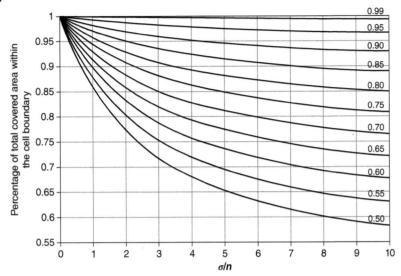

Numbers on the graphs represent the percentage of boundary coverage

Figure 2.11 Percentage of total area within a cell footprint receiving signal power above a threshold value as a function of σ/n where the BC is kept as a parameter [32].

2.3.4 Small-Scale Fading: Multipath Propagation and Doppler Effect

A major distinguishing character of the wireless mobile channel is the fact that several copies of the same radio signal arrive at the MS receiving antenna from different directions, with different amplitudes, phases, and time delays. This is the *multipath propagation* phenomenon. Depending upon phase relationship between the multipath echoes, they may add up constructively and reinforce the signal, or combine destructively to weaken the received signal. Moreover, owing to the mobile nature of the channel, the makeup of the arriving signal copies vis-à-vis amplitude, phase, time delay, and even the number of echoes in the mix, could swiftly change. As a consequence, two receivers at the proximity of each other may receive radio signals whose strengths are different by orders of magnitude. Viewing this from a different perspective, as the MS moves along its path, the phase relationship in the multipath signal mix rapidly changes causing fast variation in the envelope and phase of the received signal. This rapid fluctuation of the radio signal over a *short period of time* or *short travel distance* is called *small-scale fading* or *short-term fading* or just *fading*. The short period of time or short travel distance is defined such that the large-scale path loss due to long-term propagation, and medium-scale attenuation due to shadowing and foliage can be neglected over the considered time period and/or distance. In terms of short travel distance on the part of the MS, one is referring to distances on the order of the carrier wavelength.

In addition to multipath effect, a second problem contributing to radio signal fluctuations is *Doppler shift*. Relative motion between the transmitter and the receiver causes Doppler frequency shift. In a multipath propagation, the problem is exacerbated by the fact that various copies of the multipath signal would have different Doppler shifts that results in *Doppler spread*.

The major factors influencing small-scale fading are multipath effect, speed of the MS, motion and speed of interacting objects, and the bandwidth of the signal [4,11].

2.3.4.1 Multipath Propagation

Multipath propagation occurs when the transmitted signal has several routes to reach the receiver. The paths are created by various electromagnetic effects such as reflection, diffraction, and scattering in the radio channel. In land mobile communications systems, the LOS component may not even exist, particularly in urban environments, however, when it does exist it is usually the strongest signal component. A simple multipath scenario, consisting of a LOS ray and three non-LOS (NLOS) echoes, is shown in Figure 2.12.

When the transmitted signal is a single tone in the form of $s(t) = A \cos(2\pi f_c t)$, the received multipath signal maybe expressed as follows:

$$r(t) = \sum_k A_k \cos\left[2\pi f_c(t - d_k/c) + \phi_k\right] = \sum A_k \cos\left[2\pi f_c(t - \tau_k) + \phi_k\right]$$

$$(2.82)$$

where d_k is the length of the kth path, τ_k is the time delay associated with the kth path, and ϕ_k is the phase shift of the kth path that could also include the Doppler frequency shift.

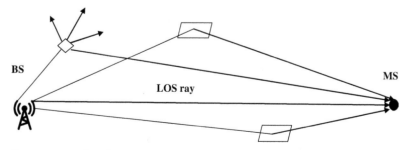

Figure 2.12 Multipath propagation with a LOS ray, two reflected echoes, and a scattered component.

In addition to being the main culprit in small-scale fading of radio signal, and as a result of it, the multipath propagation also causes *time dispersion* in the transmitted signal. Assume an extremely short pulse (approximately an impulse function) is transmitted over a multipath channel. The received signal consists of a number of pulses that arrive at the receiver with different delays and different amplitudes depending on the paths they have taken. An example is shown in Figure 2.13.

In Figure 2.13, a short pulse is transmitted at time t_0, three pulses are received at times $t_1, t_1 + \tau_{11}$, and $t_1 + \tau_{12}$. The extra delays relative to the arrival time of the first received pulse, that is, τ_{11} and τ_{12} are called *excess delays*. This reveals an important trait of multipath medium, that is, the imposition of *time spread* or *time dispersion* into the transmitted signal.

Definition 2.1 *Time dispersion is the process of stretching the transmitted signal in time in a manner that the duration of the received signal is greater than that of the transmitted signal.*

Time dispersion is the direct result of signal and its echoes taking different times to cross the channel. The testing and measurement of wireless channels using short pulses is referred to as *pulse sounding* technique.

Another attribute of multipath propagation is its dynamic behavior caused by time variation in the radio channel. As a result of such dynamism, the multipath channel characteristics varies with time. That is to say, if we repeat the pulse-sounding experiment, discussed earlier, (an example of which is shown in Figure 2.13) over and over, changes in the received pulse train will be observed in three different respects.

1) Variations in the amplitude of the individual pulses
2) Changes in relative arrival delays among the pulses
3) Change in the number of pulses in the received train

Figure 2.14 provides an example that demonstrates the *time-varying* nature of the multipath channel.

Since the channel is time varying, its impulse response is a function of two time variables. Figure 2.14 illustrates the time variation observed in a

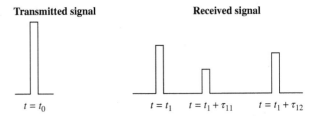

Figure 2.13 An example of multipath propagation. A short pulse is transmitted and three pulses, with different amplitudes and different time delays, are received.

Transmitted signal   **Received signal**

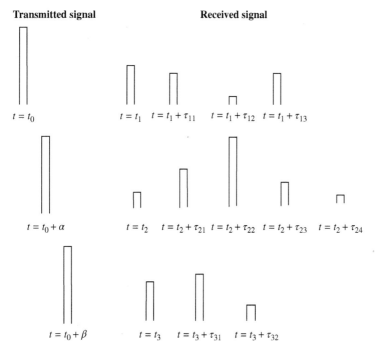

Figure 2.14 The time varying nature of a multipath channel demonstrated by the pulse sounding technique.

multipath channel and how it affects the received signal. A single pulse is transmitted through the channel at three different time instances. Since the channel characteristics vary with time, three different pulse sequences appear at the front end of the receiver. The pulse sequences are different in terms of amplitude, arrival time, excess delays, and the number of pulses in the sequence.

2.3.4.2 Double Path Example

The simplest multipath propagation scenario relates to the case in which the transmitted signal reaches the receiver through two traveling paths. A double path scenario, composed of two NLOS paths, is illustrated in Figure 2.15.

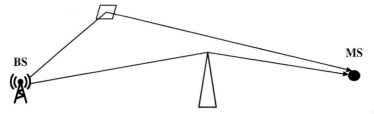

Figure 2.15 A double-NLOS-path case. The signal reaches the MS antenna through a reflected route as well as a diffracted path.

Assume that the transmitted signal is a single tone of the form of $s(t) = A\cos(2\pi f_c t)$. The received signal, according to Equation 2.82, can be written in the following form:

$$r(t) = \sum_{k=1}^{2} A_k \cos\left[2\pi f_c(t - \tau_k) + \phi_k\right] \tag{2.83}$$

The received signal is the sum of two sinusoidal waveforms of the same frequency, therefore, it has the general form of $r(t) = A_r \cos(2\pi f_c t + \phi_r)$, from Equation 2.83 it follows that

$$A_r \cos(2\pi f_c t + \phi_r) \equiv A_1 \cos\left[2\pi f_c(t - \tau_1) - \phi_1\right] + A_2 \cos\left[2\pi f_c(t - \tau_2) - \phi_2\right] \tag{2.84}$$

Our objective is to express A_r in terms of $A_1, A_2, \tau_1, \tau_2, \phi_1$, and ϕ_2. Applying the phasor concept and Euler's identity to Equation 2.84, Equation 2.85 is resulted.

$$A_r e^{j(2\pi f_c + \phi_r)} = A_1 e^{j\left[2\pi f_c(t - \tau_1) + \phi_1\right]} + A_2 e^{j\left[2\pi f_c(t - \tau_2) - \phi_2\right]}$$
$$\rightarrow A_r e^{j\phi_r} e^{j2\pi f_c t} = \left[A_1 e^{j(-2\pi f_c \tau_1 + \phi_1)} + A_2 e^{j(-2\pi f_c \tau_2 + \phi_2)}\right] e^{j2\pi f_c t} \tag{2.85}$$

Equating factors of $e^{j2\pi f_c t}$ from both sides of Equation 2.85, the desired equation that can be used to find an equation for computation of A_r is obtained.

$$A_r e^{j\phi_r} = A_1 e^{-j(2\pi f_c \tau_1 - \phi_1)} + A_2 e^{-j(2\pi f_c \tau_2 - \phi_2)}$$
$$\rightarrow A_r \cos(2\pi \phi_r) + jA_r \sin(2\pi \phi_r) = \left[A_1 \cos(2\pi f_c \tau_1 - \phi_1) + A_2 \cos(2\pi f_c \tau_2 - \phi_2)\right]$$
$$-j\left[A_1 \sin(2\pi f_c \tau_1 - \phi_1) + A_2 \sin(2\pi f_c \tau_2 - \phi_2)\right] \tag{2.86}$$

Equating the real and imaginary parts on both sides of the latter equation results in the following pair of equations.

$$\begin{cases} A_r \cos(\phi_r) = A_1 \cos(2\pi f_c \tau_1 - \phi_1) + A_2 \cos(2\pi f_c \tau_2 - \phi_2) \\ A_r \sin(\phi_r) = -A_1 \sin(2\pi f_c \tau_1 - \phi_1) - A_2 \sin(2\pi f_c \tau_2 - \phi_2) \end{cases} \tag{2.87}$$

Squaring these equations, adding them up, and using trig identities $\cos^2\theta + \sin^2\theta = 1$ and $\cos\alpha\cos\beta + \sin\alpha\sin\beta = \cos(\alpha - \beta)$ to simplify the result, leads to Equation 2.88.

$$A_r^2 = A_1^2 + A_2^2 + 2A_1 A_2 \cos\left[(2\pi f_c \tau_1 - \phi_1) - (2\pi f_c \tau_2 - \phi_2)\right] \tag{2.88}$$

Therefore, A_r is calculated from Equation 2.89.

$$A_r = \sqrt{A_1^2 + A_2^2 + 2A_1A_2 \cos\left[2\pi f_c(\tau_1 - \tau_2) - (\phi_1 - \phi_2)\right]}$$
$$= \sqrt{A_1^2 + A_2^2 + 2A_1A_2 \cos\left(2\pi f_c \Delta\tau - \Delta\phi\right)} \tag{2.89}$$

Equation 2.89 shows that the amplitude of the received signal; A_r, depends upon magnitude of A_1 and A_2 as well as delay difference $\Delta\tau$ and phase difference $\Delta\phi$, all of which are directly related to the length of the route that the two rays traverse on their way to reach the receiver antenna. It is also very interesting to observe that A_r is a function of frequency f_c as well. When the MS and the surrounding environment remain stationary, the received signal amplitude remains the same. Assuming $\Delta\phi = 0$ and $\Delta\tau = k/f_c$, the two components constructively add and $A_r = A_1 + A_2$. However, with $\Delta\phi = 0$ and $\Delta\tau = (2k + 1)/2f_c$, the echoes add destructively, as if they are to cancel each other out, and $A_r = |A_1 - A_2|$.

The analysis and equation development for double-path case can be generalized for scenarios with more than two components in a straightforward manner. However, the resulting equations become excessively long and therefore impractical. Simulation is the method of choice to handle these cases.

2.3.4.3 Doppler Shift

The Doppler shift is the change of frequency of a transmitted waveform, received by a station that is in relative motion with respect to the signal source. Assuming an unmodulated carrier signal with frequency f_0 is transmitted and the receiver is directly moving toward or away from a stationary signal source, then the apparent frequency of the received signal is determined by Equation 2.90.

$$f_{ap} = \left(\frac{c \pm v}{c}\right)f_0 = f_0 \pm \frac{v}{c}\frac{c}{\lambda_0} = f_0 \pm \frac{v}{\lambda_0} = f_0 \pm f_{d-\max} \tag{2.90}$$

Here c is the speed of electromagnetic waves, v is the speed of receiving station, and $f_d = \pm v/\lambda_0$ is the Doppler frequency shift, which is in fact the maximum Doppler shift, since the receiver is directly moving toward or away from the source.

In addition to causing a dynamic multipath propagation effect, motion in the wireless channel induces Doppler frequency shift, therefore it plays a key role in small-scale fading. When the MS is in relative motion with respect to the BS, a Doppler shift occurs in the received carrier frequency. The maximum and actual Doppler frequency shifts are given by Equations 2.90 and 2.91.

$$f_d = \frac{v}{\lambda_0}\cos\theta = f_{d-max}\cos\theta \tag{2.91}$$

Here θ is the angle between the direction of motion of the MS and the direction of the straight line connecting the MS antenna to the BS antenna, v is the speed of the mobile station, and λ_0 is the wavelength of the signal. Equation 2.91 implies that the Doppler shift can be positive or negative, depending on the value of θ. Maximum Doppler shift is corresponding to $\theta = 0$, when the velocity vector and the vector connecting MS to BS have exactly the same directions, therefore, angle between them is zero. By the same token, minimum Doppler shift corresponds to the case in which the velocity vector and vector connecting MS to BS have opposing directions, therefore, the angle between them is exactly equal to 180°.

Figure 2.16 demonstrates two cases of Doppler frequency shift. Figure 2.16a depicts a scenario in which $\theta < \pi/2$, and thus the Doppler frequency shift is positive. In Figure 2.16b, $\theta > \pi/2$ and thus the Doppler shift is negative and the apparent frequency is reduced.

Since the mobile station receives multiple copies of the same signal from different directions, the waves arriving through various paths might have different Doppler shifts. This creates what is known as *Doppler spread*, which generates a random FM noise imbedded in the received multipath signal.

2.3.4.4 Impulse Response of Multipath Channels

We have observed that MS antenna receives various multipath components and adds them up, this is a linear operation. However, the characteristics of a mobile channel change with time. Therefore, it is reasonable to assume that the multipath channel is a linear but time-varying filter. This implies that the impulse response of the multipath channel is a function of two time variables, $h(t, \tau)$. The output of such a system due to input $x(t)$ is determined by Equation 2.92

$$s(t) \rightarrow \boxed{h(t, \tau)} \rightarrow t(t)$$
$$r(t) = s(t) \cdot h(t, \tau)$$
(2.92)

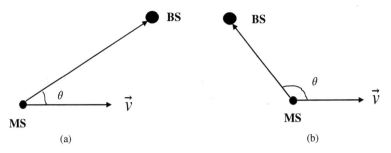

Figure 2.16 (a) Positive Doppler shift, the apparent frequency increases. (b) Negative Doppler shift, the apparent frequency decreases.

Here t is the independent time variable directly related to motion, τ is the channel multipath delay for a fixed value of t and "*" implies convolution.

Multipath channels are band-pass channels. As such, it is possible to find a complex baseband envelope representation for $h(t, \tau)$, let us denote that by $h_b(t, \tau)$, where

$$h(t, \tau) = Re\left[h_b(t, \tau)\exp\left(j2\pi f_c t\right)\right] \tag{2.93}$$

With both input and output signals being band pass, that is,

$$
\begin{aligned}
s(t) &= Re\left[s_b(t)\exp\left(j2\pi f_c t\right)\right] \\
r(t) &= Re\left[r_b(t)\exp\left(j2\pi f_c t\right)\right]
\end{aligned}
\tag{2.94}
$$

One can come up with the following input–output relationship for the system at the baseband [33].

$$r_b(t) = \frac{1}{2}s_b(t) \cdot h_b(t, \tau) \tag{2.95}$$

Here, $h_b(t, \tau), s_b(t)$, and $r_b(t)$ are the complex baseband envelopes of multipath channel impulse response, input signal, and output signal, respectively. It is often useful, for both analysis and simulation, to discretize the multipath-delay axis τ of the impulse response into equal time delay segments called *excess delay bins*. The delay corresponding to the kth bin is given by Equation 2.96. Delay discretization process is shown in Figure 2.17.

$$\tau_0 = 0, \quad \tau_1 = \Delta\tau, \quad \tau_2 = 2\Delta\tau, \quad \tau_k = k\Delta\tau, \quad \tau_{N-1} = (N-1)\Delta\tau \tag{2.96}$$

The $\tau_0 = 0$ corresponds to the delay of the first arriving component. Any number of multipath echoes received within the kth bin are represented by a single resolvable multipath component with delay equal to τ_k. This model can be used to analyze transmitted signals having bandwidths less than $1/2\Delta\tau$ [4]. Excess delay τ_k represents the delay of the kth multipath component relative to the first arriving echo. The *maximum excess delay* for this model is $\tau_{max} = n\Delta\tau$. The complex baseband impulse response for such a model is given by the following equation [31].

$$h_b(t, \tau) = \sum_{k=0}^{N-1} a_k(t, \tau)\exp\left[j2\pi f_c \tau_k(t) + \varphi_k(t, \tau)\right]\delta[\tau - \tau_k(t)] \tag{2.97}$$

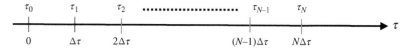

Figure 2.17 Multipath delay is discretized into delay bins.

Here $a_k(t, \tau)$ is the amplitude of the received signal of the kth multipath echo, $\tau_k(t)$ is the excess delay related to the same path, $2\pi f_c \tau_k(t)$ is the phase change due to free-space propagation, and $\varphi_k(t, \tau)$ is the phase shift due to the Doppler effect [34].

2.3.4.5 Delay Spread and Fading Modes

The term *delay spread* is used to provide a quantitative measure of a multipath channel time dispersion. This is an important wireless system parameter that is defined differently by various references. Generally speaking, delay spread is considered to be the time length of the received signal when an impulse is transmitted through the multipath channel, as shown in Figure 2.18.

This perception of *delay spread* is corresponding to τ_{max} in the discretized model of the multipath channel, discussed earlier.

Recalling the pulse sounding technique, so long as impulses (or any other type of data symbols, for that matter) are transmitted at a low enough rate, they can be easily resolved by the receiver. This is the case simply because the extensions of a single impulse are entirely received before the next impulse function is transmitted. Viewing the same phenomenon from the frequency domain perspective, under the circumstances of low data rate, all frequency components of the signal undergo fading (filtering effect) in the same fashion. This type of fading is called *flat fading*.

As data rate is increased, beyond a certain level, multipath extension of data symbols spread into adjacent symbols causing intersymbol interference (ISI). Figure 2.19 illustrates how ISI occurs. This severely distorts the signal, and without the use of a *channel equalizer* to remove ISI, the system bit error rate (BER) may become unacceptably high. It is duly noted here that in order to design a proper equalizer, the unit impulse response of the channel is needed. This can certainly be a challenge in mobile radio systems.

Viewing the ISI occurrence, illustrated in Figure 2.19, from the frequency domain point of view, it can be said that under the circumstances, different frequency components of the signal are exposed to different

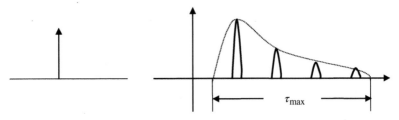

Figure 2.18 "Delay spread" defined as the time duration of an impulse response.

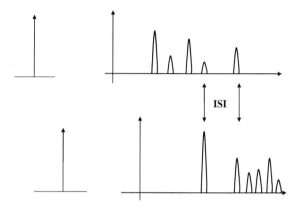

Figure 2.19 ISI occurs in a multipath channel when data rate is sufficiently high.

levels of fading, that is, different filtering effect. As such, this type of fading is known as *frequency selective fading*. Therefore, high data rates give rise to frequency-selective propagation effects. In most practical cases attempts are made to avoid frequency-selective fading conditions in the wireless channel.

2.3.4.6 Methods of Combating Frequency-Selective Fading

Consider broadband mobile cellular networks, for example AeroMACS, in which signal is transmitted at such a high rate that without any precautionary measures, the occurrence of ISI and frequency-selective fading are inevitable. If channel equalization is not desirable, what other methods are available to mitigate the effect of ISI? Suppose we reduce the size of the basic cell in the network. That implies that the MS is moved closer to the BS, while decreasing the radiated power to allow for power control in the new cellular layout. If this distance is sufficiently small, the delay spread would decrease, as delays of the multipath components are smaller when the cell size is reduced. The ISI will cease to be significant and the channel becomes approximately flat fading, thereby avoiding the need for channel equalization. In fact, in large cellular networks where the delay spread may exceed 10 μs, channel equalization is needed when data rate is as low as 64 kbit/s. While cordless communication systems inside buildings, with an excess delay spread of less than 1 μs, may still exhibit flat fading behavior at data rates exceeding 1 Mbit/s. The important conclusion is that *small cells are not just smaller, they exhibit different propagation features. Picocells* may support many megabits per second without requiring channel equalization. This is one of the major reasons that small cell technology (Picocells and Femtocells) is a key ingredient in the emerging 5G networks [35].

A second important technique for avoiding frequency-selective fading while transmitting data at high rates for broadband systems is the application of OFDM. In OFDM signaling, a wideband signal (high rate data stream) is broken down into multiple number of low rate data sequences, each of which modulates a *subcarrier*. Subcarrier signals are orthogonal over the frequency domain and each of the transmitted signal has a low enough data rate that would observe a flat fading multipath channel. Chapter 4 is entirely dedicated to the study of OFDM and OFDMA (orthogonal frequency-division multiple access).

MIMO is another technique that is chiefly applied for capacity enhancement of the radio link, but it can also be applied for coping against frequency-selective fading in dispersive channels, taking advantage of the diversity[6] that is created by the usage of multiple transmit and multiple receive antennas [36]. Normally, to mitigate the effects of frequency-selective fading, schemes that combine MIMO and OFDM are applied. Additionally, various signal processing techniques might be employed to combat this type of fading. For instance, certain modulation schemes perform more robustly in the presence of frequency-selective fading than others.

Furthermore, error correction coding (ECC), that is, channel coding, may be applied to improve the bit error rate (BER) of the received signal that has been severely hampered by the effect of intersymbol interference. However, coding alone, although reduces the effect of frequency-selective fading, cannot completely eliminate the ensued ISI [11].

Channel equalization is a direct method of combating intersymbol interference generated by frequency-selective fading. An equalizer's operation can be explained in both time domain and frequency domain. In order to design and implement an equalizer, the impulse response of the channel is needed. As our study of wireless channels' impulse response in Section 2.3.4.4 indicates, this is a dynamic function of two time variables. Consequently, the equalizer needs to be equipped with a mechanism that keeps track of variations in the channel impulse response, in other words, an *adaptive equalizer* is required, which is the challenging part of equalizer system design. References [4] and [11] provide an extensive coverage of analysis and design of adaptive equalizer for frequency-selective fading channels.

In practical wireless networks, such as LTE-based 4G (and the emerging 5G) networks, as well as in WiMAX and AeroMACS, a combination of

---

6 Diversity refers to techniques in which the same signal is received by the receiver through multiple independent channels. In mobile radio, diversity enhances the system performance of wireless link without increasing the transmitted power or required bandwidth.

techniques outlined in the previous paragraphs are applied to ensure that frequency-selective fading does not render the network performance quality below the desirable levels.

2.3.4.7 Coherence Bandwidth and Power Delay Profiles (PDPs)

When the bandwidth of the transmitted signal B_s is sufficiently narrow, or in the language of digital communication systems, the data rate is sufficiently low, all frequency components of the received signal are subject to about the same amount of attenuation, that is, the channel is a frequency flat fading one. As the transmission bandwidth increases, the frequency components at the extreme sides of the spectrum start to be attenuated differently, and the multipath channel begins to behave like a frequency-selective fading channel. The minimum transmission bandwidth at which the frequency-selective behavior of the multipath channel becomes apparent is designated by B_{co}, which is inversely related to maximum excess delay $\tau_{\max}$, that is, $B_{co} \cong k/\tau_{\max}$. This bandwidth is called *coherence bandwidth*. It must be cautioned that coherence bandwidth, very much like delay spread, does not have a universally accepted definition and equation to represent it. One reason for nonexistence of a unique formula for coherence bandwidth is that its quantity may depend upon various signal processing techniques that are used in the system, in particular, modulation formats might have direct effect on the value of coherence bandwidth. The key point, however, is that, conceptually speaking, coherence bandwidth of a multipath channel refers to the largest band of frequencies over which the channel remains flat. The significance of this concept is clear; any signal whose bandwidth is smaller than the channel coherence bandwidth would be exposed to flat fading effect. On the other hand, signals with bandwidth larger than channel coherence bandwidth would endure distortions that are associated with frequency-selective fading.

Quantitatively speaking, coherence bandwidth is the frequency separation at which the amplitude of the two frequency components become decorrelated, in the sense that the amplitudes of these components are subject to attenuation levels that are statistically uncorrelated or have very little correlation. This implies that envelope correlation coefficient $\rho(\Delta f, \Delta t)$ falls below some predetermined value. The envelope correlation coefficient is calculated from the following statistical equation.

$$\rho(\Delta f, \Delta t) = \frac{E[A_1 A_2] - E[A_1]E[A_2]}{\sqrt{\left\{E[A_1^2] - E^2[A_1]\right\}\left\{E[A_2^2] - E^2[A_2]\right\}}} \tag{2.98}$$

Where A_1 and A_2 represent the amplitudes of signals at frequencies f_1 and f_2, and at times t_1 and t_2, that is, $|f_2 - f_1| = \Delta f, |t_2 - t_1| = \Delta t$. Note that the definition and equation for $\rho(\Delta f, \Delta t)$ is identical to that of correlation coefficient of two random variables.

In order to determine some approximate expressions for coherence bandwidth, one needs to have a more appropriate definition for *delay spread*. This is accomplished by introducing the concept of *power delay profile* (PDP). PDP is the multipath signal power distribution expressed as a function of excess delay τ. PDP, therefore, may be viewed as multipath signal power spectral density (PSD) as a function of excess delay. Figure 2.20 shows a discrete and a continuous example of PDP.

We have learned, so far, that time dispersion and frequency-selective fading are manifestations of multipath propagation caused by delay spread. Excess delay is a random variable that, depending on the circumstance, may have a number of possible probability density functions (PDF). However, in most practical cases a reasonable distribution for τ is an exponential one, whose PDF is given by Equation 2.99.

$$f(\tau) = \frac{1}{2\pi\sigma_\tau} \exp\left(\frac{-\tau}{2\pi\sigma_\tau}\right) \tag{2.99}$$

Here σ_τ is what is accepted as the value of delay spread in most of the literature and is equal to the standard deviation (rms value) of excess delay computed based on power delay profile. One has to note that this standard deviation is different from the standard deviation of τ that is computed based on PDF of τ. Calculation of delay spread, that is, σ_τ, is carried out by the following two equations.

$$\begin{cases} E[\tau]_{\text{PDP}} \triangleq \dfrac{\displaystyle\int_0^\infty \tau P(\tau)d\tau}{\displaystyle\int_0^\infty P(\tau)d\tau}, & E\left[\tau^2\right]_{\text{PDP}} \triangleq \dfrac{\displaystyle\int_0^\infty \tau^2 P(\tau)d\tau}{\displaystyle\int_0^\infty P(\tau)d\tau} \\[3em] E[\tau]_{\text{PDP}} \triangleq \dfrac{\displaystyle\sum_k \tau_k P(\tau_k)}{\displaystyle\sum_k P(\tau_k)}, & E\left[\tau^2\right]_{\text{PDP}} \triangleq \dfrac{\displaystyle\sum_k \tau_k^2 P(\tau_k)}{\displaystyle\sum_k P(\tau_k)} \end{cases} \tag{2.100}$$

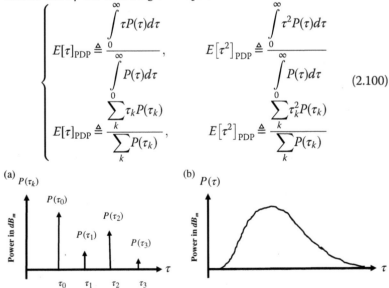

Figure 2.20 Power delay profile. (a) Discrete case. (b) Continuous case. τ is excess delay relative to the arrival time of the first echo, $\tau_0 = 0$, power values are normally expressed in dB_m.

Equation 2.100 calculates "mean value" and "mean square value" of τ based on either a continuous or a discrete PDP. The index "PDP" is used here to avoid any confusion between these expected values and the standard statistical expected values. The *delay spread* σ_τ can now be calculated by Equation 2.101.

$$\sigma_\tau = \sqrt{E[\tau^2]_{\text{PDP}} - E^2[\tau]_{\text{PDP}}} \tag{2.101}$$

One should be cautious in recognizing that delays are excess delays and are measured with respect to the arrival time of the first detectable signal component reaching the receiver, that is, $\tau_0 = 0$. Typical values of delay spread are on the order of microsecond for outdoor cellular radio channels, and in the range of nanosecond for indoor cases.

It can be shown that PDP is directly related to the baseband equivalent impulse response of the mobile channel, as shown in Equation 2.102 [4].

$$P(\tau) = |h_b(t, \tau)|^2 \tag{2.102}$$

Having calculated the delay spread, we now proceed to determine equations for coherence bandwidth. Under the assumption that excess delay has an exponential distribution (Equation 2.99), it has been shown that envelope correlation coefficient can be expressed as in Equation 2.103 [30].

$$\rho(\Delta f, \Delta t) = \frac{J_0\left(2\pi f_{d-max}\Delta t\right)}{1 + 4\pi^2 \Delta f^2 \sigma_\tau^2} \tag{2.103}$$

Here $J_0(\bullet)$ is the zeroth-order Bessel function of the first kind, f_{d-max} is the maximum Doppler shift given in Equation 2.90, and σ_τ is the delay spread. Coherence bandwidth B_{co} is equal to frequency separation Δf between two frequency components whose amplitudes are "decorrelated" (uncorrelated or lightly correlated) at the same time, that is, when $\Delta t = 0$. That is to say, coherence bandwidth is the band of frequencies over which two frequency components still have a strong amplitude correlation, meaning that they are affected by almost the same filtering effect, and thus within this band the channel remains flat. This also implies that frequency components separated by a band greater than B_{co} are subject to different filtering effect imposed by the multipath channel. Replacing $\Delta t = 0$ into Equation 2.103

$$\rho(\Delta f, 0) = \frac{1}{1 + 4\pi^2 \Delta f^2 \sigma_\tau^2} \tag{2.104}$$

The "decorrelation" is defined by the value of the correlation coefficient. According to Jakes [30], coherence bandwidth based on correlation coefficient values of 0.5 and 0.9 are as shown in Equation 2.105.

$$\begin{cases} 0.5 = \dfrac{1}{1 + 4\pi^2 \Delta f^2 \sigma_\tau^2} & \rightarrow \quad \Delta f = B_{co} = \dfrac{1}{2\pi\sigma_\tau} \\[4mm] 0.9 = \dfrac{1}{1 + 4\pi^2 \Delta f^2 \sigma_\tau^2} & \rightarrow \quad \Delta f = B_{co} = \dfrac{1}{6\pi\sigma_\tau} \end{cases} \qquad (2.105)$$

References [4] and [5] provide the following equations for coherence bandwidth based on the same values for ρ.

$$\begin{cases} \rho = 0.5 & \rightarrow \quad B_{co} \cong \dfrac{1}{5\sigma_\tau} \\[4mm] \rho = 0.9 & \rightarrow \quad B_{co} \cong \dfrac{1}{50\sigma_\tau} \end{cases} \qquad (2.106)$$

The selection of correlation coefficient based on which coherence bandwidth is defined, as was mentioned earlier, may depend on other aspects of the system. For instance, for some robust modulation schemes ρ should drop to the value 0.5 or below before the channel stops acting like a flat-fading channel, while for other modulation formats the channel behaves like frequency-selective fading one, with $\rho = 0.9$. What is shared between all four equations used for evaluation of coherence bandwidth is the inverse relationship between delay spread and coherence bandwidth. The larger the delay spread, the narrower the coherence bandwidth is. In other words, a narrow coherence bandwidth corresponds to extensive delay spread. This is the *bandwidth of the multipath channel*, not to be mistaken with the actual channel transmission bandwidth. The most widely cited equation for calculation of coherence bandwidth is the third equation in the list, that is, $B_{co} = 1/5\sigma_\tau$.

2.3.4.8 Frequency Flat Fading versus Frequency-Selective Fading

Two fundamental parameters characterizing multipath channels, namely, delay spread and coherence bandwidth, have been identified and explored. Delay spread quantifies the multipath channel time dispersion. Coherence bandwidth, on the other hand, presents a frequency domain perspective of the multipath effect and maintains that longer channel time spread results in shorter coherence bandwidth.

When the transmitted signal's period is shorter than the delay spread of the multipath channel, ISI and sever distortion follows. In other words, if the bandwidth of the transmitted signal B_{sg} is greater than the channel coherence bandwidth B_{co}, that is, $B_{sg} > B_{co}$, ISI occurs and the channel becomes frequency-selective fading. Under these circumstances, as was mentioned earlier, various frequency components (particularly extreme lower and upper frequencies) of the transmitted signal undergo different filtering effect. On the contrary, when $B_{sg} < B_{co}$, there will be no or very little ISI affecting the signal. That means all frequency components of the signal arrive at the receiver antenna with no, or insignificant, variations in distortion levels, while they are all faded in a similar fashion. Under these circumstances, the channel is frequency flat fading.

As we have mentioned earlier, preventive measures are taken to avoid frequency-selective fading conditions, as the design of the proper equalization system that can remove the effects of ISI is a formidable challenge for wireless channels.

2.3.4.9 Frequency Dispersion and Coherence Time

Delay spread and coherence bandwidth are parameters that characterize the time depressiveness of the multipath channel. However, they do not offer any insight into the time-varying nature of the channel caused by relative motion between the MS and the BS and the motion of various interacting objects in the propagation environment.

Unlike time-invariant channels, time varying channels impose *time-selective fading* on the transmitted signal. That means the channel has different filtering effect on the same frequency component of the signal at different times. Time-selective fading causes distortion due to the fact that channel characteristics, that is, channel impulse response and transfer function change while the transmitted signal is crossing the channel. In other words, the channel seen by the leading edge of the signal is different from that seen by the trailing edge.

If a signal has short-time duration such that it passes through before the channel gets a chance to change, there would be no time-selective distortion. As the signal duration increases, that is, the data rate decreases, the channel will be able to change while the signal is in flight, thereby causing time-selective distortion. The channel changes vis-à-vis motion are directly related to *Doppler spread*. Doppler spread and *coherence time* are parameters that describe the channel changes due to relative motion of MS and BS, as well as other moving objects in the transmission environment.

Doppler spread is defined as the maximum Doppler shift experienced by the signal, given by Equation 2.90 and repeated here in Equation 2.107.

$$B_{ds} \triangleq f_{d-\max} = \frac{v}{\lambda_0} = f_0 \frac{v}{c} \tag{2.107}$$

where B_{ds} is Doppler spread.

If the bandwidth of the baseband signal is much greater than B_{ds}, that is, the time duration of the signal is much less than inverse of B_{ds}, the effect of frequency depressiveness and Doppler spread is negligible at the receiver. Frequency dispersiveness or frequency spread means the reception of the same signal at different frequencies, indeed analogous to time dispersiveness in which the same signal is received at different times. When B_{ds} is much smaller than the baseband signal bandwidth, it is said that the signal experiences *slow fading*, otherwise the signal is exposed to *fast fading*.

Coherence time is roughly defined as the minimum signal duration at which frequency dispersiveness becomes appreciable. In order to quantify coherence time, we refer to envelope correlation coefficient given in Equation 2.103. The coherence time is the time separation between the same frequency components ($\Delta f = 0$) of the signal such that the components are essentially decorrelated, or the envelope correlation coefficient falls below a certain level. When $\Delta f = 0$, the envelope correlation coefficient equation is reduced to

$$\rho(0, \Delta t) = J_0^2\left(2\pi f_{d-\max}\Delta t\right) \tag{2.108}$$

If we define coherence time for time separation of the same frequency component of the signal when ρ has dropped to 0.5, we have

$$\rho(0, \Delta t) = J_0^2\left(2\pi f_{d-\max}\Delta t\right) = 0.5 \tag{2.109}$$

Solving Equation 2.109 for Δt we conclude that

$$\Delta t = T_c \cong \frac{9}{16\pi f_{d-\max}} = \frac{9}{16\pi B_{ds}} \tag{2.110}$$

Coherence time is essentially a statistical measure of the time duration over which the channel impulse response is, in effect, invariant. Rapaport [4] defines two other measures for coherence time, one corresponding to envelope correlation coefficient of 0.9, which is given by

$$T_c \cong \frac{1}{f_{d-\max}} = \frac{1}{B_{ds}} \tag{2.111}$$

It has been shown that Equation 2.111 bears an over estimation of the coherence time [4]. The second measure, which is widely accepted, is the geometric mean of the former and latter, given in Equation 2.112.

$$T_c = \sqrt{\frac{9}{16\pi f_{d-\max}} \times \frac{1}{f_{d-\max}}} \cong \frac{0.423}{B_{ds}} \tag{2.112}$$

2.3.4.10 Classification of Multipath Fading Channels

The multipath fading channels may be classified based on transmitted signal bandwidth and mobile channel characterizing parameters such as delay spread, coherence bandwidth, Doppler spread, and coherence time. Accordingly, four categories of fading channels are recognized.

1) *Frequency Flat Slow Fading Channel:* The channel is flat, which implies that all frequency components of the transmitted signal undergo essentially the same form of attenuation. This condition is present when signal bandwidth does not exceed channel coherence bandwidth. At the same time, the impulse response of the channel changes much slower than the baseband signal fluctuations, which means channel does not change (or shows insignificant variations) while the bits of transmitted signal are in flight. In other words, signal bandwidth is greater than channel Doppler spread. The frequency flat slow-fading channel condition, therefore, may be characterized by the expression given further.

$$B_{ds} < B_{sg} < B_{co} \qquad (2.113)$$

2) *Frequency Flat Fast-Fading Channel:* This is a flat channel, however, the channel impulse response changes within symbol time duration. Consequently, owing to motion in the channel, the leading edge and the trailing edge of the signal encounter different channel impulse responses. In the time domain, this means that channel coherence time is smaller than symbol period and hence time-selective fading strikes the signal. In the frequency domain, this translates into frequency dispersion. The frequency flat fast-fading channel may be conceptually identified by the following expressions.

$$B_{sg} < B_{co}, \quad B_{sg} < B_{ds} \qquad (2.114)$$

3) *Frequency-Selective Slow-Fading Channel:* Channel suffers from frequency-selective fading that is associated with the cases in which signal bandwidth exceeds the channel coherence bandwidth, or equivalently signal period is shorter than channel delay spread. As it has been argued earlier, under these circumstances, various frequency components of the signal endure different filtering effects. The channel is slow and its impulse response does not undergo significant changes while the signal is in flight. Therefore, the effects of frequency dispersion are not appreciable for this type of fading channel. In this case, the channel frequency parameters will hold the following relationships vis-à-vis signal bandwidth.

$$B_{sg} > B_{co} \quad B_{sg} > B_{ds} \qquad (2.115)$$

4) *Frequency-Selective Fast-Fading Channel:* When it happens, this channel poses the most challenging form of fading. On the one hand, the channel distorts the signal by imposing fading that is frequency selective, which means the signal bandwidth is greater than the channel coherence bandwidth. On the other hand, channel impulse response has sufficient time to alter while the bits of signal crossing the channel. Hence, in this case channel coherence time is

smaller than signal period. The corresponding characterizing expressions for this type of fading channels are provided in Equation 2.116.

$$B_{sg} > B_{co} \quad B_{sg} < B_{ds} \tag{2.116}$$

The expressions presented in Equations 2.113–2.116 are rather conceptual relationships than exact portray of fading channel conditions. After all, with the exception of signal data rate and bandwidth, all other parameters are defined by approximate equations that have not received universal acceptance.

2.3.4.11 Probabilistic Models for Frequency Flat Fading Channels

The multipath signal received at the receiving antenna may be viewed as the superposition of large number of components generated in the radio channel through propagation mechanisms of reflection, diffraction, and scattering. This scenario is commonly the case in dense urban areas and on the surface of large international airports.

2.3.4.12 Rayleigh Fading Channels

In the absence of LOS component, it can be assumed that multipath signal components are approximately independent and identically distributed. Under these circumstances the concept of the central limit theorem, or some approximate version thereof, may be invoked to conclude that the received signal is a complex Gaussian random variable [37]. This implies that the signal phase has a uniform distribution and its envelope varies in accordance with Rayleigh random variable. This type of multipath channel is said to be *Rayleigh fading channel*.

In general, a Rayleigh random variable represents the distribution of the magnitude of sum of two independent zero-mean equal-variance Gaussian random variables, like the envelope of the complex Gaussian random variable mentioned above. The PDF of Rayleigh random variable is given by Equation 2.117.

$$f_R(r) = \frac{r}{\sigma^2} e^{-r^2/2\sigma^2} u(r) \tag{2.117}$$

Here σ^2 is the variance of original Gaussian components and $u(r)$ is the unit-step function. It can be readily shown that the mean value and variance of Rayleigh random variable are given by Equations 2.118 and 2.119.

$$E(R) = \sqrt{\frac{\pi}{2}}\sigma \tag{2.118}$$

$$\sigma_R^2 = \left(\frac{4 - \pi}{2}\right)\sigma^2 \tag{2.119}$$

The CDF (integral of the PDF function) and power distribution of a Rayleigh random variable are expressed in Equations 2.120 and 2.121.

$$F_R(r) \triangleq Pr(r \leq R) = \left[1 - \exp\left(\frac{-R^2}{2\sigma^2}\right)\right] u(R) \tag{2.120}$$

$$f_P(p) = \frac{1}{2\sigma^2} \exp\left(\frac{-p}{2\sigma^2}\right) u(r) \tag{2.121}$$

Rayleigh distribution is the most widely accepted probabilistic model for the description of time variations of the received signal envelope in frequency flat fading channels. A sample of signal envelope variation versus time for Rayleigh fading channel is shown in Figure 2.21.

Figure 2.22 illustrates the distribution of the signal envelope for a Rayleigh fading channel for three values of σ.

The CDF (Equation 2.120) computes the probability that the envelope of the received signal does not exceed a given value of R. The signal power distribution, given by Equation 2.121 for Rayleigh fading channels, is an important piece of information. The communication system receivers are normally designed to operate with a minimum signal power level, that is, the receiver sensitivity. In order to provide a desired level of performance quality, however, a particular minimum power level is required, which is

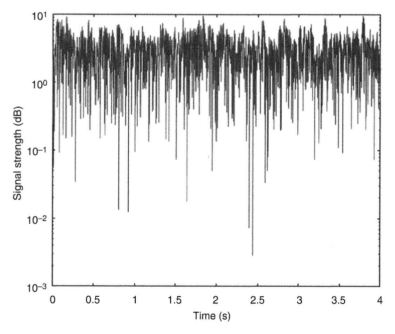

Figure 2.21 A sample function of Rayleigh fading signal envelope.

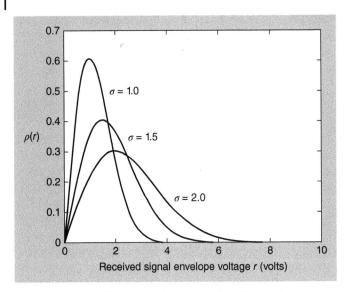

Figure 2.22 Probability density function for Rayleigh distribution.

normally greater than the receiver sensitivity, and it is usually referred to as *threshold value*. In that sense, when received signal power drops below the threshold level, the receiver is said to be going through *outage* or *fade*. The PDF of power distribution may be exploited to compute the probability of outage, which represents the fraction of time that the receiver is in fades. Assuming that the threshold power value is given by P_{TH}, and using Equation 2.121, the probability of outage for Rayleigh fading channels is calculated from Equation 2.122 given further.

$$Pr(outage) = \int_{0}^{P_{TH}} \frac{1}{2\sigma^2} \exp\left(\frac{-p}{2\sigma^2}\right) dp = 1 - \exp\left(\frac{-P_{TH}}{2\sigma^2}\right) \quad (2.122)$$

It should be noted that $2\sigma^2$ is the expected value of the total power of the signal. For instance, if we define *deep fade* to be the signal power level that is 30 dB below the mean value of the signal total power, the probability that the signal goes through deep fade can be calculated by Equation 2.122, as carried out in following manner.

$$10 \log\left(\frac{P_{TH}}{2\sigma^2}\right) = -30 \rightarrow \frac{P_{TH}}{2\sigma^2} = 0.001 \rightarrow Pr(outage) = 1 - \exp(-0.001) \cong 0.001$$

Thus, on average, the signal is lost to a deep fade 0.1% of the time. If the outage is defined as the power level of 20 dB below $2\sigma^2$, then similar computation shows that, on average, the receiver goes through outage 1%

of the time. Error correction coding with sufficient burst error correction capability can recover most of the lost bits.

The number of fades per some defined time duration, as well as the expected value of time duration of fades, are important system parameters that are used for the design of diversity schemes and error control coding formats for mitigation of the effect of multipath fading. Based on S. O. Rice's analytical work on finding joint probability density function of Rayleigh random variable and its time derivative, simple expressions have been developed for Rayleigh fading channels for computation of the expected value of the number of times that the signal crosses a particular level (level crossing rate, LCR) per second, and the expected value of time duration of the corresponding fades [38,39]. Accordingly, the LCR is calculated from Equation 2.123.

$$R_{LC} = \sqrt{2\pi} \frac{v}{\lambda} \beta e^{-\beta^2} \tag{2.123}$$

Here $\beta \triangleq R/R_{rms}$ is the normalized referenced level (with respect to *rms* value of the signal envelope), v is the speed of MS, and λ is the wavelength of the carrier wave. The expected value of fade duration is computed from Equation 2.124.

$$E\left[\tau_{fade}\right] = \frac{\left(e^{-\beta^2} - 1\right)\lambda}{\sqrt{2\pi}\beta v} \tag{2.124}$$

The expected value of fade duration combined with signal symbol rate determines the expected value of the number of data symbols lost in each fade, and thereby provides some key information needed for the design of diversity or ECC system.

2.3.4.13 Rician Fading Channels

In case one of the multipath components is much more significant than the others such that it plays a dominant role in the mix (some references maintain that the large component is nonfading and deterministic [4]), it can be shown that the fading signal envelope is distributed according to Rician probability density function given in Equation 2.125. Under these circumstances, smaller multipath components arriving from different directions are superimposed on the dominant component. Normally, if there is a strong LOS component, the channel is assumed to be Rician, such is the case in satellite mobile communication systems.

$$f_R(r) = \frac{r}{\sigma^2} e^{-\frac{r^2 + A^2}{2\sigma^2}} I_o\left(\frac{Ar}{\sigma^2}\right) u(r) \tag{2.125}$$

In this latter equation, $I_0(\bullet)$ is the zero-order modified Bessel function of the first kind, whose values may be looked up from a table, or calculated from Equation 2.126. Parameter A denotes the peak amplitude or expected value of the peak amplitude of the dominant component.

$$I_0(x) = \frac{1}{2\pi} \int_0^{2\pi} e^{x\cos\theta} d\theta \tag{2.126}$$

Figure 2.23 provides plots of Rician probability density function for $\sigma = 1$ and for four different values of parameter K. Parameter K, the "Rician parameter," is used as the third variable on the set of Rician plots more often than A. K is the ratio of power of the dominant component to the total expected value of received scattered power $2\sigma^2$, hence it is defined by Equation 2.127.

$$K = \frac{A^2}{2\sigma^2} \quad \rightarrow \quad K_{dB} = 10\log\frac{A^2}{2\sigma^2} \tag{2.127}$$

In Figure 2.23, K is expressed in direct values.

Figure 2.23 reveals that when the LOS component (or the dominant module) does not exist (or is very weak), corresponding to $K \cong 0(K \rightarrow -\infty$ dB), the Rician PDF approaches that of Rayleigh. On the other hand, as dominant component becomes stronger in the mix, that is, when $K = 10(10$ dB), the Rician PDF assumes a bell-shaped form that approximates the Gaussian probability density function.

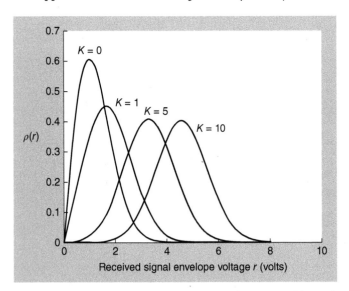

Figure 2.23 Rician probability density functions for $\sigma = 1$ and four values of K.

2.4 Challenges of Broadband Transmission over the Airport Surface Channel

In this section, we venture few general comments on intricacies and challenges of broadband signaling, over MLS spectrum, through the airport surface channel. The detailed description of airport channel characteristics over 5 GHz band is provided in Chapter 3.

Independent from the allocated spectrum over which a wireless airport communication system operates, one encounters a mixture of propagation and mobility issues that are unique to the airport surface radio channel. Specifically, the combination of mobility scenarios and propagation environments on the surface of an airport can be vastly different from one spot to another, all within a limited area. This ranges from terminal areas that house low-speed or parked stations but within a highly multipath environment, to open areas of takeoff and landing runways with high-speed aircraft movement and LOS or near LOS propagation conditions.

AeroMACS is created based on IEEE-802.16e-standared (WiMAX technology) over C-band spectrum of 5.090–5.150 GHz. Consequently, AeroMACS inherits all challenges posed by complexities of radio channels, as well as those pertaining to broadband transmission.

Regarding path loss, the signal power through large-scale propagation over Aeronautical C-band is significantly more intense than that over the legacy aeronautical VHF band, as all path loss models introduced in this chapter testify to a direct relationship between frequency and path loss.

Large-scale fading and log-normal shadowing can pose a problem, particularly in large airports (see Chapter 3) where giant aircraft (Boeing 747, Airbus A-380) are parked or move around the airport surface. The effect may manifest itself in an increase in the variance of the log-normal distribution.

Multipath fading and time dispersion due to presence of large number of reflecting and scattering objects on the airport surface can be severe in the channel, particularly in the ramp area. Doppler spread and frequency dispersion due to mobility are much wider for the airport channel, owing to high speeds that aircraft may assume on the runways when they are about to takeoff or right after landing. Thus, small-scale fading can be more severe over certain parts of the airport surface than it is for land mobile radio channels.

In addition to standard cellular network interferences, as discussed in Section 2.3.1, interference due to scarcity of spectrum, which dictates spectrum sharing and tight allocation of radio frequencies is always a challenge to reckon with. The broader issue here is to use a limited available radio spectrum to accommodate an ever-increasing number of users requiring broadband services and applications while they are on the move. AeroMACS shares its spectrum with mobile satellite systems (Globalstar). Chapters 7 and 8 will shed some light on the subject of AeroMACS interference to coallocated applications.

Noise in the form of additive white Gaussian is a concern in wideband wireless networks, as the large bandwidth essentially raises the noise floor for these networks [5].

2.5 Summary

This chapter serves to provide a brief review of cellular networking and propagation environment over which AeroMACS is presumed to operate. The background material covered in this chapter is oriented toward assisting the reader for understanding various required protocols and signal processing techniques (to be discussed in future chapters) used in the physical layer and to some extent in the MAC layer of AeroMACS networks. The chapter, also, supports the objective of making the text self-contained as much as possible.

Cellular concept and cellular network design, required for understanding of AeroMACS network architecture, is presented. Radio channel characterization, a challenging issue in the planning, development, and deployment of any wireless mobile network, is reviewed in some detail. Three categories of signal degradations associated with mobile radio channels are identified. Long-term attenuation, path loss, is a TR distance-dependent signal power loss, which is also a function of frequency band over which the network operates. Common deterministic and statistical empirical models for calculation and estimation of path loss are presented. Large-scale fading (medium-scale attenuation), represented by the log-normal shadowing model, which is essentially superimposed over the signal path loss, is explored. With large-scale path loss prediction and log-normal shadowing effect added to it, the network designer is in a position to begin to define the hardware makeup of the network, antenna gains, location and height of BS antenna towers, transmitter power, and so on. Small-scale fading, caused by multipath propagation and Doppler effect, subjects the received signal to rapid fluctuations with respect to time and displacement. In order to combat the degradation imposed by short-term fading, various radio and signal processing techniques are employed. OFDM/OFDMA, MIMO, and smart antenna, diversity, and channel coding, are examples of this sort.

There is a wealth of literature available for the reader who is interested in more rigorous treatment of the topics covered in this chapter. Section "References" lists a few of these sources [4,11,12,15,30].

References

1 W. C. Y. Lee, "Smaller Cells for Greater Performance," *IEEE Communication Magazine*, 29(11), 19–23, 1991.
2 Fujitsu, *"High-Capacity Wireless Solutions: Picocell or Femtocell?"* 2013. (Available online.)
3 V. H. Macdonald, "The Cellular Concept," *Bell System Technical Journal*, 58(1), 15–41, 1979.
4 T. S. Rappaport, *Wireless Communications: Principal and Practice*, 2nd edn, Prentice Hall, 2002.
5 B. A. Black, P.S. Dipiazza, B. A. Ferguson, D. R. Voltmer, and F. C. Berry, *Introduction to Wireless Systems*, Prentice Hall, 2008.
6 C. A. Balanis, *Antenna Theory Analysis and Design*, 3rd edn, John Wiley & Sons, Inc., New Jersey, 2005.
7 I. Katzela and M. Naghshineh, "Channel Assignment Schemes for Cellular Mobile Telecommunication Systems: A Comprehensive Survey," *IEEE Communications Surveys and Tutorials*, 3(2), 10–31, 2000.
8 D. C. Cox and D. O. Reudink, "Dynamic Channel Assignment in Two Dimension Large-Scale Mobile Radio Systems," *Bell System Technical Journal*, 51(7), 1611–1629, 1972. (Also published online by John Wiley & Sons, Inc., July 2013.)
9 K. Okada and F. Kubota, "On Dynamic Channel Assignment Strategies in Cellular Mobile Radio Systems," *IEICE Transactions on Fundamentals*, E75-A (12), 1634–1641, 1992.
10 J. R. Boucher, *Traffic System Design Handbook: Telecommunication Traffic Tables and Programs*, Wiley-IEEE Press, 1992.
11 A. F. Molisch, *Wireless Communications*, 2nd edn, John Wiley & Sons, Inc., New Jersey, 2010.
12 W. C. Y. Lee, *Mobile Communications Engineering*, McGraw-Hill, 1997.
13 D. C. Cox, "Co-Channel Interference Consideration for In Frequency Reuse Small Coverage Area Radio Systems," *IEEE Transactions On Communications*, COM-30 (1), 135–141, 1982.
14 G. L. Stuber, *Principle of Mobile Communications*, 3rd edn, Springer, 2011.
15 V. K. Garg and J. E. Wilkes, *Wireless and Personal Communications Systems*, Prentice Hall, 1996.

16 H. Huang O. Alrababi, J. Daly, D. Samardzija, C. Tran, R. Valenzuela, S. Walker, "Increasing Throughput in Cellular Networks with Higher-Order Sectorization," *Proceedings of IEEE Signals, Systems, and Computers conference*, 2010.

17 V. S. Abhayawardhana, et al., "Comparison of Empirical Propagation Path Loss Models for Fixed Wireless Access Systems," *Proceedings of IEEE Vehicular Technology Conference*, vol. 1, 73–77, 2005.

18 C-F Yang, B-C Wu, and C-J Ko, "Ray-Tracing Method for Modeling Indoor Wave Propagation and Penetration," *IEEE Transactions on Antennas and Propagation*, 46(6), 907–919, 1998.

19 E. C. Jordan and K. G. Balmain, *Electromagnetic Waves and Radiating Systems*, Prentice Hall, 1968.

20 Y. Okumura et al., "Field Strength and its variability in VHF and UHF Land Mobile Radio Service," *Review of the Electrical Communication Laboratory*, 16, 825–873, 1968.

21 M. Hata, "Empirical Formula for Propagation Loss in Land Mobile Radio Services," *IEEE Transaction On Vehicular Technology*, VT-29 (3), 317–325, 1980.

22 European Cooperation of Scientific and Technical Research EURO-COST 231, *Urban Transmission Loss Models for Mobile Radio in the 900 and 1800 MHz Bands*, Revision 2, The Hague, 1991.

23 M. Alshami, T. Arslan, J. Thompson, and A. T. Erdogan, "Frequency Analysis of Path Loss Models on WIMAX," *Proceeding of 3rd IEEE CEEC*, pp 1–6, 2011.

24 V. Erceg, et al, "An Empirically Based Path Loss Model for Wireless Channels in Suburban Environments," *IEEE Journal of Selected Areas on Communications*, 17, 1205–1211, 1999.

25 IEEE 802.16 Broadband Wireless Access Working Group, "Comparison of Propagation Path Loss Models", 2007. (Available online.)

26 ITU-R, Recommendation ITU-R P.529-3, "Prediction methods for the terrestrial land mobile service in the VHF and UHF bands," 1978-1990-1995-1999. Available online.

27 Electronic Communication Committee (ECC) within the European Conference of Postal and Telecommunications Administration (CEPT), "The Analysis of the Coexistence of FWA Cells in the 3.4–3.8 GHz band," Technical report, ECC Report 33, May 2003.

28 M. Alshami, "Evaluation of Path Loss Models at WiMAX Cell- Edge," *IEEE NTMS 4th IFIP International Conference*, Feb. 2011.

29 Y. S. Meng and Y. H. Lee, "Investigation of Foliage on Modern Wireless Communication Systems: A Review," *Progress in Electromagnetics Research*, 105, 313–332, 2010.

30 W. C. Jakes, *Microwave Mobile Communications*, IEEE Press, New Jersey, 1993.

31 S. Y. Seidel et al., "Path Loss, Scattering and Multipath Delay Statistics in Four European Cities for Digital Cellular and Microcellular Radiotelephone," *IEEE Transactions on Vehicular Technology*, 40(4), 721–730, 1991.

32 D. O. Reudink, "Properties of Mobile Radio Propagation above 400 MHz," *IEEE Transactions on Vehicular Technology*, 23(2), 1–20, 1974.

33 L. W. Couch, *Digital and Analog Communication Systems*, 8th edn, Pearson, 2013.

34 G. L. Turin et al., "A Statistical Model for Urban Multipath Propagation," *IEEE Transactions on Vehicular Technology*, VT-21, 1–9, 1972.

35 V. Jungnickel, et al., "Role of Small Cells, Coordinated Multipoint, and Massive MIMO in 5G," *IEEE Communications Magazine*, 52(5), 44–51, 2014.

36 D. Gesbert, et al., "From Theory to Practice: An Overview of MIMO Space-Time Coded Wireless Systems," *IEEE Journal on Selected Areas in Communications*, 21(3), 281–302, 2003.

37 A. Papoulis, *Probability, Random Variables and Stochastic Processes*, 4th edn, McGraw Hill, 2002.

38 R. H. Clarke, "A Statistical Theory of Mobile-Radio-Reception," *Bell System Technical Journal*, 47, 957–1000, 1968.

39 A. Abdi, K. Wills, H. A. Barger, M.-S. Alouini, and M. Kaveh,"Comparison of the Level Crossing Rate and Average Fade Duration of Rayleigh, Rice and Nakagami Fading Models with Mobile Channel Data," *Proceedings of IEEE Vehicular technology conference*, Fall, 2000.

3

Wireless Channel Characterization for the 5 GHz Band Airport Surface Area*

3.1 Introduction

This chapter discusses the characteristics of the wireless channel for the airport surface area. The chapter provides a motivation for this topic, some background on wireless channels and modeling, and specific results for the airport surface channel. Much of the airport surface-specific material here comes from a NASA project [1] that resulted in two journal papers [2,3].

3.1.1 Importance of Channel Characterization

When new radio standards and technologies are introduced for any modern application, one of the first steps involved in this complex, and often long, process is that of channel characterization. One might define channel characterization as the quantitative specification of the wireless channel. This specification usually results in a set of channel models. The use of channel models for communication system design and evaluation is widespread, and universally accepted as an important element of system optimization.

Mathematical channel characterizations provide fundamental knowledge for physical layer (PHY) waveform design and analysis. A description of the channel time variation is also important for data link layer, particularly medium access control (MAC) layer design. Before building or deploying any communication system, the use of thorough channel characterization information allows prediction and tradeoff studies that address various aspects of communication system design, such as

---

*Contributed by David W. Matolak, University of South Carolina.

AeroMACS: An IEEE 802.16 Standard-Based Technology for the Next Generation of Air Transportation Systems, First Edition. Behnam Kamali.
© 2019 the Institute of Electrical and Electronics Engineers, Inc. Published 2019 by John Wiley & Sons, Inc.

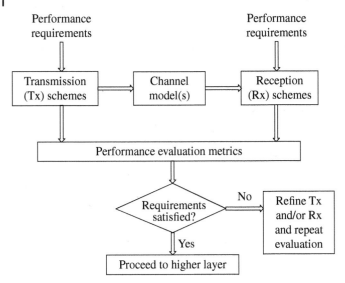

Figure 3.1 Illustration of use of channel models in evaluation of PHY Tx/Rx schemes.

communication link range, optimal channel/subchannel bandwidths, and system performance (bit error ratio, throughput, latency, etc.) for *any* potential waveform used over the channel [4]. The use of channel models to evaluate transmission and reception schemes at the PHY layer is illustrated in Figure 3.1.

Figure 3.1 can pertain to one or more simultaneously operating wireless links. In this figure, performance requirements of the communication system specify values for parameters of the transmission scheme (e.g., required bit rate, spectral characteristics), and also for the reception scheme (e.g., required packet error probability). For a selected transmission/reception scheme, the performance evaluation metrics can depend upon the channel model(s) used. If the performance evaluation outputs indicate the system will meet its requirements, then the system design can proceed on to the higher layers of the protocol stack. If the performance evaluation metrics indicate that the transmission/reception scheme will not meet requirements, then, with knowledge of the channel, appropriate remedies can be added at the transmit (Tx) or receive (Rx) or both ends of the link, and the evaluation repeated.

The physical layer performance characterization is indispensable for the design and performance prediction for higher layers in the communications protocol stack, which depend upon the physical layer for message transfer [5]. The physical layer performance directly affects the data link, MAC, and network layers, and through these, the performance of all higher layers.

Several system PHY and data link layer design parameters upon which the channel characterization has a significant effect include the following [6]:

- Modulation(s) and corresponding detection schemes [6]
- Forward error correction coding and associated interleaving schemes [7]
- Antenna characteristics, including diversity antenna parameters [8]
- Receiver processing algorithms, including those for synchronization, interference suppression, combining, and so on, all of which are adaptive [9]
- Signal bandwidths [4]
- Adaptation algorithms for resource allocation in time, frequency, and spatial domains [10]
- Physical facility siting rules [11]
- Duplexing and multiplexing schemes [9]
- Security techniques (against eavesdropping, jamming, spoofing, etc.) [12]

Accurate channel models contain mathematical descriptions that can be used for analysis, but often analysis becomes intractable, at which point evaluations can be conducted and extended via computer simulations [13,14]. Simulations are extensively used to assess and design modern communication systems. Thus, the channel model consists not only of mathematical descriptions, but also the "software implementation" of these mathematical descriptions. Ultimately, if a wireless communication system is deployed without a thorough channel characterization, the system will most certainly be suboptimal. Well-known performance limits that can arise from not accounting for channel characteristics include an irreducible channel error rate that can preclude reliable message transfer and severely limited data carrying capacity.

3.1.2 Channel Definitions

A definition for the wireless channel is the set of all parameters for all transmission paths taken by an electromagnetic signal from transmitter to receiver, in a frequency band of interest, over an area (or spatial volume) of interest. These parameters include the amplitude, phase, and delay for each path or component. In general, all parameters can be temporally or spatially varying. This is in contrast to guided wave transmission schemes (those that use wires, cables, waveguides, light guide fibers, etc.), which are largely time invariant.[1]

Strictly, one could define as many types of channel models as there are types of communication links. In practice, this is neither desirable – because of complexity – nor necessary – since many channels exhibit similar characteristics. For wireless systems, the channel is not typically under the direct or complete control of the system designer or operator. For the simplest cases, or for specific well-defined (often time-invariant) environments, the channel can be defined with high accuracy. In this case, models for the channel can be deterministic. In more complex cases, with mobility of Tx and/or Rx and/or objects in the environment, or when a model is to represent a range of environmental conditions, statistical channel models are apt. As with guided wave versus wireless channel models, another classification is deterministic versus statistical models.[2]

From the perspective of electromagnetic field theory, *any* wireless channel could be viewed as being purely deterministic, and hence channel characteristics could be calculated to any arbitrary degree of precision, at any point in space at any time – *if* one had knowledge of all geometry and electrical parameters of all objects in the environment, and *if* one could solve the field theory equations (Maxwell's equations) rapidly and accurately enough. In many settings though, particularly with mobility, the required knowledge translates to a *very* large amount of data, and hence renders this approach impractical. This motivates the use of statistical channel models. Interested readers are referred to available texts, for example, Ref. [16], for deterministic treatments of wireless channels.

There are also additional classifications of wireless channels, which include terrestrial versus aeronautical, maritime, or satellite, line of sight (LOS) versus non-LOS (NLOS), indoor versus outdoor, classifications by frequency band or primary propagation mechanism, and so on. Examples of useful texts either dedicated to wireless channel modeling or with

---

1 Guided wave channels can exhibit time variation when either anomalous events, such as accidental cutting of cables, occur or over the very long term, when cable materials degrade or their characteristics change with temperature, aging, and so on. Nonetheless, for modern communication systems that transfer messages over durations of seconds, minutes, or even hours, guided wave channels are well modeled as time invariant.

2 Note that strictly speaking, the term deterministic must be used with some caution, since in wireless settings, even the most careful design cannot account for all contingencies, atypical events may occur, and these can be treated as random. A famous example of this was when Penzias and Wilson of Bell Laboratories were first discovering the cosmic microwave background radiation [15], for which they eventually won the Nobel Prize. After carefully calibrating their system and finding themselves unable to explain results, a close inspection of their receive antenna revealed a nest of birds, whose presence altered the antenna characteristics.

multiple detailed chapters include Refs. [11] and [17–24]. Additional information, much standardized, is also available from the International Telecommunication Union (ITU) in its Recommendation series on propagation [25].

For essentially all cases, wireless channels are modeled as linear filters, and hence are characterized completely by their channel impulse response (CIR), or equivalently, their channel transfer function (CTF), the Fourier transform of the CIR. The discussion here thus focuses upon this response and its characterization.

3.1.3 Airport Surface Area Channel

The airport surface area is defined here as all outdoor area on airport property. This includes runways, taxiways, areas near gates, maintenance areas, and all areas in between. As is well known, this area is a dynamic environment where airline activities such as baggage handling, fueling, and catering take place throughout the day and night, while aircraft are taking off and landing, taxiing, pushing into and pulling out of gates, while airport security vehicles and other ground vehicles are moving about. Figure 3.2, from Ref. [2], shows a photograph taken from the air traffic control tower (ATCT) at JFK International Airport, illustrating features in the airport surface environment.

The airport surface area (ASA) channel is defined as the channel between the antenna at the ground site and the antenna on some mobile

Figure 3.2 View of some large airport features at John F. Kennedy International Airport, taken from air traffic control tower (ATCT) [2].

device located on the airport surface. The ground site antenna is often atop the ATCT, but may also be located on an airport building roof, or on a small tower on the ASA itself. The mobile devices on the airport surface may be aircraft, ground vehicles, or carried by persons. Most of our results pertain to the channel with the ground-site antenna at the ATCT, and the mobile device contained within a vehicle on the ASA; some results with the ground-site antenna located at an airport field site on the ASA are also provided. The ASA channel is a terrestrial, point-to-multipoint channel with some features in common with other terrestrial mobile channels; the distinguishing features are those unique to the ASA, for example, large metallic aircraft and an open area often containing very large buildings on its perimeter.

Worth mentioning is that airports are of various sizes, and in Refs [2] and [3], we created a three-level classification of airports as small, medium, and large. Large airports are busy, with many large jet airplanes (e.g., 747's and 777's), and at least several hundred aircraft arriving and departing per day, typically 80 or more during a busy hour [26]. The diameter of most large airports is typically no more than 5 km. Examples of large airports include JFK and Miami International Airport (MIA). The *medium* airport class has much in common with the large airports. Medium-sized airports have buildings on the airport surface, but these are not as large or as numerous as in the large airport class. These airports also typically do serve the largest jets. An example of medium-sized airport is Cleveland Hopkins International Airport (CLE). The medium airports are still significantly bigger and busier than small, general aviation (GA) airports, which may have a single runway and only one building structure.

Several distinct propagation channel regions exist within most airports. As with terrestrial channel models, this includes LOS and NLOS regions. Many terrestrial channel models are also specific to an environment type [6], for example, urban, suburban, or rural. Defining such region types objectively can be difficult, and can often lead to further specification into subclasses [27].

In Refs [2] and [3], we classified ASA propagation regions into three types: LOS-Open (LOS-O), NLOS-Specular (NLOS-S), and NLOS. The LOS-O areas are those clearly visible from the ATCT (or ground-site antenna), with no significant scattering objects nearby, for example, runways and portions of taxiways. The NLOS-S regions are those in between the other two and exhibit *mostly* NLOS conditions, but have a distinct specular, first-arriving component in the CIR, in addition to lower energy multipath components (MPCs). An example of NLOS-S region would be ground vehicle lanes near terminal buildings, where a significant diffracted signal component is received. The NLOS regions are those that have a completely obstructed LOS to the ATCT. These regions are near airport gates or behind large airport buildings.

Aircraft and ground vehicles may traverse all three types of regions as they move about the ASA. This has consequences for statistical channel models that will be addressed subsequently. A final comment on the ASA channel regards the spatial distribution of MPCs: scattering is almost *never* isotropic about the mobile terminal.

3.2 Statistical Channel Characterization Overview

In this section, we briefly describe wireless channel characteristics that are often modeled statistically. This includes a description of key channel parameters. As background, we first describe the CIR and CTF.

3.2.1 The Channel Impulse Response and Transfer Function

A general form for the multipath CIR is as follows:

$$h(\tau, t) = \sum_{k=0}^{L(t)-1} \alpha_k(t)\exp\{j[\omega_{D,k}(t)\{t - \tau_k(t)\} - \omega_c\tau_k(t)]\}\delta(\tau - \tau_k(t))$$

$$= \sum_{k=0}^{L(t)-1} \alpha_k(t)\exp\{j\phi_k(t)\}\delta(\tau - \tau_k(t)) \tag{3.1}$$

Here $h(\tau, t)$ is the output of the channel at time t due to an impulse input at time $t - \tau$, in complex baseband form. This CIR is in the form of "discrete impulses" via the Dirac deltas, and hence presumes infinite bandwidth. Nevertheless, the equation is often a good approximation, and the impulses $\delta(\tau - \tau_k(t))$ can be replaced by pulses of finite duration. In both forms of Equation 3.1, functions of time are indicated only by parentheses, for example, $\tau_k(t)$. The interpretation of Equation 3.1 is that the channel imposes specific discrete attenuations (α_k), phase shifts (ϕ_k), and delays (τ_k) upon any signal transmitted. This approximation is typically very good for signal bandwidths of several tens of MHz or more, but is not accurate for very wideband signals such as ultrawideband (UWB) [28]. Other types of channels, such as HF troposcatter channels, are better represented by a continuous function of both τ and t. The CIR in Equation 3.1 represents the response between two antennas; multi-antenna channels can be represented as vectors or matrices with entry responses of the form given in Equation 3.1.

The remaining parameters in Equation 3.1 are the radian carrier frequency $\omega_c = 2\pi f_c$, which can of course also be a function of time, and the kth resolved Doppler frequency $\omega_{Dk} = 2\pi f_{D,k}$ with $f_{D,k}(t) = v(t)f_c$ $\cos[\theta_k(t)]/c$. Here, $v(t)$ is relative velocity between Tx and Rx, $\theta_k(t)$ is the aggregate phase angle of all components arriving in the kth delay "bin," and c is the propagation velocity, well approximated by the speed of light. The delay bin width is approximately equal to the reciprocal of the signal bandwidth – components separated in delay by an amount smaller than the bin width are "unresolvable." The kth resolved component in (3.1) thus often consists of multiple terms "subcomponents" from potentially different spatial angles $\theta_{k,i}$. We do not address "spatial" channel modeling, for example Refs [29,30], here, other than brief comments in a subsequent section.

The CTF corresponding to Equation 3.1 is

$$H(f,t) = \sum_{k=0}^{L(t)-1} \alpha_k(t)\exp\{j[\omega_{D,k}(t)\{t - \tau_k(t)\}]\}\exp[-j\omega_c\tau_k(t)]e^{-j2\pi f\tau_k(t)}$$

(3.2)

where the frequency dependence is expressed by the final exponential term. The second exponential can change significantly with small changes in delay $\tau_k(t)$ when f_c is large. This second term typically dominates the small-scale fading variation, as f_c is usually *much* larger than $f_{D,k}$. As an example, for ASA applications, if the carrier frequency is 5 GHz, and relative velocity is 60 m/s (roughly 135 miles/h), $f_{D,\max} = 1$ kHz $\ll f_c$.

In Figure 3.3, we illustrate a conceptual CIR (magnitude). This diagram illustrates variation (fading) in time t and variation of impulse energy ($\sim\alpha_k^2$) with delay τ. This type of CIR is useful in analysis, simulations, or hardware in terms of the common tapped-delay line (TDL) model. The TDL is a linear, finite IR filter, as shown in Figure 3.4.[3] In Figure 3.4, the input signal is $s(t)$ the output signal is $y(t)$, and other parameters are as defined for Equation 3.1.

3.2.2 Statistical Channel Characteristics

Essentially, any of the CIR parameters, other than the carrier frequency (unless it is "randomized" by transmission) and the propagation velocity, can be modeled statistically. Typically, the MPC amplitudes α_k, phase shifts ϕ_k (and spatial angles of arrival θ_k embedded within these phases), and delays τ_k are modeled as random. Also, modeled as random are more "composite" channel features such as attenuation, delay spread, and

---

3 Implicit in Figure 3.3 is the concept of multiple timescales: the short-term "delay" scale (τ) and the longer-term timescale over which the channel's parameters evolve. For most channels considered to be "slowly fading," the CIR is viewed as "decaying" or ending in delay τ long before any of the components change in time t appreciably. This will be true for ASA channels.

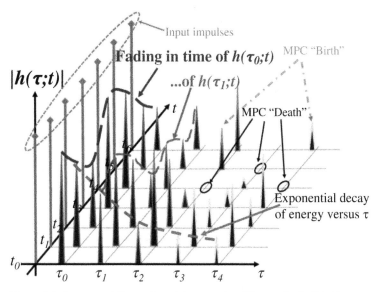

Figure 3.3 Conceptual illustration of time-varying CIR magnitude $|h(\tau,t)|$.

Doppler spread, which arise from statistics of the CIR. We briefly address these next.

Attenuation is typically quantified as the loss in power of a transmitted signal. This can be expressed, in dB, as $20 \log10[|h(\tau,t)|]$, and the models are known as path loss models. The well-known, deterministic, free-space model (the simplest possible wireless channel path loss model) is given by $L_{fs} = 20 \log(4\pi d/\lambda)$, with d the link distance and λ the wavelength. Modern path loss models are most often specified, in dB, as

$$L(d) = L_0 + 10\nu \log(d/d_{\min}) + X, \qquad d_{\min} \le d \le d_{\max} \qquad (3.3)$$

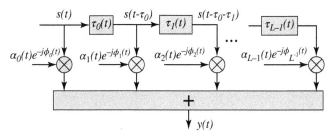

Figure 3.4 General form of wideband TDL model corresponding to CIR in Equation 3.1.

where L_0 is a constant value of attenuation at distance d_{min}, ν is the path loss exponent, and X is a zero mean Gaussian random variable with standard deviation σ_X. Parameters L_0, ν, and σ_X are most often determined empirically for a given setting. In the model of (3.3), the only statistical parameter is σ_X.

Delay spread is a measure of the width, in delay τ, of the CIR. There are multiple measures for this width, with the most common being the root mean square delay spread (RMS-DS), denoted σ_τ [6]. Other metrics include the delay window and delay interval [31]. The RMS-DS can be computed as follows

$$\sigma_\tau = \sqrt{\frac{\sum_{k=1}^{L-1} \alpha_k^2 \tau_k^2}{P} - \mu_\tau^2}$$

$$\sigma_\tau = \sqrt{\frac{\sum_{k=0}^{L-1} \alpha_k^2 \tau_k^2}{P} - \mu_\tau^2}$$

(3.4)

where α_k and τ_k are as defined in (3.1), and μ_τ is the mean energy delay, given by

$$\mu_\tau = \frac{\sum_{k=1}^{L-1} \alpha_k^2 \tau_k}{P}$$

(3.5)

The term $P = \sum_{k=0}^{L-1} \alpha_k^2$ in (3.4) and (3.5) is the total power in the power delay profile (PDP). Both (3.4) and (3.5) can be computed for either an average CIR over a set of data, or for a single CIR. In the latter case, the parameters in (3.4) and (3.5) are termed the *instantaneous* RMS-DS [32] and instantaneous mean energy delay, respectively.

The computation of the average CIR or average PDP was first addressed in the classic paper by Bello [33]. In this work, the CIR (and consequently, the CTF as well) was treated as a random process, and multiple correlation functions were defined to characterize its second-order statistics. Details of this stochastic treatment are beyond the scope of this chapter, and can be found in Refs [6] and [33], for example. Here, we briefly describe the correlation functions of the CIR and CTF, and how these give rise to common statistical parameters such as the RMS-DS.

The CIR correlation function[4] is

$$R_{hh}(\tau_1, \tau_2, t_1, t_2) = E[h(\tau_1, t_1)h^*(\tau_2, t_2)]$$

(3.6)

---

4 Bello focused on correlation and not covariance functions because of many cases of interest, for example, NLOS conditions, the channel is considered to be zero mean. Also, for stationary or wide-sense stationary assumptions, the correlation and covariance differ only by a constant.

where the asterisk denotes complex conjugation. The CTF correlation function is analogously defined:

$$R_{HH}(f_1, f_2, t_1, t_2) = E[H(f_1, t_1)H^*(f_2, t_2)] \qquad (3.7)$$

The functions R_{hh} and R_{HH} are related through the double Fourier transform, that is,

$$R_{HH}(f_1, f_2, t_1, t_2) = \int\int R_{hh}(\tau_1, \tau_2, t_1, t_2) e^{-j2\pi f_1 \tau_1} e^{-j2\pi f_2 \tau_2} d\tau_1 d\tau_2 \qquad (3.8)$$

Since four-dimensional functions are complicated to work with, simplifying assumptions are often made. The first is uncorrelated scattering (US), which means that MPCs at different values of delay are uncorrelated. This reduces R_{hh} to $R_{hh}(\tau, t_1, t_2)\delta(\tau - \tau_1)$, and R_{HH} to $R_{HH}(\Delta f, t_1, t_2)$, with $\Delta f = |f_1 - f_2|$. The second assumption is wide-sense stationarity (WSS) in time, which results in replacing t_1 and t_2 in both R_{hh} and R_{HH} with $\Delta t = |t_1 - t_2|$. The resulting WSSUS correlation functions are then $R_{hh}(\tau, \Delta t)\delta(\tau - \tau_1)$ and $R_{HH}(\Delta f, \Delta t)$, with the latter termed the spaced-frequency/spaced-time correlation function. When $\Delta t = 0$, R_{hh} becomes $R_{hh}(\tau)$, and this is the average PDP. The RMS-DS is standard deviation of $R_{hh}(\tau)$.

From the Fourier relationships, with $\Delta t = 0$, the standard deviation of $R_{HH}(\Delta f)$ is termed the correlation or coherence bandwidth, B_{co}. An approximate relationship is $\sigma_\tau \sim 1/B_{co}$ and more conservatively, as discussed in Chapter 2, $\sigma_\tau \sim 1/(5B_{co})$ [34]. These two parameters – the RMS-DS and correlation bandwidth – are widely used measures of delay dispersion and frequency correlation, respectively.

The correlation functions also allow us to quantify the channel's time variation. Specifically, setting $\Delta f = 0$, the standard deviation of $R_{HH}(\Delta t)$ is termed the coherence time T_c, roughly the time over which the channel's statistics remain constant. From the Fourier relations again, the coherence time is reciprocally related to the RMS Doppler spread B_{ds}: $T_c \sim 1/B_{ds}$. The Doppler spread's physical interpretation is the range of Doppler shifts impressed upon the transmitted signal due to platform velocity and the various angles of arrival of the MPCs.

3.2.3 Common Channel Parameters and Statistics

Recall from our characterization of mobile radio channel that four related parameters, σ_τ, B_{co}, T_c and B_{ds}, are widely used to quantify the channel's effect upon signals. Along with the path loss model parameters, particularly the path loss exponent ν and the standard deviation σ_X, these parameters provide the communication engineer with vital information for assessing channel effects upon transmitted signals.

Since attenuation can differ significantly with the presence or absence of a LOS component in the channel, the probability of LOS present [35], in a given environment, is sometimes provided as an additional channel parameter. This can be of use in assessing the statistics of attenuation over an area.

Other channel statistics often estimated and reported are the number of MPCs, MPC fading distributions, correlations among MPCs, and probabilities of MPC occurrence and duration. For estimation of all these MPC statistics, one generally collects measurement data (or, sometimes, simulation data). The correlation functions $R_{hh}(\tau)$ and $R_{HH}(\Delta f, \Delta t)$ can be estimated, but strictly these require the WSSUS assumption, which will not pertain for long durations or very wide frequency spans. Hence, recent work (see, for example, Ref. [36] and references therein) has investigated the estimation of the channel stationarity time or stationarity distance (SD). With knowledge of the approximate value of SD, one can proceed to estimate the correlation functions and parameters. For the NLOS-S and NLOS ASA regions, for typical vehicle velocities, the channel can be statistically *non*stationary over fairly short durations (tens to hundreds of milliseconds).

Finally, spatial parameters of wireless channels are also of interest in modern communication systems that employ multiple antennas at either Tx or Rx or both (MIMO). With antenna arrays at one or both link ends, spatial correlations among antenna elements can be estimated. The spatial distribution of received power, often expressed as a function of azimuth angle, is also often reported. Additional spatial statistics can be found in the literature, for example, Ref. [30].

3.3 Channel Effects and Signaling

In this section we connect more directly the channel characteristics and the signals transmitted over the channel. A discussion of fading is followed by the relations among the various channel parameters and communication system signaling parameters.

3.3.1 Small-Scale and Large-Scale Fading

Fading, as was reviewed in Chapter 2, is used broadly as a term for signal amplitude variation. In some cases and some frequency bands (e.g., ionospheric refraction at HF), such variation is caused by variation in the propagation medium itself. Although inhomogeneities in tropospheric properties do occur in terrestrial channels, for frequency bands below about 10 GHz and link ranges on the order of 10 km and smaller, these medium-induced variations are far less significant than those due to obstruction, mobility, and multipath propagation. Thus, the discussion here focuses on these effects.

For terrestrial channels, fading is often classified according to its rate of variation, that is, slow or fast, but this classification is often imprecise when used to refer to LOS obstruction (shadowing) and multipath fading, respectively. Another common and more accurate classification is small-scale versus large-scale fading. In short, small-scale fading is due to the constructive and destructive addition of the multipath components, the $\alpha e^{j\varphi}$ terms in (3.1). This fading occurs on spatial scales on the order of a wavelength. Large-scale fading, often called shadowing, is attributable to obstruction or blockage of the LOS component by an obstacle that is large with respect to a wavelength, for example, hundreds of wavelengths.

In the majority of terrestrial channels, as we have seen in Chapter 2, large-scale fading is modeled as having a log-normal distribution: The attenuation in dB is well modeled as Gaussian. This is manifested in the path loss model of (3.3) by the variable X, which again is zero mean with standard deviation σ_X.

Small-scale fading pertains to the CIR amplitude $|h(\tau, t)|$, or when multiple subcomponents are present, to each of the resolvable terms in the CIR sum of (3.1). When the number of components (or subcomponents) is large, the central limit theorem yields amplitude fading that can be either Rician or Rayleigh distributed, depending on whether or not the CIR has a dominant component. Other distributions, such as the Nakagami, Weibull, and K-distributions, are also sometimes employed [37]. Descriptions of all these distributions are widely available in the literature.

3.3.2 Channel Parameters and Signaling Relations

As implied by some of the channel parameter names such as delay spread and Doppler spread, the effects of the channel on transmitted signals can be inferred. A distortionless channel is one that has a CIR of the form

$$h(\tau, t) = \alpha(t)\delta(\tau - \tau_0(t)) \qquad (3.9)$$

where the rate of change of amplitude $\alpha(t)$ and delay $\tau_0(t)$ is very *slow* with respect to the transmitted signal symbol duration T_s; in the strict sense both α and τ_0 should be essentially constant over the transmission duration. In the frequency domain, the distortionless channel has a CTF that has constant gain and linear phase over the band of the transmitted signal. These relations clearly show that the channel's characteristics (distortionless or not) depend upon the transmitted signal characteristics. A distortionless channel does not alter the transmitted signal waveform shape.

Table 3.1 Example channel parameters and corresponding signal/system parameters they affect.

Channel parameters	Affected signal/system design parameters
Multipath delay spread σ_τ and coherence bandwidth B_{co}	Signal bandwidth B_{sg} and symbol rate R_s subcarrier bandwidths; equalization method
Channel attenuation α	Transmit power P_t, link ranges, modulation/detection, and FEC type
Doppler spread B_{ds} and coherence time T_c	Data block or packet size, FEC type and strength, transceiver adaptation rates, duplexing method
Spatial correlation ρ_s and temporal correlation ρ_t	Diversity method, FEC type, multiplexing method

When more than one component is present in the CIR, the channel can still be distortionless if all the relative[5] delays $\tau_k \ll T_s$. A distortionless channel is also known as a frequency-flat or frequency nonselective channel, in reference to the shape of the CTF. In contrast, when the range of relative delays or the maximum MPC delay τ_{L-1} is on the order of, or larger than, the symbol duration, the channel is said to be dispersive, and transmitted pulses undergo distortion, and their shape is changed. Such channels are said to be frequency selective. For assessing average channel effects, often one compares the RMS-DS σ_τ to the symbol time T_s for declaring the presence or absence of distortion. Hence, for distortionless transmission we require $\tau_k \ll T_s$, or expressed in the equivalent form, $B_{co} \gg 1/T_s = R_s$ with R_s being the symbol rate.

In terms of time variation, when the symbol time $T_s \ll T_c$, the coherence time, the channel is said to be slowly fading. Equivalently, fading is slow when the Doppler spread $B_{ds} \ll R_s$.

For both frequency selectivity and fading rate, the previous relations must be viewed as guidelines or approximations – there is no sharp transition between distorting and nondistorting or between slow and fast fading. These effects occur on a continuum.

Table 3.1 lists a number of important channel parameters and the signal design parameters they directly affect. The signal design parameters refer mostly to the physical and data link layers.

---

5 The relative delay of the kth component is $\tau_k - \tau_0$. Often the bulk propagation delay of the first arriving component τ_0 is set to zero for convenience.

3.4 Measured Airport Surface Area Channels

In this section, we describe results of a measurement campaign for characterizing the ASA channel. Detailed results appear in Refs [1–3].

3.4.1 Measurement Description and Example Results

Measurements were made at two large airports, JFK and MIA, one medium airport, CLE, and three small GA airports, Tamiami Airport (TA) in Miami, FL, Ohio University Airport (OU) in Albany, OH, and the Burke Lakefront Airport (BL) in Cleveland, OH. The measurements employed a spread spectrum wireless channel sounder, consisting of a transmitter unit and a receiver unit. Transmit power was 2 W, center frequency 5.15 GHz, and signal bandwidth approximately 50 MHz. The channel sounder gathered PDPs (and MPC phases) at a selectable rate from 2–60 PDPs/s.

Omnidirectional monopole antennas were used at both Tx and Rx for most measurements, but some measurements employed directional horn antennas to increase range. The PDPs were recorded, and two thresholds were employed: The first threshold removed all components below 25 dB from the strongest component; the second threshold was used to remove noise spikes that could be mistakenly interpreted as MPCs, with a false interpretation probability of less than or equal to 0.001. Detailed descriptions and specifications for the equipment and test procedures can be found in Refs. [1] and [2].

The mobile unit receiver was transported in a ground vehicle that moved at velocities ranging from 0 to approximately 50 km/h. Figure 3.5 shows an aerial photograph of the Miami airport [2], with numbered test points shown. For this airport, the mobile receiver moved between the numbered test points in numerical order. Similar procedures were followed at the other airports.

In total, over 35,000 PDPs were taken in the various airports, with Tx located at the ATCT. An additional 15,000 PDPs were taken with the Tx at airport field sites and in point-to-point link tests. Table 3.2, from Ref. [2], lists the total number of PDPs taken in each case. To separate the NLOS-S from the NLOS regions, an RMS-DS threshold was employed. This threshold also appears in Table 3.2. For RMS-DS values $\sigma_\tau < \sigma_{\tau 1}$, the channel was declared LOS, for RMS-DS values $\sigma_{\tau 1} < \sigma_\tau < \sigma_{\tau 2}$, the channel was classified as NLOS-S. When $\sigma_\tau > \sigma_{\tau 2}$, NLOS conditions pertained. (Note that visual inspection of the path from Tx to Rx was also used to confirm presence/absence of LOS).

Table 3.3 summarizes the measured results for RMS-DS. Statistics here include minimum, mean, and maximum values, and pertain to the

Figure 3.5 Aerial view of Miami International Airport, showing numbered test locations [2]. (Reproduced with permission of IEEE.)

instantaneous RMS-DS. Corresponding coherence bandwidth estimates appear in Table 3.4.

We also provide some summary information on the ASA channel MPC (also termed "tap," with reference to the TDL model) amplitude statistics. The tap or MPC associated with the first arriving signal generally has the least amount of fading; often this is a LOS or diffracted component. Hence, it is well modeled by the Rician distribution, characterized by the Rician "K-factor," the ratio of power in the dominant component to that in the scattered components that make up this first tap.

For different channel bandwidths, the channel model has a different number of taps, so in Table 3.5, we provide values for the range of the Weibull distribution "shape factor" parameter[6] β, across *all* remaining taps (other than the first). Also of note is that for the NLOS setting, the first tap is generally not strong, so the Rician characterization does not apply. For the NLOS setting, we hence have no first-tap K-factor, and the Weibull β parameter applies to all taps. In the point-to-point case, the range of K-factors for the first tap applies here only to the "boresight" antenna alignment (see Ref. [1]).

---

6 This shape factor is analogous to the Rician K-factor: $\beta = 2$ constitutes Rayleigh fading, and smaller values of β mean more severe fading, larger values mean less severe fading.

Table 3.2 Number of measured PDPs for each propagation region [2,3].

Airport	Total number of PDPs				
	Mobile			Field site transmit	
	NLOS $(\sigma_{\tau 2})$	NLOS-S $(\sigma_{\tau 1})$	LOS-O	NLOS $(\sigma_{\tau 2})$	NLOS-S $(\sigma_{\tau 1})$
JFK	6693 (800)	7103	—	7272 (800)	2240
MIA	6299 (1000)	5950	—	909 (1000)	1408
CLE	1332 (500)	852 (125)	443	—	—
OU	—	1108	—	—	—
BL	—	652 (125)	256	—	—
TA	2248 (500)	2955	—	—	—

RMS-DS values (σ_τ) in ns.

Table 3.3 Summary of measured RMS-DS values for three settings [2,3].

Airport	RMS-DS (ns) [min; mean; max]					
	Mobile			Point to point	Field site transmit	
	NLOS	NLOS-S	LOS-O	LOS-O	NLOS	NLOS-S
JFK	[800; 1469; 2456]	[21.4; 311; 798.7]	—	—	[802; 1475; 2433]	[5.8; 317.3; 799.5]
MIA	[1000; 1513; 2415]	[23.1; 459; 999.9]	—	[5.6; 163; 249]	[1000; 1625; 2451]	[8; 443; 997]
CLE	[500; 1206; 2472]	[125; 295; 499]	[14; 65; 124]	[1; 18.12; 202]	—	—
OU	—	[14; 293; 2416]	—	—	—	—
BL	—	[126; 429; 2427]	[5; 44; 124]	—	—	—
TA	[502; 1390; 2404]	[15; 256; 499]	—	—	—	—

Table 3.4 Summary of computed coherence bandwidth (B_c) values for three settings [2,3].

Airport	B_c (MHz) for correlation of [0.9; 0.5; 0.2]					
	Mobile			Point to point	Field site transmit	
	NLOS	NLOS-S	LOS-O	LOS-O	NLOS	NLOS-S
JFK	[NA; 1.56; 18.8]	[4.7; 15.2; 21.5]	—	—	[NA; 1.56; 18.8]	[4.3; 14.8; 16.8]
MIA	[NA; 0.78; 14.8]	[1.56; 15.2; 22.3]	—	[15.2; 50; 50]	[NA; 1.16; 13.6]	[4.3; 11.7; 16.8]
CLE	[NA; 10.1; 21.4]	[5.9; 17.1; 22]	[6.6; 17.6; 22.6]	[4.7; 12; 22.2]	—	—
OU	—	[3.9; 12.5; 21.4]	—	—	—	—
BL	—	[3.5; 13.2; 19.6]	[6.6; 16; 21.4]	—	—	—
TA	[NA; 9; 16]	[3.9; 14.1; 20.8]	—	—	—	—

Table 3.5 Summary of computed Rician K-factor values and range of Weibull β-factors for three settings [2,3].

Airport	K-factors (dB); and range of β-factors (min; max)					
	Mobile			Point to point	Field site transmit	
	NLOS	NLOS-S	LOS-O	LOS-O	NLOS	NLOS-S
JFK	(1.42; 2.09)	10.1 (1.67; 2.18)	—	—	(1.52; 2.1)	10.8 (1.58; 2.05)
MIA	(1.51; 2.13)	9.3 (1.66; 2.04)	—	(23–25)	(1.74; 2.8)	8.4 (1.5; 1.73)
CLE	(1.59; 2.25)	9.5 (1.77; 1.89)	13.1 (1.69; 2.2)	(14.1–15)	—	—
OU	—	8.9 (1.75; 2.36)	—	—	—	—
BL	—	12.5 (1.7; 2.73)	14.5 (1.88; 2.4)	—	—	—
TA	(1.4; 1.87)	11 (1.54; 2.02)	—	—	—	—

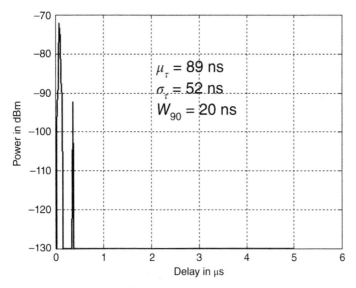

Figure 3.6 Example PDP for CLE, LOS-O region [1].

Examples of PDPs are shown in Figures 3.6–3.8. These plots illustrate typical results in the three different ASA propagation regions. The text on the figures provides some delay statistics, where W_{90} denotes the 90% energy CIR delay window [25]. The LOS-O channel in Figure 3.6 shows one additional MPC due to a reflection from a large building. Figure 3.7, for the NLOS-S region, shows at least eight MPCs, and the PDP in Figure 3.8, for the NLOS region, has the largest number of MPCs (at least 30). Additional results, including distributions of RMS-DS and coherence bandwidth, appear in Refs [1–3].

3.4.2 Path Loss Results

We employed the log-distance formulation to model path loss. For the NLOS-S regions, for path loss in dB, the following formula was obtained for link distances up to $2\,\mathrm{km}$

$$PL_{\mathrm{NLOS\text{-}S}}(d) = 103 + 10(2.3)\log_{10}(d/d_0) + X \qquad (3.10)$$

where the reference distance is $d_0 = 462$ m. The path loss exponent is thus $\nu = 2.3$. The variable X is a zero-mean Gaussian random variable with standard deviation 5.3 dB. The constant attenuation $L_0 = 103\,\mathrm{dB}$ is very close to the free-space value of approximately 100 dB.

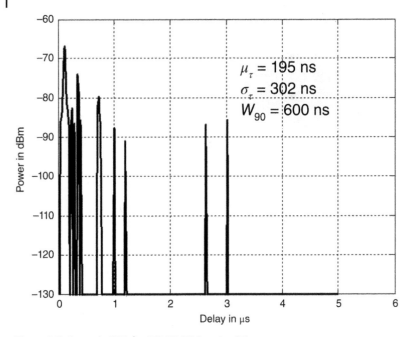

Figure 3.7 Example PDP for JFK, NLOS-S region [1].

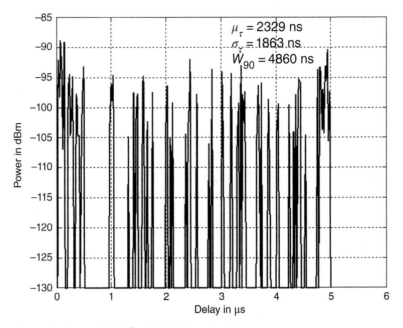

Figure 3.8 Example PDP for JFK, NLOS region [1].

The path loss exponent for the LOS-O areas was found to be essentially that of free space (i.e., $\nu = 2$). The large difference in the relative heights of the transmitting and receiving antennas, and the relatively short distances yield values for free space path loss that are within a few dB of those for the common "two-ray" model [6]. Because of problems with one of the measurement equipment GPS receivers, insufficient data was gathered for the path loss modeling for NLOS regions.

3.5 Airport Surface Area Channel Models

The models derived for the three ASA regions are in the form of tapped delay lines. Models for multiple values of channel bandwidth were constructed, with the widest channel bandwidth being 50 MHz. Models for narrower bandwidths are obtained from the 50 MHz models by combining MPCs (taps), for example, a 25 MHz model is obtained by combining two adjacent MPCs in the 50 MHz model.

Also introduced in Refs [1–3] was the concept of MPC persistence. For each MPC in the CIR of (3.1), a random process $z_k(t)$ multiplies $\alpha_k(t)$. The process $z_k(t)$ takes values either 0 or 1, hence indicating the absence or presence of the kth MPC, respectively. Such processes are also sometimes called "birth–death" processes (Figure 3.3). First-order homogeneous discrete Markov chains were employed to model the MPC persistence processes. For each MPC, the process is characterized by the steady-state probability vector $SS = [P_0 \ P_1]^T$ (superscript T denotes transpose), and the two by two transition matrix TS. Element $P_0 = 1 - P_1$ denotes $Pr[z_k = 0]$ and analogously for P_1. One example of TS matrix for tap # 3 for the NLOS region for a large airport is given in (3.11), along with the corresponding SS vector in (3.12). Each element P_{ij} in TS is defined as the probability of going from state i to state j, with $P_{i0} + P_{i1} = 1$ for $i = 0$ and 1.

$$TS_3^{(\text{NLOS})} = \begin{bmatrix} P_{00} & P_{01} \\ P_{10} & P_{11} \end{bmatrix} = \begin{bmatrix} 0.4259 & 0.5741 \\ 0.1965 & 0.8035 \end{bmatrix} \tag{3.11}$$

$$SS_3^{(\text{NLOS})} = \begin{bmatrix} P_0 \\ P_1 \end{bmatrix} = \begin{bmatrix} 0.2551 \\ 0.7449 \end{bmatrix} \tag{3.12}$$

Thus, this particular MPC is present for approximately 74% of the time, and when present, the probability of staying present is approximately 80%. Based upon curve fits to the persistence process data, MPC probability of occurrence $Pr[z_k = 1]$ versus delay was also derived [2].

For all MPCs, amplitude distributions were fit using the Weibull probability density function (pdf). This pdf is defined as

$$p_w(x) = \frac{\beta}{a^\beta} x^{\beta-1} \exp\left[-\left(\frac{x}{a}\right)^\beta\right] \tag{3.13}$$

where again, β is a shape factor that determines fading severity, $a = \sqrt{\Omega/\Gamma[(2/\beta)+1]}$ is a scale parameter with $\Omega = E(x^2)$ and Γ the gamma function.

The channel models in the following sections thus consist of tables of MPC amplitude statistics[7] and Markov chain persistence parameters. Since the AeroMACS system will employ a channel bandwidth of 5 MHz, only the 5 MHz models are presented here. Models for 10 MHz bandwidths appear in Refs [2] and [3], and models for the full 50 MHz bandwidth appear in Ref. [1]. Correlations among the various MPCs were also estimated; interested readers can turn to Refs [2] and [3] for details.

3.5.1 Large/Medium-Sized Airports

The large airport channel parameters for the NLOS and NLOS-S regions appear in Table 3.6. For the medium-sized airports, Table 3.7 contains the parameters.

3.5.2 Small Airports

Delay dispersion (RMS-DS) statistics for small airport channels appear in Table 3.8. Small airport channel parameters are provided in Table 3.9.

3.6 Summary

In this chapter, we began with a discussion of the importance of channel modeling, some channel definitions, and a description of the airport surface area channel. This was followed by an overview of statistical channel characterization, in which we described the channel impulse response and transfer function, which completely characterize the channel. This led to a discussion on statistical channel characteristics, including path loss variation and the correlation functions of the channel impulse response and transfer function, which are quantified using

---

7 For all MPCs other than the first, phase was well modeled as uniform on [0,2π].

Table 3.6 Channel parameters for 5 MHz channels: large airport [2].

Tap index k	Energy	Weibull shape factor (β_k)	$P_{1,k}$	$P_{00,k}$	$P_{11,k}$
		NLOS			
1	0.6496	2.28	1.0000	NA	1.0000
2	0.0789	1.65	0.9172	0.1748	0.9255
3	0.0579	2.0	0.8699	0.2003	0.8804
4	0.0516	1.63	0.8445	0.2543	0.8626
5	0.0440	2.0	0.8213	0.2678	0.8407
6	0.0437	1.67	0.8093	0.2571	0.8249
7	0.0390	2.0	0.7926	0.2737	0.8098
8	0.0352	2.0	0.7657	0.2957	0.7847
		NLOS-S			
1	0.9503	3.5	1.0000	NA	1.0000
2	0.0356	1.47	0.6941	0.5312	0.7935
3	0.0142	1.58	0.5196	0.6942	0.7173

Table 3.7 Channel parameters for 5 MHz channels: medium airport [2].

Tap index k	Energy	Weibull shape factor (β_k)	$P_{1,k}$	$P_{00,k}$	$P_{11,k}$
		NLOS			
1	0.8309	1.64	1.0000	NA	1.0000
2	0.0683	1.32	0.8086	0.3018	0.8356
3	0.0444	1.34	0.7379	0.3454	0.7684
4	0.0297	1.40	0.6974	0.3989	0.7389
5	0.0267	1.36	0.6578	0.4710	0.7257
		NLOS-S			
1	0.9728	4.04	1.0000	NA	1.0000
2	0.0272	1.55	0.6011	0.5532	0.7028
		LOS-O			
1	0.9893	5.83	1.0000	NA	1.0000
2	0.0107	1.59	0.4458	0.6667	0.5830

Table 3.8 Summary of measured RMS-DS values for three airports [3].

Airport	RMS-DS (ns) [min; mean; max]		
	NLOS	**NLOS-S**	**LOS-O**
bl	—	[126; 429; 2,427]	[5; 44; 124]
OU	—	[14; 293; 2,416]	—
ta	[502; 1,390; 2,404]	[15; 256; 499]	—

Table 3.9 Channel parameters for 5 MHz channels: small airport [3].

Tap index k	Energy	Weibull shape factor (β_k)	$P_{1,k}$	$P_{00,k}$	$P_{11,k}$
		NLOS			
1	0.730	2.12	1.0000	NA	1.0000
2	0.098	1.57	0.9162	0.2426	0.9306
3	0.064	1.49	0.8704	0.2824	0.8935
4	0.040	1.67	0.8004	0.3168	0.8300
5	0.037	1.68	0.7935	0.2608	0.8081
6	0.029	2	0.7639	0.3515	0.8000
		NLOS-S			
1	0.959	4.22	1.0000	NA	1.0000
2	0.041	1.41		0.6028	0.6996	0.8016
		LOS-O			
1	0.986	5.15	1.0000	NA	1.0000
2	0.014	1.58	0.5092	0.7744	0.7810

metrics in the delay, frequency, time, and Doppler domains. Some discussion of these metrics – RMS delay spread, coherence bandwidth, coherence time, and Doppler spread – was presented, and these metrics were connected to transmission signaling parameters along with a brief discussion on fading. An extensive airport surface area channel measurement campaign was summarized. Example measurement results for RMS delay spread, coherence bandwidth, and small-scale fading Rician K-factors were provided. Detailed airport surface area channel models, in the form of tapped-delay lines, were also presented.

References

1 D. W. Matolak, "Wireless Channel Characterization in the 5 GHz Microwave Landing System Extension Band for Airport Surface Areas," Final Project Report for NASA ACAST Project, Grant Number NNC04GB45G, May 2006.

2 D. W. Matolak, I. Sen, and W. Xiong, "The 5 GHz Airport Surface Area Channel: Part I, Measurement and Modeling Results for Large Airports," *IEEE Transactions on Vehicular Technology*, 57(4), 2014–2026, 2008.

3 I. Sen and D. W. Matolak, "The 5 GHz Airport Surface Area Channel: Part II, Measurement and Modeling Results for Small Airports," *IEEE Transaction on Vehicular Technology*, 57(4), 2027–2035, 2008.

4 J. G. Proakis, *Digital Communications*, 2nd edn, McGraw-Hill, New York, NY, 1989.

5 W. Stallings, *Data and Computer Communications*, 7th edn, Prentice Hall, Upper Saddle River, NJ, 2004.

6 G. Stuber, *Principles of Mobile Communications*, Kluwer Academic Publishers, Norwell, MA, 1996.

7 S. G. Wilson, *Digital Modulation and Coding*, Prentice Hall, Upper Saddle River, NJ, 1996.

8 H. V. Poor, G. W. Wornell, (eds.), *Wireless Communications: Signal Processing Perspectives*, Prentice Hall, Upper Saddle River, NJ, 1998.

9 A. F. Molisch (ed.), *Wideband Wireless Digital Communications*, Prentice Hall, Upper Saddle River, NJ, 2001.

10 L.-L. Yang and L. Hanzo, "Multicarrier DS-CDMA: A Multiple Access Scheme for Ubiquitous Broadband Wireless Communications," *IEEE Communications Magazine*, 41(10), 116–124, 2003.

11 J. D. Parsons, *The Mobile Radio Propagation Channel*, 2nd edn, John Wiley & Sons, Inc., New York, NY, 2000.

12 D. J. Torrieri, *Principles of Secure Communication Systems*, 2nd edn, Artech House, Boston, MA, 1992.

13 M. C. Jeruchim, P. Balaban, and K. S. Shanmugan, *Simulation of Communication Systems: Modeling, Methodology, and Techniques*, 2nd edn, Kluwer Academic Press, Boston, MA, 2000.

14 W. H. Tranter, K. S. Shanmugan, T. S. Rappaport, and K. L. Kosbar, *Principles of Communication System Simulation with Wireless Applications*, Prentice Hall, Upper Saddle River, NJ, 2004.

15 S. Weinberg, *The First Three Minutes: A Modern View of the Origin of the Universe*, Basic Books, New York, NY, 1993.

16 C. A. Levis, J. T. Johnson, and F. L. Teixeira, *Radiowave Propagation: Physics and Applications*, John Wiley & Sons, Inc., Hoboken, NJ, 2010.

17 H. L. Bertoni, *Radio Propagation for Modern Wireless Systems*, Prentice Hall, Upper Saddle River, NJ, 2000.

18 A. Saakian, *Radio Wave Propagation Fundamentals*, Artech House, Norwood, MA, 2011.

19 F. Perez Fontan and P. Marino Espineira, *Modeling the Wireless Propagation Channel*, John Wiley & Sons, Inc., Hoboken, NJ, 2008.

20 M. Patzold, *Mobile Fading Channels*, John Wiley & Sons, Inc., Hoboken, NJ, 2002.

21 S. Salous, *Radio Propagation Measurements and Channel Modeling*, John Wiley & Sons, Inc., Hoboken, NJ, 2013.

22 F. Hlawatsch and G. Matz (eds.), *Wireless Communications Over Rapidly Time-Varying Channels*, Academic Press, Oxford, UK, 2011.

23 W. Jakes (ed.), *Microwave Mobile Communications*, IEEE Press, New York, NY, 1994.

24 A. G. Kanatas and A. D. Panagopoulos, *Radio Wave Propagation and Channel Modeling for Earth–Space Systems*, CRC Press, London, UK, 2016.

25 International Telecommunications Union, Radio Communications Sector, 2016. Available at http://www.itu.int/pub/R-REC.

26 Federal Aviation Administration, Benchmark Report, October 2004. Available at http://www.faa.gov/events/benchmarks/.

27 A. F. Molisch, H. Asplund, R. Heddergott, M. Steinbauer, and T. Zwick, "The COST259 Directional Channel Model – Part I: Overview and Methodology," *IEEE Transactions on Wireless Communications*, 5(12), 3421–3433, 2006.

28 R. C. Qiu, "A Study of the Ultra-Wideband Wireless Propagation Channel and Optimum UWB Receiver Design," *IEEE Journal on Selected Areas Communications*, 20(9), 1628–1637, 2002.

29 D. Gesbert, H. Bolcskei, D. A. Gore, and A. J. Paulraj, "Outdoor MIMO Wireless Channels: Models and Performance Prediction," *IEEE Transactions on Communications*, 50(12), 1926–1934, 2002.

30 B. Clerkxx and C. Oestges, *MIMO Wireless Networks*, 2nd edn, Academic Press, Oxford, UK, 2013.

31 International Telecommunications Union, "Multipath Propagation and Parameterization of Its Characteristics," Recommendation ITU-R P.1407-5, September 2013.

32 A. F. Molisch and M. Steinbauer, "Condensed Parameters for Characterizing Wideband Mobile Radio Channels," *International Journal on Wireless Information Networks*, 6(3), 133–154, 1999.

33 P. A. Bello, "Characterization of randomly time-variant linear channels," *IEEE Trans. Commun. Syst.*, CS-11(4), 360–393, 1963.

34 B. H. Fleury, "An Uncertainty Relation for WSS Processes and Its Application to WSSUS Systems," *IEEE Transactions on Communications*, 44(12), 1632–1634, 1996.

35 J. Jarvelainen, S. L. H. Nguyen, K. Haneda, R. Naderpour, and U. T. Virk, "Evaluation of Millimeter-Wave Line-of-Sight Probability With Point Cloud Data," *IEEE Wireless Communications Letters*, 5(3), 228–231, 2016.

36 D. W. Matolak and R. Sun, "Air-Ground Channel Characterization for Unmanned Aircraft Systems—Part I: Methods, Measurements, and Results for Over-water Settings," *IEEE Transactions on Vehicular Technology*, 66(1), 24–44, 2016.

37 M. K. Simon and M. S. Aloini, *Digital Communication over Fading Channels*, John Wiley & Sons, Inc., New York, NY, 2000.

4

Orthogonal Frequency-Division Multiplexing and Multiple Access

4.1 Introduction

Orthogonal frequency-division multiple access (OFDMA) is a network access technology employed in a number of contemporary wireless telecommunication networks, including IEEE 802.16-Std.-based networks such as WiMAX and AeroMACS. The parent technical concept of OFDMA, that is, orthogonal frequency-division multiplexing (OFDM), is a multicarrier modulation (MCM) scheme, as well as a multiplexing technique. In fact, OFDM is a special case of MCM in which frequency-division multiplexing (FDM) is applied to a large number of subcarriers with two distinguishing attributes. First, unlike standard FDM, the spectrum of the created secondary channels[1] mutually overlap, leading to a higher spectral efficiency for OFDM. Furthermore, the secondary channels exhibit orthogonality in the frequency domain, meaning that at the central frequency of each secondary channel (subcarrier frequency), the spectrum of all other secondary channels experiences a zero-crossing. In this manner, a single high-rate traffic data is split into N parallel low-rate data streams, each of which modulate a subcarrier and are transmitted with a bandwidth efficiency that is much higher than that of the conventional FDM systems.

Before elaborating on OFDM and OFDMA, which are essentially methods of spectral resource sharing amongst a large number of network clients, it is important to distinguish the difference between the two related concepts of *multiplexing* and *multiple access*. Multiplexing means the simultaneous transmission of multitude of signals using a single transmission medium. On the other hand, when a multitude of users randomly access a network and request service, the method of resource sharing that is offered by the network is referred to as multiple access technique.

---

1 The term "subchannel" is reserved to be used in conjunction with OFDMA, instead the phrase "secondary channel" is used here.

AeroMACS: An IEEE 802.16 Standard-Based Technology for the Next Generation of Air Transportation Systems, First Edition. Behnam Kamali.
© 2019 the Institute of Electrical and Electronics Engineers, Inc. Published 2019 by John Wiley & Sons, Inc.

A practical breakthrough was transpired in OFDM technology in early 1970s when Weinstein and Ebert applied discrete Fourier transform (DFT) to OFDM for the implementation of a part of the system modulation and demodulation process [1,2]. Soon, advances in digital signal processing (DSP) and very large scale integrated (VLSI) circuit technologies in the 1980s removed practical obstacles, such as hardware and computational complexities, for the implementation of OFDM and coded OFDM (COFDM) systems. In the 1980s, OFDM/COFDM were applied to high-speed telephone modems and digital mobile communications systems. In the 1990s, COFDM was adopted for a number of broadband applications that include HDSL (high-bit-rate digital subscriber links) and ADSL (asymmetric digital subscriber links) [3], digital audio broadcasting (DAB), digital video broadcasting, and HDTV [4]. In late 1990s and early 2000s, the high-speed wireless LAN; Wi-Fi (IEEE 802.11a standard), was developed based on OFDM [5]. In the first two decades of the twenty-first century, the fourth-generation wireless cellular (4G), WiMAX, and Aero-MACS networks were designed and developed based on OFDM and OFDMA [6]. It is widely accepted that the combination MIMO–OFDM is one of the most promising techniques to support future high-performance/high data rate applications in broadband wireless networks. Insofar as application of OFDM and OFDMA in the emerging fifth-generation (5G) cellular networks is concerned, although "the jury is still out on that," there are strong indications that OFDM modulation/multiplexing and OFDMA access technology are serious contenders for application in 5G networks. However, 5G promises to deliver the gigabit experience to the mobile user, which requires a capacity increase of up to three orders of magnitude with respect to current long-term evolution (LTE) systems. Consequently, there is a widespread agreement that 5G ambitious goals may have to be realized through a combination of multilayer innovative techniques [7].

In Chapter 2 we explored the problems that plague the multipath channel. Frequency selective fading that filters various spectral components of the signal differently, thereby causing severe distortion, was identified as the most challenging concern. It was also mentioned that one of the methods of counteracting frequency selective fading in dispersive channels is the application of OFDM. In this chapter, we examine OFDM and its virtues in combating time and frequency dispersive channel distortions.

4.2 Fundamental Principles of OFDM Signaling

OFDM is a spectrally efficient and a robust transmission technique for broadband digital communication systems that is particularly effective against many channel adversities, including frequency selective fading and impulse noise. Perhaps, the key motivation behind the widespread adoption of OFMD signaling in mobile communication is its ability to convert a wideband frequency selective fading channel into a series of approximately flat fading narrowband channels. In other words, OFDM is effective in removing the frequency-dependent aspect of fading without requiring sophisticated equalization procedures; however, it should be noted that OFDM does not eliminate the fading effect itself.

A major advantage of OFDM over conventional FDM is its spectral efficiency. Figure 4.1 shows the division of spectrum of conventional FDM into the secondary channels, and the same for the OFDM scheme. The figure clearly demonstrates the spectral efficiency advantages offered by

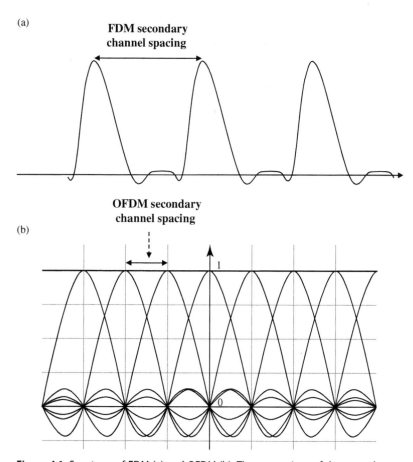

Figure 4.1 Spectrum of FDM (a) and OFDM (b). The comparison of the secondary channel spacing demonstrates the spectral efficiency advantage of OFDM.

OFDM. First, as Figure 4.1 illustrates, the OFDM secondary channels' spectral spacing is much narrower than that of conventional FDM; enabling, for a given bandwidth allocation, the creation of many more subcarriers and secondary channels for OFDM. Second, the OFDM secondary channels not only spectrally overlap, but are also orthogonal.

4.2.1 Parallel Transmission, Orthogonal Multiplexing, Guard Time, and Cyclic Extension

In conventional serial transmission systems, the data bits (or symbols) modulate a single carrier waveform, where the modulated signal occupies the entire channel bandwidth. In contrast to this, parallel signaling partitions the traffic data into N symbol (or bit) streams each with a symbol duration of T_s. Each symbol stream modulates a subcarrier; the sum of these waveforms constitutes a multicarrier (MCM) signal. Clearly, each of the subsignals occupy a portion of the available bandwidth. When the frequencies of neighboring subcarriers are spaced exactly by reciprocal of the symbol rate, that is, $1/T_s$ Hz, it can be readily shown that they exhibit orthogonality over the symbol interval. In other words, tone frequencies are of the form k/T_s where k is an integer. These tones are orthogonal simply because of the following computation.

$$\frac{1}{T_s} \int_0^{T_s} \cos(2\pi k_1 t/T_s)\cos(2\pi k_2 t/T_s)dt = \begin{cases} 1 & \text{if } k_1 = k_2 \\ 0 & \text{if } k_1 \neq k_2 \end{cases} \quad (4.1)$$

Under this orthogonality circumstance, the MCM signal constitutes an OFDM waveform.

More often than not, OFDM is accompanied by channel coding. The resulting signaling technique is referred to as Coded-OFDM; COFDM.

Assume that the input bit stream to the OFDM modulator is given by sequence $\{a_k\}$. This sequence is converted into N parallel symbol streams, which is represented by $\{S_k\}$. Each of the streams modulate a subcarrier of the OFDM system. Note that unless the subcarrier modulations are binary, $\{S_k\}$ is a complex sequence. As Equation 4.1 indicates, by selecting the frequencies of subcarriers as multiple integer of the inverse of the actual symbol duration, orthogonality of secondary channels is guaranteed. Subcarrier waveforms orthogonality in OFDM signaling can be interpreted from two different perspectives:

1) In the time domain, orthogonality implies that individual signals may be separated through correlation processes without imposing inter-symbol interference on each other. This is an important feature of OFDM signaling.

2) From the frequency domain point of view, orthogonality of OFDM subsignals implies that the signals satisfy the first Nyquist criterion in the Fourier domain. In other words, when the spectrum of one subsignal is at its maximum value the spectra of all other subsignals cross the frequency axis. A direct consequence of this property is that the spectrum of OFDM signal becomes almost a flat one. This is shown in Figure 4.1.

4.2.1.1 Cyclic Prefix and Guard Time

Since OFDM, through parallel transmission, spreads out the effect of frequency selective fading over a number of secondary channels, it can essentially convert the channel into a series of approximately flat fading channels, and thus mitigating the effect of ISI to a great extent. To eliminate ISI in the channel almost completely, a *guard time* is added to each OFDM symbol. In other words, the guard time is inserted to eliminate the possible residual ISI caused by frequency selective fading channel, most of which has already been removed by channel split process that is inherent in OFDM signaling. Guard time is normally selected to be larger than the channel delay spread (denoted by σ_τ as per Chapter 2), or larger than maximum channel delay spread (σ_{max}), which can also be considered as the time duration of the multipath channel impulse response. Should the guard interval be composed of merely baseline, the subcarrier signals would not remain orthogonal, which would give rise to intercarrier (actually intersubcarrier) interference (ICI)[2] problem. To circumvent this problem, the OFDM signal is cyclically extended within the guard interval, this is referred to as cyclic prefix (CP), since the guard band is normally added to the beginning of the OFDM symbol. The time duration of the CP or guard band is denoted by T_{cp}. Note that the guard time is not a part of useful signal duration and will be discarded after the detection process. The relationship between useful symbol period, guard interval, and transmission time duration T is given by $T = T_s + T_{cp}$. The ratio of the guard interval T_{cp} to T_s (useful symbol duration), T_{cp}/T_s, which reflects the bandwidth overhead required for the guard time is, by and large, application dependent. For IEEE 802.16e Std.-based networks (WiMAX and AeroMACS), T_{cp}/T_s may be selected from values $1/2$, $1/4$, $1/8$, and $1/16$. In classical OFDM applications, however, since the insertion of guard interval reduces the data throughput, T_{cp}/T_s is usually selected to be less than $1/4$, which includes the ratio of $1/32$ as well [8].

---

2 In the interest of preventing ambiguity, the following two terms are defined. ISI refers to intersymbol interference between various OFDM signals. The term ICI (intercarrier interference) is used to denote cross talk or interference between subcarrier signals within a single OFDM waveform.

The proper selection of CP time duration maintains subcarrier orthogonality, ensures that the delayed echoes of the OFDM signal have integer number of cycles within the transmission time duration (essentially converts linear convolution to cyclic convolution), therefore, multipath signals with delays less than guard time do not cause intersymbol interference, ISI. Thus, the CP allows the complete reception of one OFDM symbol before the next symbol is received. The insertion of CP, also, increases the receiver tolerance to symbol time synchronization error [4].

4.2.2 Fourier Transform-Based OFDM Signal

For radio communications OFDM creates a doubly modulated signal. First, the parallel streams of data modulate the subcarriers whose frequencies are determined by $k \cdot f_s$, where $f_s = 1/T_s$. The subcarrier signals are then added and multiplexed to the RF signal of carrier frequency f_c. Thus, the OFDM symbol, in the mth transmission interval, may be expressed by

$$
\begin{aligned}
s_m(t) &= \sum_{k=0}^{N-1} Re\big[S_{(m-1)N+k} \exp\big(j2\pi\{kf_s + f_c\}t\big)\big] \quad (m-1)T \le t \le mT \\
&= 0 \qquad\qquad\qquad\qquad\qquad\qquad\qquad\qquad\qquad otherwise
\end{aligned}
$$

$$(4.2)$$

Hence, the OFDM general signal that extends from $-\infty$ to $+\infty$ can be written as follows.

$$
s_{OFDM}(t) = \sum_{m=-\infty}^{\infty} \sum_{k=0}^{N-1} Re\left[S_{(m-1)N+k}\Pi\left\{\frac{t - (2m-1)T/2}{T}\right\}\exp\big(j2\pi\{kf_s + f_c\}t\big)\right]
$$

$$(4.3)$$

Here $\Pi\left(\dfrac{t - t_0}{T}\right)$ is a rectangular function that is equal to 1 over $t_0 - T/2 \le t \le t_0 + T/2$ and it is equal to zero outside of this interval. The equivalent complex baseband signal over the fundamental OFDM signal interval ($m = 1$) derived from Equation 4.2 is given in Equation 4.4.

$$
s(t) = \sum_{k=0}^{N-1} S_k \exp\big(j2\pi kf_s t\big) \quad for \quad 0 \le t \le T \tag{4.4}
$$

Clearly, $s(t)$ is the inverse Fourier transform of N complex symbols $\{S_k\}$. Taking samples of this signal that are apart by T_s/N provides the result given in Equation 4.5.

$$
s(nT_s) = \sum_{k=0}^{N-1} S_k \exp\big(j2\pi kf_s nT_s/N\big) \tag{4.5}
$$

A comparison is now made between this latter equation and the standard inverse Fourier transform equation for N samples of a general signal $x(t)$, as given in Equation 4.6.

$$x(nT_s) = \frac{1}{N} \sum_{k=0}^{N-1} X\left(\frac{k}{NT_s}\right) \exp(j2\pi nk/N) \tag{4.6}$$

Ignoring the constant factor of $1/N$ in front of Equation 4.6, since $S_k s$ are actually samples of Fourier transform of $s(t)$, the two equations are identical under the following condition:

$$\frac{f_s T_s}{N} = \frac{1}{N} \quad \rightarrow \quad f_s = \frac{1}{T_s} \tag{4.7}$$

This is in fact the same condition that is required for orthogonality. Therefore, one consequence of maintaining orthogonality is that the OFDM signal can be defined by Fourier transform [9]. In summary, the complex symbols $\{S_k\}$ are Fourier transform of the modulated signal, which is intuitively obvious since these symbols will be multiplied by their corresponding complex exponential functions to realize the modulation process. To put Equation 4.5 into standard inverse discrete Fourier transform form, by making use of Equation 4.7, and as it is customary, dropping T_s from the equation, that is, letting $s(nT_s) = s_n$, we conclude

$$s_n = \sum_{k=0}^{N-1} S_k \exp(j2\pi kn/N) \tag{4.8}$$

Equation 4.8 implies that the discrete version of the OFDM signal may be generated by computing the IDFT[3] of the complex input symbols at the transmitter, and conversely the complex symbols are obtained (or estimated) through DFT calculation of the down-converted signal at the receiver. In practice, for sufficiently large value of N, these computations are implemented using fast Fourier transform (FFT) algorithms. If N is not a power of 2, the symbols are padded with zeros such that total number of symbols is a power of 2. It is also noted that the number of subcarriers is equal to the size of FFT. That is to say, for instance, a FFT size of 512 corresponds to 512 available subcarriers.

4.2.3 Windowing, Filtering, and Formation of OFDM Signal

The OFDM signal formed by performing IDFT and adding guard interval is equivalent to a number of unfiltered modulated (BPSK, QPSK, QAM,

---

3 As it was mentioned earlier, the standard equation for IDFT contains a factor of $1/N$ in front of the summation sign. The effect of this factor can be reflected in the computation of DFT at the receiver side.

etc.) signals whose out-of-band spectrum falls off rather slowly. To force the spectrum to decay more rapidly around the boundary and outside of the allocated band, in order to comply with regulatory spectrum masking, two techniques of *windowing* and *filtering* are available. These two techniques may use the same mathematical equations, for example, raised cosine, root raised cosine, and so on. However, windowing refers to multiplication in the time domain, or equivalently convolution in the frequency domain; whereas filtering entails convolution in the time domain and multiplication in the frequency domain. As such computational complexity of their implementation is different. In general, the implementation of a window, in the context of DSP realization, is less complex than that of a filter with the same characteristics.

In practice, the OFDM signal is generated according to the following algorithmic order.

1) The complex symbols are padded with sufficient zeros to get proper number of samples for IFFT computation.
2) The computation of IFFT sample values is carried out.
3) The samples of IFFT are inserted at the start of the OFDM symbol, and the first samples of IFFT are appended to the end of the OFDM symbol (CP).
4) The symbol is then multiplied by a raised cosine window, or filtered by a raised cosine filter.

The OFDM signal so generated will be added to the previous OFDM signal with a delay of $T = T_s + T_{cp}$ [4].

Time–Frequency Resource in OFDM Signaling

It is insightful to view OFDM bandwidth resource from the perspective of time–frequency framework, as shown in Figure 4.2.

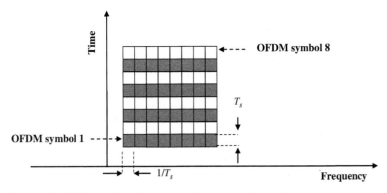

Figure 4.2 OFDM symbols illustrated in frequency–time grid.

Figure 4.2 illustrates a frequency–time resource grid for eight OFDM symbols. Each small square represents the basic resource unit, which is a subcarrier (tone) signal made available at certain point in time. The available bandwidth, in this example, is divided into eight secondary channels. The frequency spacing separating adjacent subcarriers is equal to the inverse of OFDM symbol time duration T_s, which guarantees subcarrier orthogonality.

4.2.4 OFDM System Implementation

In practice, the OFDM transmitter and receiver are implemented as a digital system using FFT algorithm, however, in the interest of gaining more insight, we examine the analog implementation of the OFDM transmitter circuit. Figure 4.3 illustrates a block diagram for an OFDM transmitter system in the analog domain. The serial stream of traffic data symbols, which might originate from a single source or a multiple of sources, are primarily converted into N parallel data streams, each of which modulates a sinusoidal subcarrier signal. The subcarrier signals, as shown in Figure 4.3, are generated by a bank of oscillator circuits. The subcarrier signals (secondary signals) are added next. Guard time, through cyclic extension, is arranged for the resulting signal. The signal is then multiplied by a, perhaps, raised cosine windowing function. The last phase is the RF modulation of the OFDM signal, which is the process of up-converting the signal to any required radio frequency band.

The oscillation frequencies of the bank of oscillators shown in the circuit diagram of Figure 4.3 are determined based on a total available OFDM channel bandwidth B_T, the starting frequency of the band f_0, and the number of required subcarriers N, as shown in Figure 4.4.

The available bandwidth is divided into N bands to create N secondary OFDM channels, therefore, $B_T/N = 1/T_s$. The frequency of tones (subcarriers) are, hence, given by Equation 4.9.

$$
\begin{cases}
f_1 = f_0 + \dfrac{B_T}{2N} = f_0 + \dfrac{1}{2T_s} \\[2mm]
f_2 = f_0 + \dfrac{3B_T}{2N} = f_0 + \dfrac{3}{2T_S} \\[2mm]
\vdots \quad \vdots \qquad \vdots \qquad \quad \vdots \qquad \vdots \\[2mm]
f_N = f_0 + \dfrac{(2N-1)B_T}{2N} = f_0 + \dfrac{(2N-1)}{2T_s}
\end{cases}
\tag{4.9}
$$

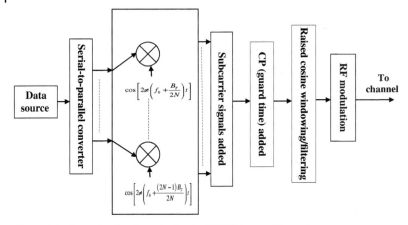

Figure 4.3 An analog implementation for OFDM transmitter system.

At the OFDM receiver side, following signal detection, what essentially takes place is that all signal processing operations performed at the transmitter are reversed.

4.2.5 Choice of Modulation Schemes for OFDM

Modulation schemes for OFDM signals are normally selected from the families of MPSK and QAM digital modulation techniques. In general, a modulation scheme is selected for a given application based on evaluation of various performance measures of the technique, and how closely it matches the characteristics of the transmission channel, and the requirements set forth by the user [10]. Important criteria for this selection are bandwidth efficiency and spectral behavior, power/error performance (BER), noise and interference immunity, and hardware/software complexity of the modulator and detector schemes. For OFDM, the choice of modulation schemes for various subchannels is dictated by a compromise between data rate requirements and transmission robustness. An interesting feature of OFDM is that different modulation techniques may be exploited for different subcarriers. This would facilitate layered services in a wireless network such as WiMAX and AeroMACS [11].

Figure 4.4 Dividing the OFDM bandwidth into N secondary channels (subcarriers).

4.2.6 OFDM Systems Design: How the Key Parameters Are Selected

The selection of OFDM parameters involves a process of compromise between available resources, user's requirements, and channel characteristics. The primary resource available to the system designer is the available channel bandwidth. The user requirements are normally expressed in terms of data rate and data reliability, that is, bit error rate (BER). Channel statistics such as noise PSD for AWGN (additive white Gaussian noise) channels, or fading characteristics as well as delay spread and coherence bandwidth for frequency selective fading channels are needed as a starter for the design of OFDM system. Normally, available bandwidth, data throughput, and tolerable delay spread are given. As a rule of thumb, the guard interval should be selected two to four times longer than delay spread, where the exact value depends on modulation and coding schemes applied in the design [4]. The choice of useful signal duration is a critical one. To minimize SNR loss caused by the guard time, it is desirable to have the useful signal duration much longer than the guard interval. However, longer useful symbol duration results in an increase in the number of subcarriers and smaller carrier spacing, which implies more sensitivity to phase noise and carrier frequency error, as well as larger size FFT algorithm [12]. In practice, how closely two subcarriers can be placed depends on tolerable phase noise and carrier offset. For mobile reception of OFDM signal, the subcarrier spacing must be large enough to make the Doppler shift negligible. The useful symbol duration should be selected such that channel impulse response does not change during symbol time period [13].

4.3 Coded Orthogonal Frequency-Division Multiplexing: COFDM

Channel coding techniques are generally used in flat fading channels. In the context of OFDM, this implies that the channel coding is an appropriate signal processing technique to be used in conjunction with subcarriers of OFDM signal for the purpose of mitigation of fading effect [11].

Several coding schemes such as convolutional codes [14], TCM codes [11], Reed–Solomon (RS) codes [15], and Turbo and LDPC (low-density parity check) codes [16] have been proposed and studied for this purpose. In modern wireless networks, such as WiMAX, Aero-MACS, and LTE-based 4G and 5G, however, modulation and coding schemes are grouped together and selected adaptively based on the channel conditions. The corresponding protocol is known as adaptive modulation coding (AMC). AMC will be explored in Chapters 5 and 7.

4.3.1 Motivation

In mobile radio, one has to often deal with channels that exhibit a highly fluctuating frequency response over the available signal bandwidth, that is, the frequency selective fading phenomena. As was mentioned earlier, the key motivation behind the widespread adoption of OFMD signaling in mobile communication applications is its ability to convert a wideband frequency selective fading channels into a series of flat fading narrowband secondary channels by selecting the secondary channel bandwidth smaller than the coherence bandwidth of the multipath channel [13]. In other words, OFDM is effective in removing the frequency selective filtering effect of the fading channel; however, it does not eliminate the signal fading itself. Nevertheless, once OFDM converts a wideband frequency selective channel into a set of narrowband flat fading channels, coding techniques may be applied to combat fading in the frequency-flat channels. Consequently, channel coding are appropriate signal processing techniques to be applied to secondary channels, created by subcarriers of OFDM signal, to mitigate the effect of fading [11].

4.3.2 System-Level Functional Block Diagram of a Fourier-Based COFDM

In this section, a system-level characterization of a COFDM is presented. Figure 4.5 illustrates a functional block of a Fourier transform-based COFDM system. The traffic data bits originating from various sources may have to be compressed. Such is the case in digital audio and digital video broadcasting, and a number of other applications. The source coding block in Figure 4.5 accounts for both noiseless and noisy compression coding. The compressed data bits are next encoded into channel code words that may be complemented by a time domain interleaving process whose function is to break down long bursts of detection errors to shorter bursts or even convert them to random errors. The serial encoded bits are then converted to parallel bits, which is then partitioned into N symbols. A symbol in this context means a defined number of bits clustered together. Each symbol is mapped to a constellation according to the modulation format used for the corresponding subcarrier. These symbols are in fact the frequency domain representation of the OFDM signal. Thus, in order to generate the corresponding time-domain-modulated signal an IFFT computation block is required. Consequently, the modulator of the OFDM signal consists of two blocks: symbol mapping

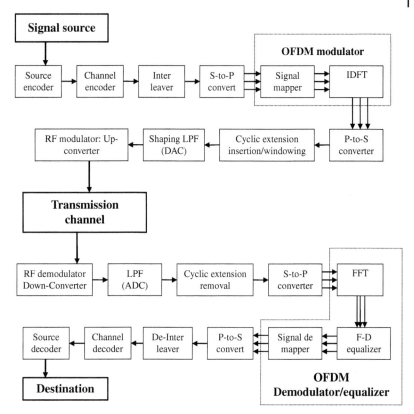

Figure 4.5 A general functional block diagram of Fourier transform-based COFDM system.

and IFFT computer. These two blocks are equivalent to the conventional digital modulator, which in this case requires a bank of N product modulators and N sinusoidal oscillators, as shown in Figure 4.2. In order to prepare the signal for transmission through the physical channel, the parallel signal is first transferred to serial, guard interval is added, and the windowing process is carried out. This provides a series of samples that are applied to a LPF (DAC) for conversion into analog OFDM waveforms. For some applications, such as cable broadcast, the signal at this stage can be amplified and delivered to the transmission channel. For wireless and radio transmission, the signal, however, is up-converted to a radio frequency signal, as shown in Figure 4.5. At the receiver side, as illustrated in Figure 4.5, the signal processing operations performed at the transmitter are reversed one by one.

4.3.3 Some Classical Applications of COFDM

Early applications of MCM and OFDM techniques have been in HF military communications systems [17]. The COFDM idea was conceived in the 1950s, however, widespread applications emerged in the 1980s–2010s. For OFDM/COFDM to come to technological prominence, a period of 30 years was needed in order to overcome the practical difficulties that existed in their hardware implementation. The ability to define the COFDM signal in the Fourier domain, the advances made in modulation, channel coding, equalization techniques, and digital signal processing in the 1980s, and the ease with which software implementation is carried out in DSP processors, or dedicated VLSI circuits, has made OFDM an attractive choice for many applications, particularly in wireless telecommunication systems. Early on in the 1980s, OFDM Technology gave birth to a new generation of telephone modems of data rates above 14 kbps which were then considered as "ultrahigh-speed voiceband data modems" [18]. Trellis coded modulation and OFDM were among techniques that made these developments possible [19]. We now address few forerunner applications of COFDM in wireless networks whose evolved versions have become standard signaling for WMAN (wireless metropolitan area network), such as WiMAX and AeroMACS, and WWAN (wireless wideband area networks) of the twenty-first century.

4.3.3.1 COFDM Applied in Digital Audio Broadcasting (DAB)

One of the early applications of COFDM in radio broadcasting is in DAB for mobile users within the framework of EURIKA 147 Project in Europe [14]. The DAB standard was initiated as a European research project in the 1980s and it was launched for the first time by the Norwegian Broadcasting corporation in June 1995. BBC (British Broadcasting Company) launched its first DAB digital radio broadcasts in September 1995. BBC had provided DAB coverage for over 60% of the UK population as of 1998. By mid-1998 more than 150 million people around the world were within the coverage of a DAB transmitter [20]. An upgraded version of DAB was released in February 2007, which is designated as DAB+.

OFDM and rate-compatible punctured convolutional coding (RCPC) form the core of COFDM signaling applied in DAB and DAB+. A major reason for the selection of OFDM modulation is the possibility to use *single-frequency network*. In a single-frequency network, the user receives several copies of the same signal from a number of transmitters with different delays corresponding to the distance between the receiver and the transmitters. As long as the time difference between the arrivals of two signals is less than OFDM guard interval, no ISI or ICI will occur. The presence of two or more time-shifted copies of the signal provides a diversity advantage in the network, in the sense that the probability that the sum of the signals has an unacceptably low power because of propagation, shadowing, and flat fading losses is much lower than that of an individual signal [4].

The DAB signal occupies a bandwidth of 1.536 MHz with a gross transmission rate of 2.3 Mbit/s. The net data rate, however, vary between 0.6 and 1.8 Mbit/s depending on the convolutional code rate, which is application dependent. The individual carriers are modulated using differential QPSK (DQPSK). To protect subcarriers from flat fading effect, convolutional code of constraint length 7 is applied. To allow for both equal and unequal error protection, the code rate varies from $1/3$ to $3/4$, which gives rise to an ensemble code rate average of $1/2$. DAB operates on four different transmission modes, using four different sets of OFDM parameters, as shown in Table 4.1. Modes I through III are optimized for performance in different operating frequency ranges, while mode IV is designed to provide larger area coverage at the expense of more sensitivity to Doppler shift.

The DAB waveform consists of a number of audio signals, sampled at 48 kHz with an input resolution of up to 22 bits that can provide CD quality audio output [20].

4.3.3.2 COFDM Applied in Wireless LAN (Wi-Fi): The IEEE 802.11 Standard

Since early 1990s, wireless local area networks (WLAN) designed and developed over so-called ISM bands (industrial, scientific, medical band over 0.9, 2.4, and 5 GHz) have been around. In June 1997, the IEEE approved an international interoperability standard: the IEEE 802.11 Standards. In July 1998, the IEEE 802.11 working group adopted OFDM signaling for their new 5 GHz standard that supports data rates 6, 9, 12, 18, 24, 36, 48, and 54 Mbps. To accommodate various supported

Table 4.1 COFDM parameters for DAB operating modes.

Parameter	Mode I	Mode II	Mode III	Mode IV
Nominal frequency	<375 MHz	<1.5 GHz	<3 GHz	<1.5 GHz
No. of subcarriers	1536	386	192	768
Useful symbol duration	1 ms	250 μs	125 μs	500 μs
Guard interval	246 μs	62 μs	31 μs	123 μs
Subcarrier modulation	DQPSK	DQPSK	DQPSK	DQPSK
Channel coding	RCPC	RCPC	RCPC	RCPC

Table 4.2 Data rate-dependent parameters of COFDM in IEEE 802.11 Standard.

Data rate M bits/s	Modulation format	Convolutional code rate	Coded bits per OFDM symbol	Data bits per OFDM symbol
6	BPSK	1/2	48	24
9	BPSK	3/4	48	36
12	QPSK	1/2	96	48
18	QPSK	3/4	96	72
24	16-QAM	1/2	192	96
36	16-QAM	3/4	192	144
48	64-QAM	2/3	288	192
54	64-QAM	3/4	288	216

data rates, the standard allows for the use of four different modulation schemes, namely, BPSK, QPSK, 16-QAM, and 64-QAM for subcarriers. The selected channel coding scheme is convolutional coding with variable rate that is dependent on the data rate. Thus, the choice of modulation parameters and channel coding rate depend on the selected data rate and is set according to Table 4.2 [21].

Table 4.3 lists other important timing parameters of COFDM signaling used in the IEEE 802.11 standard. Chapter 5 reveals that the same COFDM ideas that are applied in IEEE 802.11 standards are carried over to IEEE 802.16 standards and WiMAX networks.

A key parameter that predominantly affects the selection of other OFDM parameters is the guard interval of 800 ns. Clearly, this guard interval creates robustness against delay spread up to several hundred nanoseconds, depending on the coding rate and the applied modulation scheme. Practically speaking, this implies that the signaling is robust enough for the system to be used in any indoor environment, including large factory buildings. It can be also used in outdoor environments, although directional antennas may be needed to reduce the delay spread.

To limit the SNR effect of the guard interval to 1 dB, the symbol duration is selected to be 4 µs. The useful symbol duration, therefore, as shown in Table 4.3, is 3.2 µs, which is equal to the inverse of subcarriers frequency spacing, 0.3125 MHz. By using 48 subcarriers and applying various modulation scheme provided in the Standard, uncoded data rates of 12–72 Mbps are achievable. In addition to 48 data subcarriers, each OFDM symbol contains 4 pilot subcarriers that is used to track the residual carrier frequency offset remaining after the initial carrier frequency correction. To protect the signal against deep fade, a ½ rate,

Table 4.3 Timing parameters of COFDM signaling in IEEE 802.11 Std.

No. of data subcarriers	No. of pilot subcarriers	Subcarrier frequency spacing	Guard interval duration	Symbol interval	Useful symbol duration
48	4	0.3125 MHz	800 ns	4 µs	3.2 µs

constraint 7, convolutional code is used. The code rate can be adjusted to 2/3 and ¾ through the process of puncturing. This provides coded data rates in the range of 6–54 Mbps [5].

4.4 Performance of Channel Coding in OFDM Networks

As it was pointed out earlier, one of the major reasons for adoption of OFDM in many wireless applications is that OFDM signaling is capable of rendering a wideband frequency selective fading channel to a series of narrowband channels that suffer only from flat fading. To combat the effects of deep fading that may occur in the channel, various channel coding schemes have been applied or proposed. Rate compatible punctured convolutional codes alone or concatenated with Reed–Solomon codes have been extensively applied in early OFDM systems [5,16,22–25]. Later, turbo codes [16,26,27], low-density parity check (LDPC) codes [28] were proposed in the literature. Convolutional and Turbo convolutional codes combined with several schemes of modulation in AMC protocols are universally applied in standard-based wireless networks [29].

Using computer simulation, Arijon and Farrell studied the performance improvement in OFDM systems when RS codes and concatenated RS–RS codes are employed. The simulation models were developed for multipath fading channels in both urban and suburban areas. The results indicated a performance improvement with single RS codes is proportional to RS code word length, while code rate is kept fixed. Concatenated RS–RS code provided much more performance improvement. In the latter part of the study, it was also concluded that in the case of RS–RS concatenated coding, the use of a powerful code as outer and a less powerful code as inner code provides a better result than the other way around, the result that is well known in RS-convolutional code concatenation. The final concluding remark made are the claims that RS–RS concatenated code can remove most of the fading effect in the multipath channel [15].

The channel coding technique applied in DVB-T and DVS-S is a concatenation of a RCPC code as the inner code and a shortened (204, 188) RS code as the outer code. The RS code is the shortened version of systematic (255, 239) code that has a random error correcting capability of eight words. The mother code of the inner convolutional code is of code rate ½ and 64 states. The generator polynomials for the mother code are $G_1 = (171)_{OCT}$ $G_2 = (133)_{OCT}$. The system allows for a range of punctured convolutional codes of equivalent rates of 2/3, 3/4, 5/6, and 7/8, which facilitates the selection of the most appropriate error correction coding scheme for a given service. Table 4.4 provides the results of simulated performance of various modulation/coding schemes for nonhierarchical transmission in DVB-T standard for 8 MHz channels. Perfect channel estimation without phase noise has been assumed. The table provides two sets of data. First, the required CNR (carrier-to-noise ratio) for BER 2×10^{-4} after the Viterbi decoder for all combinations of coding rates and modulation formats are given. Then, the net bit rates after the Reed–Solomon decoder for four different relative values of guard interval time are listed. Note that the figures are approximate simulation values [25].

Table 4.5 provides the same information as Table 4.4 for hierarchical transmission of QPSK and nonuniform 16-QAM [4].

Turbo-coded-based OFDM system for terrestrial video broadcasting have been investigated and proposed. In particular, the feasibility of application of turbo codes and the potential performance improvements have been examined. It has been demonstrated, through simulation, that the turbo code provides in excess of 5 dB SNR performance improvement over conventional convolutional codes. This in turn improves bit error rate and video quality [16].

The parameter ϑ used in Table 4.5 is the cyclic prefix time duration (guard time).

Later, a new block turbo-coded-OFDM (BTC-OFDM) was developed for WWANs to provide performance improvement compared to that used in the IEEE 802.11a standard. This scheme utilizes diversity in both time and frequency domains to achieve more robustness against frequency and time fading [27] (see Chapter 5 for definition and classification of diversity).

COFDM signaling with iterative-decoded LDPC codes have been studied for time-varying wireless channels [28]. A space-time coded (STC) OFDM–MIMO–LDPC signaling technique has been investigated for correlated frequency and time selective fading channels. It is shown that the product of the time-selectivity order and the frequency-selectivity order is a key parameter to characterize the outage capacity of the correlated fading channel. It has been demonstrated that compared

Table 4.4 C/N required for various coding/modulation combinations for the BER $= 2 \times 10^{-4}$ for nonhierarchical transmission at the output of the Viterbi decoder.

Modulation format	Code rate	C/N Gaussian (dB)	C/N Rayleigh (dB)	C/N Rician (dB)	Bit rate (Mb/s) For $\vartheta = 1/4$	Bit rate (Mb/s) For $\vartheta = 1/8$	Bit rate (Mb/s) For $\vartheta = 1/16$	Bit rate (Mb/s) For $\vartheta = 1/32$
QPSK	½	3.1	3.6	5.4	4.98	5.53	5.85	6.03
QPSK	2/3	4.9	5.7	8.4	6.64	7.37	7.81	8.04
QPSK	¾	5.9	6.8	10.7	7.46	8.29	8.78	9.05
QPSK	5/6	6.9	8.0	13.1	8.29	9.22	9.76	10.05
QPSK	7/8	7.7	8.7	16.3	8.71	9.68	10.25	10.56
16-QAM	½	8.8	9.6	11.2	9.95	11.06	11.71	12.06
16-QAM	2/3	11.1	11.6	14.2	13.27	14.75	15.61	16.09
16-QAM	¾	12.5	13.0	16.7	14.93	16.59	17.56	18.10
16-QAM	5/6	13.5	14.4	19.3	16.59	18.43	19.52	20.11
16-QAM	7/8	13.9	15.0	22.8	17.42	19.35	20.49	21.11
64-QAM	½	14.4	14.7	16.0	14.93	16.59	17.56	18.10
64-QAM	2/3	16.5	17.1	19.3	19.91	22.12	23.42	24.13
64-QAM	¾	18.0	18.6	21.7	22.39	24.88	26.35	27.14
64-QAM	5/6	19.3	20.0	25.3	24.88	27.65	29.27	30.16
64-QAM	7/8	20.1	21.0	27.9	26.13	29.03	30.74	31.67

Also shown is bit rate for different values of guard intervals.

with the conventional space-time trellis code (STTC), the LDPC-based STC can significantly improve the system performance by exploiting both the spatial diversity and the selective fading diversity in wireless channels [30].

4.5 Orthogonal Frequency-Division Multiple Access: OFDMA

Radio spectrum is one of the key resources for development of wireless communication systems. However, each wireless application is allocated with a finite band of frequencies within the available spectrum. This defines a fundamental limitation on this key resource. In order to accommodate multiple mobile devices that wish to get connected to the network and communicate simultaneously, it is necessary to set up a

Table 4.5 Required C/N for hierarchical transmission to achieve BER $= 2 \times 10^{-4}$ at the output of the Viterbi decoder.

Modulation	Code rate	C/N Gaussian (dB)	C/N Rayleigh (dB)	C/N Rician (dB)	Bit rate (Mb/s) For $\vartheta = 1/4$	Bit rate (Mb/s) For $\vartheta = 1/8$	Bit rate (Mb/s) For $\vartheta = 1/16$	Bit rate (Mb/s) For $\vartheta = 1/32$
QPSK	½	4.8	5.4	6.9	4.98	5.53	5.85	6.03
QPSK	2/3	7.1	7.7	9.8	6.64	7.37	7.81	8.04
QPSK	¾	8.4	9.0	11.9	7.46	8.29	8.78	9.05
NU 16-QAM	½	13.0	13.3	14.9	4.98	5.53	5.85	6.03
NU 16-QAM	2/3	15.1	15.3	17.9	6.64	7.37	7.81	8.04
NU 16-QAM	¾	16.3	16.9	20.0	7.46	8.29	8.78	9.05
NU 16-QAM	5/6	16.9	17.8	22.4	8.29	9.22	9.76	10.05
NU 16-QAM	7/8	17.9	18.7	24.1	8.71	9.68	10.25	10.56

Also shown is the bit rate at the output of the RS decoder.

method of resource sharing, that is, a *multiple access technique*. In *homogeneous networks*, a single access technology is applied throughout the network, however, the *heterogeneous networks* are designed to support multitude of access technologies.

A related issue is *duplexing*, which addresses how uplink (MS to BS transmission) and downlink (BS to MS transmission) communications are managed. In the old analog relay-selector-based telephone systems, a single analog channel (perhaps a twisted wire) was utilized to transmit a two-way communication between both sides of the conversation, as well as accommodating for control and signaling needs. Wireless communications, however, requires two separate radio channels for uplink and downlink, and in many cases dedicated radio channels for control and signaling. Two-way communication through radio channels may be conducted by transmitting and receiving via different channels, or if the same channel is to be used the time has to be interleaved for transmission and reception, this leads to two distinct methods of duplexing that are in common use. Frequency domain (division) duplexing (FDD), in which the uplink and downlink data are transmitted through two different radio channels with wide spectral separation, in order to

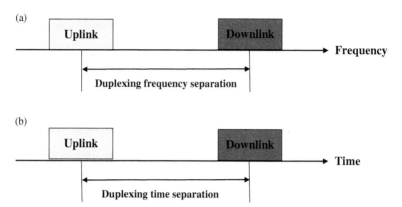

Figure 4.6 Duplexing methods: FDD (a) and TDD (b).

avoid interference. The second method is time domain (division) duplexing (TDD), wherein uplink and downlink signals are separated by time, that is, they are transmitted at different times. Figure 4.6 illustrates the FDD and TDD concepts.

Depending on the application, multiple access technology, as well as other network parameters, a particular duplexing method may be preferred.

4.5.1 Multiple Access Technologies: FDMA, TDMA, CDMA, and OFDMA

The oldest, and conceptually simplest, method of multiple access is realized when the available spectrum is divided into N nonoverlapping bands of frequencies to form radio channels. Each user is provided with a single transmission channel, and as such, a multitude of users may simultaneously communicate through the network. This configuration is called *frequency-division multiple access* (*FDMA*). Few notes are in order at this point. First and foremost, it is noted that the allocated bandwidth for radio channels offered by FDMA is, theoretically, equal to the total available bandwidth divided by N. Second, as it was alluded to earlier, wireless communications with FDMA (or any other access technique, for that matter) requires a duplex channel to support transmission on both uplink and downlink. Barring other technical concerns, FDD seems to be a natural duplexing choice for FDMA technique. FDMA–FDD technology is employed in the 1G analog cellular networks, where analog FM is used to exclusively carry voice and audio. Finally, although conceptually channel orthogonality has been guaranteed by dividing the available spectrum to N nonoverlapping bands, adjacent channel interference may occur due to imperfect channel filter design and

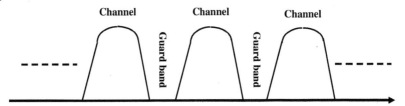

Figure 4.7 FDMA spectral division into radio channels and guard bands.

behavior, as well as radio propagation anomalies. In order to mitigate the effect of this sort of interference, it is necessary to leave a guard band between adjacent channel spectra. This is illustrated in Figure 4.7.

Despite its simplicity, FDMA suffers from a number of shortfalls, the most important of which is spectral inefficiency. This is, primarily, due to the required guard bands, as shown in Figure 4.7, which essentially constitute spectral overhead and reduce the number of available radio channels in the network. Moreover, each FDMA channel can support only a single mobile device. Furthermore, when an occupied FDMA channel is in idle state, it cannot be utilized by other users and essentially becomes a wasted resource. For more extensive coverage on FDMA and its shortfalls, see Refs [31] and [32].

Radio resources may be delivered to the user through time slots, that is, diving time into short periods and assigning each user with these time periods in a cyclic or random fashion for single-carrier transmission and reception of traffic data. This is the essence of *time division multiple access (TDMA)*. Packet radio networks, in essence, apply TDMA technique in which time slots become available randomly. However, in many TDMA systems, the user is allocated with regular and cyclical time slots during which she may transmit or receive data in the burst construct. Outside of the user dedicated time slots no transmission occurs, thus with TDMA, the user experiences noncontinuous communications. Clearly, TDMA technology is well suited for digital transmission and digital modulation schemes. In the context of cellular networks and on the uplink, TDMA is an access method in which MSs transmit to the BS intermittently using the same radio channel but with the timing of their transmissions so arranged that the bursts coming from various MSs do not overlap when they arrive at the BS antenna but arrive in a predefined sequence. A similar arrangement is made for the downlink. When N uniform time slots are available, one round of traffic data and the associated overhead bits collected from the corresponding N users along with some required added operating bits is called a *TDMA frame*. The structure of a TDMA frame, as described by Rappaport [31], is shown in Figure 4.8.

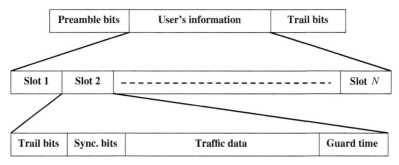

Figure 4.8 A structure for TDMA-TDD frame [31].

The preamble section of the frame contains information related to address and synchronization that both the BS and the MS can use for identification purposes. We note that with TDMA, the natural duplexing choice is TDD, although TDMA may employ FDD as well. If we assume that the underlying technology is TDMA–TDD, then the time slots are alternatively used for uplink and downlink transmissions. For instance, odd slots might be used for uplink transmission and even slots for downlink communication. The generic term of "traffic data" is used to imply that user's information may be of any type, for example, voice, audio, video, data, and map.

One major advantage of TDMA over FDMA is that, within the provided time slots, the user can occupy larger bandwidth, theoretically, even the entire available spectrum. Thus, traffic data may be transmitted at high rate, however, equalizers are needed to cope with frequency selective fading induced ISI. The wide bandwidth, also, allows the user to take advantage of frequency diversity that is available within the allocated spectrum. Moreover, the sensitivity against FM noise created by Doppler spread is reduced. No frequency guard band is needed in TDMA since each user is allowed to use the entire available bandwidth, however, time slots need to be separated by temporal guard band [32]. A major challenge with TDMA, like any other digital communications system, is the synchronization process and its associated high overhead. It is imperative that each TDMA data burst is routed to its proper destinations, this is ensured by proper synchronization between the receiver and the transmitter.

Spread spectrum modulation schemes, such as direct sequence spread spectrum (DSSS) and frequency hopping (FH), are distinguished from other forms of modulation in that they generate broadband signals with bandwidths several orders of magnitude larger than that of the baseband signal. The broadband modulated signal is generated by a pseudonoise (PN) code (also known as spreading codes) that is independent from the baseband signal. Normally, a user's signal occupies the entire allocated spectrum, and yet coexist with other users' signals that also occupy the entire available bandwidth. The PN codes are generated by linear feedback shift register systems [33]. (*Code division multiple access*) (CDMA) is a network access method that is based on spread spectrum concept. Cellular CDMA networks are developed, by and large, based on DSSS technique in which the digital data is multiplied by a fast fluctuating binary PN code to produce wideband signal. At the receiver side, the desired signal is reproduced by correlating the received signal with the spreading code of the user, which is known to the receiver. Each user is provided with a unique PN code, and shares the entire bandwidth with other users. Ideally, signal orthogonality is guaranteed if the spreading codes used for various users are all mutually orthogonal. However, there are practical limitations in producing PN codes that are perfectly orthogonal. With practical PN codes, after signal correlation, the effect of other users' signal on the received signal appears as a low power wideband noise. As the number of users in the channel increases, the noise floor rises and eventually degrades the performance, which in turn limits the number of users that defines the capacity of CDMA network. CDMA offers a number of unique features. In CDMA, many users can share the entire bandwidth without posing significant interference to one another. Network access requires virtually no management procedure, that is, the user can access the network and start communicating. One conspicuous advantage of CDMA, over FDMA and TDMA networks, is the fact that in CDMA the network performance gracefully degrades as the number of users in the network increases, whereas in FDMA and TDMA, the maximum number of users is a set value, and additional users will be blocked from accessing the network. One rough definition of the capacity of a multiple access system is the number of users that can be simultaneously accommodated with a given bandwidth, transmitted power, and error performance. The answer to the question of which one of these three access technologies performs better in the context of cellular networks, and in terms of offered capacity, is still a matter of dispute. Nonetheless, some have claimed that for a fixed available spectrum, CDMA capacity is larger than that of FDMA by at least an order of magnitude [34].

Earlier in this chapter, OFDM was introduced as a modulation scheme, as well as a multiplexing method. The main virtue of the standard OFDM for radio communications, as has been mentioned before, is that it may be used to convert a wideband frequency selective fading channel to a number of narrowband flat channels. Furthermore, OFDM technology can also be used to form an alternative multiple access technology. The idea is to expand OFDM concept (a point-to-point communication link) in a way that her resources, shown in Figure 4.2, can be shared among many users. There are a number of ideas for organizing an OFDMA system. One choice is combining OFDM with TDMA, that is to say, allowing a user to occupy the entire channel bandwidth and utilize all subcarriers for a time period equal to one, two (or any integer multiple of) OFDM symbol duration, before the next user takes over the channel. A second idea is to use packet radio protocols in which every user transmits a complete packet using OFDM modulation. The methods of users accessing the channel may be controlled by any of the packet radio access approaches [32]. An alternative method is to divide the set of subcarriers into a number of subsets that are called *subchannels* and interface each subchannel to a single user, thereby providing multiple access. This is the scheme that is commonly referred to as OFDMA and is applied in WiMAX and AeroMACS networks. The incentive behind OFDMA's widespread applications is discussed in the next section and more technical details are provided in Chapters 5 and 7.

4.5.2 Incentives behind Widespread Applications of OFDMA in Wireless Networks

The development of radio technologies has been a rapid ongoing process over the past four decades, addressing the ever-increasing demand on capacity, outreach and coverage, QoS, and more recently, various broadband applications. Advanced signal processing techniques and their associated algorithms on multiple access procedures, modulation and coding schemes, smart antenna systems, radio resource supervision, mobility management, and so on, are employed to meet the challenges of providing high-quality broadband services in large volumes. In this section, we concentrate on OFDMA, an access technology that is widely employed in modern standard-based wireless networks.

As will be discussed more thoroughly in Chapter 5, the IEEE 802.16 standard [35] provides specifications for WMAN air interface for fixed, nomadic, portable, and mobile broadband wireless access systems. In particular, the standard includes requirements for high data rate LOS and NLOS operation in frequencies below 11 GHz for both licensed and license-free bands. OFDM and OFDMA technologies, for their superior performance in combating frequency selective fading distortion, are recommended by the Standard to be applied for operation in sub 11 GHz NLOS applications.

In addition to superior performance in multipath channels, there are a number of other reasons that OFDMA is widely accepted in wireless networks. In the narrowband case, signal fading reduces the link range, therefore additional power budget is needed to ensure a required BER at the receiver is achieved. For instance, it has been reported that to bring down the probability of bit error below a nominal level of 10^{-3} for a given application, an additional link budget of about 30 dB is required [36]. This is the case since the entire signal is modulated on a single fixed carrier. In a broadband OFDM system, however, a subchannel consisting of a number of subcarriers is used to transport the traffic data. And in case one subcarrier experiences deep fade, chances are that a number of other subcarriers, in the subchannel, do not experience the same. Under these circumstances combining OFDMA with channel coding schemes enables the recovery of the lost data symbols, and can essentially provide significant fade-margin improvement. This is a natural *frequency diversity* provided by OFDMA. Additionally, OFDMA easily lends itself to other forms of diversity techniques for further performance improvements. Smart and multiple antenna system, which represents spatial diversity, signal repetition; a form of time diversity, and STC; a combination of time and spatial diversity, are particularly well suited with OFDMA. The combination of frequency, time, and space diversity methods results in significant performance improvement in fading channels [37].

The Shannon channel capacity theorem sets an upper limit on how fast data can be reliably transmitted through a given channel. This limit is called channel capacity and it is expressed by Equation 4.10 for Gaussian channels.

$$C_{ch} = B_{ch} \log_2 \left[1 + \frac{S}{N} \right] \tag{4.10}$$

Where C_{ch} is channel capacity, B_{ch} is channel bandwidth, and S/N is signal-to-noise ratio. Channel capacity theorem states that as long as information data rate is below channel capacity, reliable and, theoretically, error-free communication is possible. This theorem is germane to data rate over a channel with a single transmitting antenna and a single receiving antenna. However, this limit does not set a data rate restriction for reliable communications for systems in which multiple antennas are used at either the transmitting side or the receiving side, or both. In fact, each combination of transmit and receive antennas can be considered as a separate channel. Under these circumstance, the overall effective channel capacity and system throughput will significantly enhance almost proportional to the number of transmitter–receiver combinations that form independent channels. This is the crux of performance enhancement achieved in multipath channels when MIMO configurations are applied. OFDMA readily supports MIMO.

4.5.3 Subchannelization and Symbol Structure

The Fourier transform-based OFDM divides the allocated spectrum into secondary orthogonal channels, whose center frequencies are corresponding to the subcarriers. When these subcarriers are organized for the purpose of providing multiple access, they are classified into three types.

- *Data subcarriers* are strictly used for transportation of traffic data. The process of *subchannelization* is the creation of subchannels by clustering several subcarriers and assigning each subchannel to a user. These subcarrier clusters are all data subcarriers.
- *Pilot subcarriers* are used for channel measurement and estimation, for various algorithm used in the wireless network, for instance, in AMC algorithms. Pilot subcarrier are also used for synchronization.
- *Null subcarriers* consist of guard band subcarriers and DC subcarrier, and are used for prevention of energy leakage from one symbol to the adjacent symbols. The DC subcarrier is corresponding to the center of the RF bandwidth and is not used for data transmission.

OFDMA symbols are constructed by arranging all subcarriers of the underlying OFDM symbols into data, pilot, and null subcarriers. There are three possibilities in creating subchannels.

1) One choice is clustering adjacent subcarriers to form subchannels. The selection of adjacent subcarriers can be conducted adaptively by the *scheduler.* In this manner subcarriers with high signal-to-interference-plus-noise ratio (SINR) are selected and subcarriers with low SINR are avoided, therefore a "loading gain" is achieved with this arrangement. Another advantage of this *adjacent subcarrier allocation* (ASA) method is that, since the subcarriers assigned to subchannel are adjacent, they are correlated and may be used for channel estimation rather readily. This approach requires a relatively stable channel condition, therefore, it is suitable for low-speed mobile and nomadic applications [37]. In this scheme, every subcarrier is assigned to a subchannel, and therefore, to some user. This facilitates *multiuser diversity.* Figure 4.9 illustrates OFDMA symbol for adjacent subcarrier allocation.

2) A second method of subcarrier permutation for construction of subchannels is to form subsets of subcarriers that are "equally spaced" from each other and assign them to various subchannels, this is known as *evenly distributed subcarrier* (EDS) allocation. The major advantage of this method is the high degree of frequency diversity that it provides. There is no multiuser diversity in this scheme. Figure 4.10 portrays an example of evenly distributed subcarrier allocation.

3) Another alternative for molding subchannels is the assignment of randomly distributed subcarriers (RDS) into a subchannel. In other words, a subchannel is formed by selection of subcarriers that are randomly distributed across the allocated spectrum. Generally

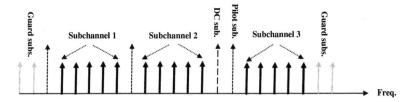

Figure 4.9 An example of OFDMA symbol structure for ASA scheme.

speaking, this method has the same advantages and shortfalls as EDS technique. The main advantages of this method, however, are maximization of frequency diversity and randomization of adjacent cell interference, thereby averaging the intercell interference [32]. This is the most suitable approach for mobile environment where the channel characteristic changes rapidly. This is the method of choice for WiMAX and AeroMACS networks. Figure 4.11 illustrates an example for RDS scheme.

4.5.4 Permutation Modes for Configuration of Subchannels

There are a number of permutation modes for assigning subcarriers into subchannels. Partial usage of subchannels (PUSC), full usage of subchannels (FUSC), and adaptive modulation coding (AMC) are the most common modes that are applied in WiMAX and AeroMACS networks. PUSC and FUSC use EDS and RDS, and AMC uses ASA. Other available permutation modes are optional PUSC (OPUSC), optional FUSC (OFUSC), tile usage of subchannels 1 (TUSC1), and tile usage of subcarrier 2 (TUSC2). Some of these modes are exclusively used in DL, some are applied in both DL and UL [38]. In FUSC, which is used only for DL, all subchannels are allocated to the BS transmitter. One *slot* of FUSC DL is composed of one subchannel in a single OFDM symbol. In contrast to FUSC, in PUSC permutation, which is applied in both DL and UL, a

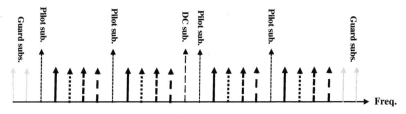

Figure 4.10 A generic example of OFDMA symbol structure for EDS scheme. Subcarriers assigned to the same subchannel are dashed similarly.

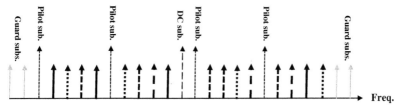

Figure 4.11 A generic example of OFDMA symbol structure for RDS scheme. Subcarriers assigned to the same subchannel are dashed similarly.

subset of subchannels are allocated to a transmitter, this subset is referred to as a *segment*. When the cell is sectored, a segment is assigned to a sector transmitter. The other subchannels are divided into two segments each of which is allocated to the transmitter of each of the remaining sectors [39]. ASA offers better channel estimation capabilities and readily allows for adaptive modulation coding (AMC). Reference [39] provides a detailed account of subchannel formations for FUSC, PUSC DL, PUSC UL, ASA UL, and ASA DL, consistent with IEEE 802.16-2004 and IEEE 802.16e standards.

4.5.4.1 The Peak-to-Average Power Ratio Problem

It is widely recognized that high *peak-to-average power ratio* (PAPR) is a drawback for OFDM signaling. PAPR is essentially a measure of power fluctuations of a signal over time. When a received RF signal suffers from high PAPR, power amplifiers required for the signal amplification need to have a large linear dynamic range. In case the power of the signal extends beyond the linear range, signal undergoes a nonlinear clipping distortion. It has been shown that in OFDMA UL, PAPR in RDS is significantly larger than that of ASA that represents an advantage of ASA over RDS [40].

4.6 Scalable OFDMA (SOFDMA)

The original OFDMA access technology for WMAN, as expressed in IEEE 802.16-2004-Std for applications in fixed WiMAX networks, was set for FFT size of 2k, that is, for 2048 subcarriers. The number of tones was to remain fixed regardless of the size of the available bandwidth. Problem arises when bandwidth is small and mobility gets involved. For instance, when bandwidth is 1.25 MHz, with 2048 subcarriers, the tone spacing is about 610 Hz. This makes the network very sensitive to impairments caused by mobility, in particular by the Doppler effect. It has been shown that the performance achieved with FFT size of 2k for high bandwidth (around 20 MHz and greater) are very good, however, degradation becomes apparent as allocated bandwidth is lowered [41]. In order to address this physical layer problem and to provide the same performance quality across all desired bandwidths, instead of fixed number of tones, fixed tone spacing was proposed and introduced in IEEE 802.16-2005e-Std. This is the central idea of scalable orthogonal frequency-division multiple access (SOFDMA) [35]. In other words, scalability in the context of IEEE 802.16 standards (WiMAX and AeroMACS) means that the size of FFT and the number of OFDM subcarriers are changed proportional to the size of available bandwidth, while the tone spacing is kept fixed.

The fundamental features of SOFDMA may be outlined as follows [37,42].

1) Subcarrier spacing is fixed and independent from the size of the allocated bandwidth.
2) The size of FFT and the number of subcarriers scale with bandwidth.
3) The number of subcarriers per subchannel is fixed and independent from bandwidth.
4) The number of subchannels scales with bandwidth. In other words, the larger the allocated bandwidth, the more subchannels are offered by the network.
5) The capacity of each subchannel remains fixed and independent from bandwidth.
6) Subchannels represent the smallest bandwidth allocation units, and they remain fixed as bandwidth changes.
7) SOFDMA supports advanced signal processing techniques for NLOS propagation, such as AMC, MIMO in both DL and UL, hybrid automatic repeat request (H-ARQ), and a number of diversity techniques.

4.6.1 How to Select the OFDMA Basic Parameters vis-à-vis Scalability

Preliminary step in the design of any OFDMA-based system is the characterization of the underlining wireless link and a rough estimation of key parameters of the multipath channel, leading to determination of FFT size, number of subcarriers, and the process of subchannelization. The number of subcarriers per network bandwidth that sets the subcarrier spacing presents a trade-off between protection against multipath and Doppler effect, on the one hand, and system complexity and cost, on the other hand. In particular, Doppler spread, coherence time, delay spread, and coherence bandwidth (please refer to Chapter 2 for definition of these parameters) of the multipath channel are needed. In order to optimize protection against multipath and Doppler spread, while system complexity is kept at a reasonable level, the Wide-Sense Stationary Uncorrelated Scattering (WSSUS)[4] model is used [42]. We embark on computation of two key WSSUS model parameters, namely, coherence time and coherence bandwidth of the channel, the approach is the one introduced in Ref. [42] and variable values such as speed and frequency are consistent with those of C-band AeroMACS.

The takeoff speed of commercial jet aircrafts, depending on headwind and the type of aircraft, is typically in the range of 150–180 mph (240–285 km/h), roughly about 66–80 m/s. Core AeroMACS band of operation is allocated over 5.091–5.150 GHz (additional bands may become available for Aero-MACS), corresponding to wavelength range of 0.05893–0.05825 m. From Equation (2.106), the worst-case Doppler spread can be calculated.

$$B_{ds} = \frac{v}{\lambda} = \frac{80}{0.05825} \cong 1{,}373\,\text{Hz} \tag{4.11}$$

Using Equation (2.111), coherence time can be computed

$$T_c \cong \frac{0.423}{B_{ds}} = \frac{0.423}{1373} \cong 3.08 \times 10^{-4} \tag{4.12}$$

As it was pointed out in Chapter 2, coherence time represents a time duration within which the channel may be assumed to remain stationary. Thus, Equation 4.12 implies that signals whose rate is greater than $1/T_c = 1/3.08 \times 10^{-4} \cong 3{,}247$ bps can pass through the channel without suffering from significant time selective distortion. The second implication of Equation 4.12 is for channel estimation and equalization an update rate of approximately 3,247 Hz is required.

An important design parameter, for OFDMA-based systems, is the subcarrier spacing that may be roughly calculated based on the notion that secondary channels (subcarriers) are required to have flat fading characteristics under the worst-case delay spread. This means that the subcarrier spacing needs be equal to worst case (smallest) coherence bandwidth. The outdoor delay spreads for cellular networks are typically on the order of few to 10 μs. Following the computation carried out in Ref. [42] and assuming a delay spread of 20 μs (to be on the safe side) and using the first part of Equation (2.105) with $\rho = 0.5$, it follows that

$$B_{ch} = \frac{1}{5\sigma_\tau} = \frac{1}{5 \cdot 20 \cdot 10^{-6}} = 10000\,\text{Hz} = 10\,\text{kHz} \tag{4.13}$$

---

4 WSSUS is a stochastic-process-based modeling that is commonly used for multipath channel under the assumptions that channel correlation function is time invariant, and scatterers with different delays are uncorrelated. These assumptions are reasonable for short-term variations of radio channel [43,44].

Equation 4.13 implies that for delay spread values of up to 20 μs, the subcarrier channels with 10 kHz bandwidth behave like frequency flat fading channels.

Another concern related to selection of subcarrier spacing is phase noise. It has been shown that the phase noise impairment intensifies as subcarrier spacing becomes narrower. Without FFT size scaling, as the system bandwidth decreases, so does subcarrier spectral spacing, and therefore the negative impact of phase noise intensifies.

Considering all key parameters of a given wireless channel, as outlined above, potential channel impairments, and the allocated bandwidth, one can select a FFT size and subcarrier spacing for a given OFDMA scenario.

4.6.2 Options in Scaling

The network bandwidths supported by the IEEE 802.16 standards for applications in WiMAX and its off springs, such as AeroMACS, are 20 MHz, 10 MHz, 5 MHz, 2.5 MHz, and 1.25 MHz. As scalability is applied, FFT size and most other OFDM parameters are scaled with the allocated bandwidth. Table 4.6 summarizes OFDMA key parameters, as well as subcarrier characterization and allocation, as the bandwidth and FFT size are scaled, for DL channel.

Table 4.6 OFDMA DL scalability parameters and subcarrier allocation.

Parameter/BW (MHz)	≥ 20	≥ 10	≥ 5	≥ 2.5	≥ 1.25
FFT size	2048	1024	516	256	128
Sampling frequency (Ms/s): f_s	22.857	11.429	5.714	2.857	1.429
Sampling factor: η_{sam}	8/7	8/7	8/7	8/7	8/7
Subcarrier spacing (kHz)	$\cong 11.16$	$\cong 11.16$	$\cong 11.16$	$\cong 11.16$	$\cong 11.16$
T_s(μs)	89.6	89.6	89.6	89.6	89.6
$T_{cp} = T_s/8$ (μs)	11.2	11.2	11.2	11.2	11.2
$T = T_s + T_{cp}$ (μs)	100.8	100.8	100.8	100.8	100.8
No. of guard and DC subcarriers	345	173	86	43	21
No. of pilot subcarriers	167	83	42	21	11
No. of data subcarriers	1536	768	384	192	96
No. of data subcarriers per subchannel	48	48	48	48	48
No. of subchannels	32	16	8	4	2

Few notes are in order regarding Table 4.6. First, the oversampling by a factor of 8/7 is observed, this is represented by a parameter designated as sampling factor and shown by η_{sam} in the table. This is specified in the IEEE 802.16 standards as a global constant. As Table 4.6 indicates, the subscriber spacing remains fixed for all cases which is a direct result, and one of the main objectives of scaling. Other constants in the table are the number of data subcarriers allocated for each subchannel that has a fixed value of 48, useful symbol time duration T_s, cyclic prefix time duration (guard time) T_{cp}, and total symbol time duration T. All other parameters are scaled with bandwidth. A similar table can be set up for OFDMA UL as well that, although not identical, would have a lot of commonalities with Table 4.6.

4.7 Summary

In this chapter, fundamental principles of OFDM signaling are reviewed. The salient features of coded OFDMA (COFDM), which represent an effective technique for signal transmission over frequency selective fading channels, are explored. The main virtue of OFDM is its ability to convert a wideband frequency selective fading channel to a series of narrowband flat fading channels. The secondary channels may use different transmit power levels and modulation schemes. While OFDM can remove the frequency selective aspect of a fading channel, error control coding techniques can complement the function of OFDM in combating and eliminating the fading effect itself. With all the flexibility and efficiency that COFDM (harbinger of AMC techniques applied in WiMAX and AeroMACS) offers, the hardware/software complexity is not a challenging issue, since fast Fourier transform devices may be employed for the implementation of the modulator and demodulator in the system. Application of COFDM in digital audio broadcast (DAB) and in IEEE 802.11 standard-based wireless LAN (Wi-Fi) is discussed in some detail. Comments are made on performance of various channel coding schemes when they are applied in OFDM networks.

OFDMA is a multiple access technique that offers significant advantages for broadband wireless transmission over its rival technologies FDMA, TDMA, and CDMA. Primary reason for this preeminence is the ability to combat against frequency selective filtering effect of the multipath channel. In addition, OFDMA inherits all of salient properties of OFDM; bandwidth efficiency, ease of implementation through the application of FFT algorithms, its natural parallelism in the frequency domain that allows for independent per-subcarrier processing, and so on. The process of subchannelization is described and the division of subcarriers into three categories of guard and DC subcarriers, pilot subcarriers, and data subcarriers, is explained. When the neighboring subcarriers are clustered together to form subchannels, adjacent subcarrier allocation (ASA) is resulted, which is a configuration particularly suitable for application of AMC. Distributed subcarrier allocations are divided into evenly distributed subcarrier (EDS) arrangement and randomly distributed subcarrier (RDS) format. Several subchannel permutations are at disposal of the network designer. Partial usage of subchannels (PUSC), in which only a subset of subchannels are assigned to the transmitter, full usage of subchannels (FUSC), and adaptive modulation coding (AMC) are the most common forms of these configurations that are applied in WiMAX and AeroMACS networks. More detailed information on this subject are offered in Chapters 5 and 7.

When considering OFDMA for broadband wireless access, the number of subcarriers, for an allocated network bandwidth, represents a trade-off between robustness against multipath channel hindrances, on the one hand, and network complexity and cost as well as Doppler spread, on the other hand. When the number of subcarriers is increased, symbol time duration may be raised that leads to a better immunity against frequency selective filtering. On the other hand, increasing the number of sub-carriers leads to requirement of more intense signal processing, larger PAPR, and higher sensitivity to Doppler spread and phase noise, due to narrower subcarrier spectral spacing. It has been shown that the optimum subcarrier spacing is about 11 kHz [42]. Consequently, with wider allocated bandwidth and fixed subcarrier spacing, one has to scale other key parameters of the OFMDA, such as FFT size (number of subcarriers). This is the main idea in scalable OFDMA (SOFDMA). SOFDMA is the foundation on which the physical layer protocols of mobile WiMAX and AeroMACS networks are built.

References

1 J. Salz and S. B. Weinstein, "Fourier Transform Communications System," presented at the *Computer Machinery Conference*, Pine Mountain, GA, Oct. 1969.

2 S. B. Weinstein and P. M. Ebert, "Data Transmission by Frequency Division Multiplexing Using the Discrete Fourier Transform," *IEEE Transactions in Communications Technology*, COM-19 (5), 1971.

3 P. S. Chow, J. C. Tu, and J. M. Cioffi, "Performance Evaluation of a Multichannel Transceiver System for ADSL and VHDSL Services," *IEEE Journal of Selected Arias in Communications*, SAC-9, 6, 1991.

4 R. Van Nee and R. Prasad, *OFDM for Wireless Multimedia Communications*, Artech House, 2000.

5 R. Van Nee et al., "New High Rate Wireless LAN Standards," *IEEE Communications Magazine*, 37(12), 82–88, 1999.

6 J. Skold, S. Parkvall, and E. Dahlman, *4G: LTE/LTE-Advanced for Mobile Broadband*, 2nd edn, Academic Press, 2014.

7 P. Banelli, et al., "Modulation Formats and Waveforms for 5G Networks: Who Will Be the Heir of OFDM?: An Overview of Alternative Modulation Schemes for Improved Spectral Efficiency," *IEEE Signal Processing Magazine*, 31(6), 80–93, 2014.

8 Y. Wu and W. Y. Zou, "Orthogonal Frequency Division Multiplexing: A Multi-Carrier Modulation Scheme," *IEEE Transactions on Consumer Electronics*, 41(3), 1995.

9 P. Shelswell, "The COFDM Modulation System: The Heart of Digital Audio Broadcasting," *Electronics and Communication Engineering Journal*, 7(3), 127–136, 1995.

10 B. Kamali and R. L. Brinkley, "Reed-Solomon Coding and GMSK Modulation for Digital Cellular Packet Data Systems," *Proceeding of 1997 IEEE Pacific Rim Conference on Communications, Computer, and Signal Processing*, vol. 2, pp. 745–748, 1997.

11 W. Y. Zou and Y. Wu, "COFDM: An Overview," *IEEE Transactions on Broadcasting*, 4(1), 1–8, 1995.

12 T. Pollet, M. Van Blade, and M. Moeneclaey, "BER Sensitivity of OFDM Systems to Carrier Frequency Offset and Wiener Phase Noise," *IEEE Transactions on Communications*, 43 (2/3/4), 191–193, 1995.

13 H. Sari et al., "Transmission Techniques for Digital Terrestrial TV Broadcasting," *IEEE Communications Magazine*, 33(2), 100–109, 1995.

14 B. Lefloch, R. Halbert-Lassalle, and D. Castelain, "Digital Sound Broadcasting to Mobile Receivers," *IEEE Transactions on Consumer Electronics*, 35, (3), 493–503, 1989.

15 I. M. Arijon and P. G. Farrell, "Performance of an OFDM System in Frequency Selective Channels Using Reed–Solomon Coding Schemes," *IEE Colloquium on Multipath Countermeasures*, 6/1–6/7, 1996.

16 C-S. Lee, T. Keller, and L. Hanzo, "OFDM-Based Turbo-Coded Hierarchical and Non-Hierarchical Terrestrial Mobile Digital Video broadcasting," *IEEE Transactions on Broadcasting*, 46(1), 2000.

17 M. S. Zimmerman and A. L. Kirsch, "The AN/GSC-10 (KATHRYN) Variable Rate Data Modem for HF Radio," *IEEE Transactions on Communications Technology*, COM-15, 197–204, 1967.

18 B. Hirosaki "A 19.2 kbps Voiceband Data Modem Based on Orthogonally Multiplexed QAM Techniques," *Proceedings of IEEE ICC'85*, Chicago IL, June, 1985.

19 G. Ungerboeck, "Channel Coding with Multilevel/Phase Signals," *IEEE Transactions on Information Theory*, IT 28, 55–67, 1982.

20 W. H. W. Tuttlebee, and D. A. Hawkins, "Consumer Digital Radio: from Conception to Reality," *Electronics and Communication Engineering Journal*, 10(6), 263–276, 1998.

21 IEEE Standard 802.11a, "Part 11: Wireless LAN Medium Access Control (MAC) and Physical Layer (PHY) Specifications; Amendment 1: High-Speed Physical Layer in the 5 GHz Band," November 2000.

22 U. Reimers, "Digital Video Broadcasting," *IEEE Communications Magazine*, 36(6), 104–110, 1998.

23 U. Reimers, "DVB-T: the COFDM-Based System for Terrestrial Television," *Electronics and Communication Engineering Journal*, 9(1), 28–32, 1997.

24 S. O'Leary, "Hierarchical Transmission and COFDM Systems," *IEEE Transactions on Broadcasting*, 43(2), 166–174, 1997.

25 ETSI, "Digital Video Broadcasting (DVB): Framing Structure, Channel Coding, and Modulation for Digital Terrestrial Television," EN 300 744 V1.4.1, 2001.

26 F. Said and A. H. Aghvami, "Two Dimensional Pilot Assisted Channel Estimation for Turbo Coded OFDM Systems," *IEE Colloquium on Turbo Codes in Digital Broadcasting*, November 1999.

27 M. Torabi and M. R. Soleymani, "Turbo Coded OFDM for Wireless Local Area Network," *Proceedings Of IEEE CCECE'02*, May 2000.

28 G. A. Al-Rawi et al., "An Iterative Receiver for Coded OFDM Systems Over Time-Varying Wireless Channels," *Proc. IEEE ICC'03* vol. 5, 2003.

29 R. Zhang and L. Cai, "Joint AMC and Packet Fragmentation for Error Control Over Fading Channels" *IEEE Transactions on Vehicular Technology*, 59(6), 3070–3080, 2010.

30 B. Lu, X. Wang, and K. R. Narayanan, "LDPC-Based Space-Time Coded OFDM Systems Over Correlated Fading Channels: Performance Analysis and Receiver Design," *IEEE Transactions on Communications*, 50(1), 74–88, 2002.

31 T. S. Rappaport, *Wireless Communications: Principal and Practice*, 2nd edn, Prentice Hall, 2002.

32 A.F. Molisch, *Wireless Communications*, 2nd edn, John Wiley & Sons, Inc., New Jersey, 2011.

33 A. Goldsmith, *Wireless Communications*, Cambridge University Press, 2007.

34 A. J. Viterbi, *CDMA: Principles of Spread Spectrum Communication*, Addison Wesley, 1995.

35 IEEE 802.16–2009 Standards: Local and Metropolitan Area Networks, Part 16: Air Interface for Broadband Wireless Access Systems, 2009.

36 IEEE 802.16 Broadband Wireless Access Working Group, Channel Models for Fixed Wireless Applications. Available at http://ieee802.org/16, IEEE, New York, 2002.

37 V. Bykovnikov, *The Advantages of SOFDMA for WiMAX*, Intel Corporation, 2005.

38 K. Etemad, H. Yaghoobi, and M. Olfat, Overview of Mobile WiMAX Air Interface in Release 1.0. *WiMAX Technology and Network Evolution*, Wiley-IEEE Press, 2010.

39 M. Maqbool, M. Coupechoux, and P. Godlewski, "Subcarrier Permutation Types in IEEE 802.16e," Telecom ParisTech Research Report, April 12, 2008.

40 J. Lee, X. Wu, and R. Laroia, *OFDMA Mobile Broadband Communications: A Systematic Approach*, Cambridge University Press, 2013.

41 IEEE C802.16d-04_47, "Applying Scalability for the OFDMA PHY Layer," 2004.

42 H. Yaghoobi, "Scalable OFDMA Physical Layer in IEEE 802.16 Wireless MAN," *Intel Technology Journal*, 8(3), 201–212, 2004.

43 P. Bello, "Characterization of Randomly Time-Variant Linear Channels," *IEEE Transactions on Communications*, 11(4), 360–393, 1963.

44 J. S. Sadowsky and V. Kafedziski, "On the Correlation and Scattering Function of WSSUS Channel for Mobile Communications," *IEEE Transactions on Vehicular Technology*, 47(1), 270–282, 1998.

5

The IEEE 802.16 Standards and the WiMAX Technology

The philosophy behind the development of any international technical standard is to establish a platform through which rapid and global deployment of innovative, cost-effective, interoperable, multivendor products is facilitated vis-à-vis an underlying technology or a class of technologies. In the case of IEEE 802.16 standards, there is an additional objective: the promotion of competition between broadband wireless access (BWA) technology for metropolitan area network (MAN) on the one hand, and more traditional wireline MAN networks such as cable networks on the other hand. The term *broadband*, according to ITU terminology, implies data rates greater than 1.5 Mbits/s.

In 1999, the IEEE Standards Board formed the IEEE 802.16 Working Group, tasked with developing and publishing air interface standards for wireless metropolitan area networks. The original IEEE 802.16 standard was developed in 2001 and published in 2002, defining air interface specifications for a fixed BWA alternative to wired metropolitan area access networks, such as fiber optic systems, coaxial cable networks driven by cable modems, and digital subscriber line (DSL). In 2004, several earlier versions of IEEE 802.16 regulations and standards were updated and combined and subsequently published as a full standard under the title of *IEEE 802.16-2004 standards* for fixed broadband wireless access. In December 2005, a new amendment to the standard, referred to as *IEEE 802.16e*, was ratified by the IEEE, which supports mobility. IEEE 802.16e adopts OFDMA for improved multipath performance for NLOS environment. Furthermore, IEEE 802.16e Amendment supports SOFDMA over the physical layer that enables optimum performance in channel bandwidths of integer multiples of 1.25, 1.5, and 1.75 MHz, ranging from 1.25 to 20 MHz. The next version of the standard is *IEEE 802.16-2009*, which was later amended by *IEEE 802.16j-2009*.

AeroMACS: An IEEE 802.16 Standard-Based Technology for the Next Generation of Air Transportation Systems, First Edition. Behnam Kamali.
© 2019 the Institute of Electrical and Electronics Engineers, Inc. Published 2019 by John Wiley & Sons, Inc.

The IEEE 802.16j represents an alteration to the IEEE 802.16e standards in which relay-based multihop network capability is incorporated. This amendment specifies PHY and MAC layer enhancements to IEEE 802.16-2009, enabling the operation of relay stations over licensed bands without requiring any modifications in the SS or the MS systems, and with full backward compatibility with IEEE 802.16-2004 and IEEE 802.16e. The more recent version of the standard is IEEE 802.16-2012, which supports provisions for machine-to-machine communications. IEEE 802.16-2012 is a rollup of many previous 802.16 standards, as such it supersedes IEEE 802.16-2009, IEEE 802.16j, and IEEE 802.16m. More recently, IEEE 802.16-2013, provisioning higher reliability, has been approved and published in March 2013.

Worldwide Interoperability for Microwave Access (WiMAX) is a wireless IEEE 802.16-standard-based broadband access technology for wireless metropolitan area network (WirelessMAN) capable of providing wireless access to entities of various sizes: from a complex of buildings in college campuses to a whole city, and even beyond. WiMAX is often referred to as commercialized IEEE 802.16 WirelessMAN, in wireless communications circles. WiMAX was originally intended as a communication standard for fixed wireless access operating over 11–60 GHz bands. However, since 2005, when mobility was introduced into the technology, it gradually shifted its focus toward competing with wideband wireless cellular networks of 3G and 4G, operating over sub 11 GHz bands. WiMAX is the wireless parent technology of Aeronautical Mobile Airport Surface Communications system (AeroMACS).

In this chapter, a brief overview of IEEE 802.16e and its later version IEEE 802.16-2009 as well as IEEE 802.16j-2009 standards are presented. These are the parts of IEEE 802.16 standards that are of interest to us for the study of AeroMACS networks. Mobile WiMAX technologies, driven from IEEE 802.16e standards, is reviewed with the objective of providing technical background information for AeroMACS networks. WiMAX Forum's recent release on system profiles and certification profiles is also discussed.

5.1 Introduction to the IEEE 802.16 Standards for Wireless MAN Networks

The IEEE 802.16 is a family of standards, defining air interface specifications for broadband wireless metropolitan area networks. The original version of these series of standards was published in October 2002 for fixed broadband access. The AeroMACS technology has been developed based on a later version of the Standard, that is, IEEE 802.16-2009, which was later amended by IEEE 802.16j-2009. Few important remarks regarding these standards are in order at the outset.

- The IEEE 802.16 standards only specify the air interface protocols for the wireless network. In other words, it standardizes protocols related to physical layer (PHY) and media access control (MAC) layer, exclusively. The standards do not cover higher layer protocols and network architecture. However, communication between two entities requires several additional layers of networking; therefore, in order to define an end-to-end communication link, PHY and MAC protocols must be complemented by higher layer protocols, such as IP (Internet Protocols), Ethernet, and so on.
- The IEEE 802.16 standards are based on cellular communications framework. In fact, the IEEE 802.16 network architecture is quite similar to that of cellular telephone networks (see Chapter 2 for extensive description of cellular networks).
- The air interface associated with the IEEE 802.16 standards are designed based on a common MAC protocol but with the flexibility that allows PHY layer protocols to evolve as the standards are adopted for different parts of radio spectrum and according to various associated regulations.
- The IEEE 802.16 MAC layer supports the point-to-multipoint (PMP) and mesh architectures.
- The IEEE 802.16 standards contain a large number of mandatory and optional protocols, to the extent that it is very difficult, if not impossible, to manufacture cost-effective devices and systems that can cover all the protocols.

The IEEE Standard 802.16.1™-2012 entitled *IEEE Standard for WirelessMAN-Advanced Air Interface for Broadband Wireless Access Systems* is a more recent version of the standard that is designated as "IMT[1]-Advanced" by the International Telecommunication Union – Radiocommunication Sector (ITU-R) [1].

The WirelessMAN, when it was published for the first time in 2002, introduced a new paradigm on high data rate wireless access: connecting homes, businesses, government entities, research and educational campuses, and so on, to core communication networks worldwide. Economically speaking, this wireless technology is more cost-effective than the wireline and cable alternatives; for the mere fact that it has the capability to cover a broad geographical area without high expenses associated with

---

1 IMT-Advanced is the International Mobile Telecommunications-Advanced standard requirements issued by the ITU-R in 2008 for 4G mobile phone and Internet access service.

development of cable infrastructure. The main features of IEEE 802.16 standards are briefly discussed in the following paragraphs.

Operating frequencies for IEEE 802.16 have been divided into two bands. This division is commensurate with the wireless channel propagation modes. First is 2 – 11 GHz spectrum for both licensed and license-free bands. This band provides a physical environment where LOS propagation may or may not exist, and significant multipath effect is observed. Thus, over this band it is assumed that signal propagates through the channel by a combination of LOS, NLOS, and near LOS modes. The ability to support near-LOS and NLOS scenarios requires additional PHY functionalities that include advanced power management techniques, interference mitigation, the use of multiple antennas, and so on [1]. It should be noted that although license-free bands (IMS bands) within this range have the same physical characteristics as other parts of this spectra, they suffer from two drawbacks. First, these bands are shared by many users; therefore, there exists an interference and a coexistence issue over license-exempt spectra. Second, regulations governing these bands limit the output power of the transmitters.

The second operating band for IEEE 802.16 standards is located across 10–66 GHz, over which the LOS propagation is deemed a practical necessity.[2] With this assumption, single-carrier (SC) modulation is naturally selected in the physical layer for this band, and for this reason it is designated as "WirelessMAN – SC™" [3]. Our focus is on 2–11 GHz band within which WiMAX and AeroMACS networks operate.

In accordance with the general cellular configuration, each "cell" covered by IEEE 802.16 consists of a single BS and several SSs and MSs, where the BS provides connectivity with the core network. The data rates supported by IEEE 802.16 have been reported over a rather wide range. A reasonable rate is 10 Mb/s;however, figures as high as 100 Mb/s have been cited in Ref. [4]. The theoretical maximum data rate over 10 – 66 MHz is in excess of 120 Mb/s.

The key ingredients of the IEEE 802.16-2009 PHY layer are OFDMA and SOFDMA, discussed in Chapter 4. Multiple access is provided through subchannelization and by assigning one or more subchannels to every user. In CDMA, different qualities of service and data rates can be provided to different users by assigning different number of spreading codes to each user. In OFDMA, same is accomplished by assigning different number of subchannels to the client.

---

2 In the frequency range of 10–66 MHz, the wavelengths are rather short and signal path loss is significant; therefore, a line of sight (LOS) path between the transmit and receive antennae is required. Under these circumstances, multipath effect is negligible [2].

The IEEE 802.16 core MAC common part sublayer (CPS) provides functionalities such as system access, bandwidth allocation, connection establishment, and connection maintenance. QoS is applied to the transmission and scheduling of data over the PHY. The MAC also contains a separate security sublayer providing authentication, secure key exchange, and encryption [1].

5.2 The Evolution and Characterization of IEEE 802.16 Standards

The IEEE-802.16 standard is a large suite of wireless broadband standards composed and authored by the IEEE standard groups. As was mentioned earlier, the IEEE 802.16 standards were originally developed to define specifications for a fixed broadband wireless alternative to wired access networks. The publication of the original version of the standard, IEEE Standard 802.16-2001 in April 2002 [5], defining the air interface specification for WirelessMAN, ushered in a new era of wireless metropolitan area networking. The 2001 standard was initially designed to operate over 10 − 66 GHz band, where extensive spectrum was available at the time. Shortly after that, an amendment to the standard was published, extending the air interface support to the lower frequency band of 2 − 11 GHz for both license and license-exempt bands [6]. Inclusion of this spectrum offered a potential opportunity for many home and small business users to access core networks at a reduced cost, albeit at lower rates [3]. The IEEE Standard 802.16-2001 and the follow-up amendment were the precursor of a complete standard suite for fixed wireless MAN that was published in 2004.

5.2.1 IEEE 802.16-2004 Standard

In 2004, IEEE 802.16-2001 standard and several other previous versions of IEEE 802 standards, regulations, and amendments to standards were updated and combined and published under the title of IEEE 802.16-2004 standards. This is a full WMAN standard for fixed nodes. The original design of PHY layer of IEEE 802.16-2004 was motivated by the need to operate under the NLOS condition for fixed nodes. Three air interface specifications are provided. The first scheme allows for a single carrier modulation, which is often referred to as Wireless MAN-SC. Wireless MAN-OFDM is the second scheme that offers a TDMA 256-subcarrier OFDM. The third is a 2024-subcarrier OFDM scheme in which the multiple access methodology is no longer TDMA but rather OFDMA, that is, multiple access is provided by subchannelization of the OFDM subcarriers and assigning different subchannels to different users [3].

5.2.2 IEEE 802.16e-2005 Standard

In December 2005, a new amendment to the standard, referred to as IEEE 802.16e-2005, was ratified by the IEEE, which supports mobility. IEEE 802.16e adopts OFDMA for improved multipath performance for NLOS environment. In addition, IEEE 802.16e Amendment supports SOFDMA at the PHY layer that enables optimum performance in channel bandwidths of integer multiple of 1.25 MHz, ranging from 1.25 to 20 MHz. Furthermore, this standard incorporates many state-of-the-art communication and radio transmission techniques such as MIMO and SAS, Turbo and low-density parity check (LDPC) coding with iterative decoding, advanced (adaptive) coding and modulation (AMC) schemes to maximize the network throughput, Hybrid ARQ (HARQ) system to provide high data reliability, and on.

5.2.3 IEEE 802.16-2009 Standard

The next version of the standard is IEEE 802.16-2009, which was amended by IEEE 802.16j-2009 (the most recent variation is IEEE 802.16-2012, with an amendment for machine-to-machine communications). Like its predecessor, this standard exclusively specifies MAC and PHY layers protocols. However, this standard combines fixed and mobile PMP BWA systems while providing multiple services. The MAC layer is structured to support a flexible PHY layer, that is, multiple physical layer specifications, each suited to a particular operational environment, are supported. The standard is a revision of IEEEE 802.16-2004 standard and consolidates material from IEEE Std 802.16e-2005, IEEE 802.16-2004/Cor1-2005, IEEE 802.16f^{TM}-2005, and IEEE Std 802.16g^{TM}-2007, along with additional maintenance items and enhancements to the management information base specifications. It should be noted that this revision supersedes[3] and therefore makes IEEE 802.16-2004 standard obsolete [2].

5.2.4 IEEE 802.16j Amendment

The IEEE 802.16j represents an amendment to the IEEE 802.16-2009 standard in which relay-based multihop network capability is incorporated. This amendment specifies PHY and MAC layer enhancements to

---

3 A superseded standard is a standard that has been replaced by a more current and approved standard.

IEEE 802.16-2009, enabling the operation of relay stations (RS) over licensed bands without requiring any modifications in subscriber stations and with full backward compatibility with IEEE 802.16-2004 and IEEE 802.16e. Referring to IEEE 802.16j project approval, the IEEE 802.16's Relay Task Group states the following:

"The approval of this project aims to enable exploitation of such advantages by adding appropriate relay functionality to IEEE Std 802.16 through the proposed amendment. The multihop relay is a promising solution to expand coverage and to enhance throughput and system capacity for IEEE 802.16 systems. It is expected that the complexity of relay stations will be considerably less than the complexity of legacy IEEE 802.16 base stations. The gains in coverage and throughput can be leveraged to reduce total deployment cost for a given system performance requirement and thereby improve the economic viability of IEEE 802.16 systems. Relay functionality enables rapid deployment and reduces the cost of system operation. These advantages will expand the market opportunity for broadband wireless access. This project aims to enable exploitation of such advantages by adding appropriate relay functionality to IEEE Std 802.16 through the proposed amendment" [7].

Thus, in relay-augmented wireless networks, small form-factor relays are used to support the expansion of the base station coverage and/or improve the network capacity, without any change in the radio resource distribution arrangement of the network. Typically, the relay-based configuration is rolled out in the early stages of network development to provide coverage to a larger area at lower cost than that of providing additional base stations [8]. In Chapter 8, IEEE 802.16j-based multihop relay systems and their potential applications in AeroMACS networks are studied in detail.

5.2.5 The Structure of a WirelessMAN Cell

As was mentioned earlier, the IEEE 802.16j amendment to IEEE 802.16-2009 introduces the multihop relay as an additional network entity that enables the extension of radio outreach and may be used to enhance system capacity in an access network. In our description of AeroMACS networks in the upcoming AeroMACS chapters, a general assumption is that the IEEE 802.16j-based relay stations are an integral part of these networks. With that in mind, an IEEE-802.16-2009-based BWA typical cell includes a single BS, and a number of SSs, MSs, and RSs. The base station provides connection between the SS/MS/RS on the one hand, and

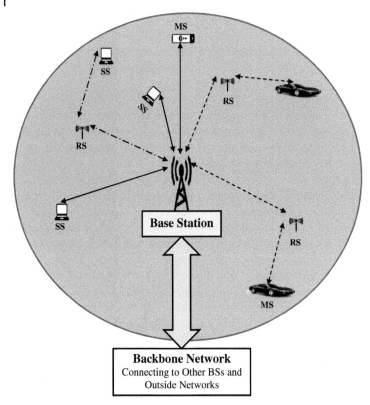

Figure 5.1 An IEEE 802.16-2009 BWA system fortified with relay stations.

other BSs and a core network that links to the outside networks such as the Internet and PSTN, on the other hand. A general structure[4] of this network is shown in Figure 5.1.

As Figure 5.1 illustrates a typical "cell" of a WirelessMAN (IEEE 802.16-2009 amended by IEEE 802.16j) network consists of a BS, a number stationary stations (SS), mobile stations (MS), and RS. The RSs, by and large, are assumed to be stationary, although mobile RSs are also incorporated in the IEEE 802.16j amendment. The BS communicates with SSs and MSs directly or through a one or more intermediary relay stations. It should be pointed out that it is possible that the direct channel (DL; BS to MS/SS) may be established through one or more RSs; that is, multihop connection, while the reverse channel (UL, MS/SS to BS) could

---

4 The phrase "network structure," or "network composition," is used here to avoid the use of the term "architecture," as architecture in network jargons is usually related to protocols and suite of protocols.

be a direct link, that is, a single hop link. Multihop relays are placed in the cell for a number of different reasons, the most important of which is the extension of radio outreach into areas that are severely shadowed by natural or man-made obstacles such as tall buildings and hilltops.

In Chapter 8 potential applications and usage scenarios of multihop relays are given an extensive coverage. In the context of AeroMACS networks, stationary stations include airport terminals, security nodes, airline offices, airport authorities' offices, and of course the control tower, and so on. Examples of mobile stations on an airport surface are aircrafts, fuel and catering trucks, airport surface buses, security vehicles, and so on.

The WirelessMAN network operates on the basis of PMP architecture. The BS uses omnidirectional or wide-beam antenna, and if the cell is sectorized, the BS uses multiple identical antennae or a single antenna that is capable of handling multiple independent sectors. The BS is connected to the backbone network, which is the communication mechanism that connects the BS to other BSs in the greater WirelessMAN network, as well as to outside networks such as the Internet and PSTN. This connection might be provided by fixed wireless links or wireline cable connections. All other links in the networks are wireless.

5.2.6 Protocol Reference Model (PRM) for the IEEE 802.16-2009 Standard

The IEEE 802.16-2009 reference model is a logical representation of the network architecture and protocol structure proposed by the standard. Since the standard defines the air interface specifications, only two protocol layers are included in the Protocol Reference Model, namely, the PHY and MAC layers. Figure 5.2 illustrates the PRM for this standard. As Figure 5.2 shows, the MAC layer consists of three sublayers:

1) *The Service-Specific Convergence Sublayer* (*CS*): The main task of this sublayer is to take network layer (most probably IP) data and transform or translate that into a format called MAC service data units (SDUs) that can be accepted and understood by the MAC CPS. The port of data entry into CS is called CS service access point (SAP). The MAC CPS data reception port is called MAC SAP. Classifying external network SDUs and associating them to the proper MAC service flow identifier (SFID) and connection identifier (CID) is also incorporated within CS.

2) *CPS*: This is the core of MAC protocols that performs key functionalities of system access, connection establishment and maintenance, bandwidth management and allocation, mobility support, modulation and coding combination selection, automatic repeat request system

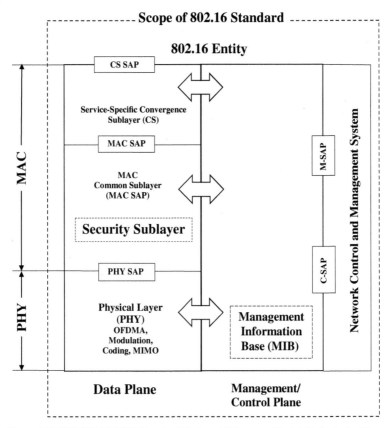

Figure 5.2 IEEE 802.16-2009 Standard Protocol Reference Model. (Adopted with permission from Ref. [2].)

support, call admission control, and provision of multicast and broadcast services. Quality of service is also applied by MAC CPS to transmission and scheduling of data over the PHY layer.

3) *Security Sublayer*: MAC provides a separate security sublayer for authentication, including device authentication and user authentication, data encryption (integrity and confidentiality), and secure key exchange [2].

In this paragraph and the next, a brief description of DL and UL communications for IEEE 802.16 (amended by IEEE 802.16j) is provided. The DL of WirelessMAN operates based on PMP transmission format, where the BS transmits signal to SS/MS/RS, as shown Figure 5.1. The IEEE 802.16 operates with a BS as the central node with a sectorized antenna, capable of simultaneously operating multiple independent sectors. Within a given sector and a frequency channel, all SS/MS/RS receive the same signal, in other words, this is a broadcast transmission. Since the BS is the only node that transmits over DL, it does not have to coordinate transmission with any other station in the cell, except for the division of the time slots in a possible overall TDD scheme, where time slots are divided into DL and UL slots. The SS/MS/RSs check the CIDs in the received data, packaged into the form of "Protocol Data Units, PDUs," and retain only those PDUs that are addressed to them. This is the way the BS communicates with each individual user [2]. In addition to each individually addressed messages, broadcast and multicast messages may also be transmitted by the BS, wherein the BS communicates to all active receivers or to just a selected number of users, respectively.

On the UL, SS/MS/RSs share the link on a demand basis. Users may demand various services, and depending upon class service utilized, the SS/MS/RS may be issued permission to continue transmitting signal. It is also possible that the BS grants the permission for continual transmission to the SS/MS/RS, upon the receipt of a request from them. Generally speaking, within each sector, SS/MS/RSs adhere to a transmission protocol that controls the contention between users and tailors the service for each user for delay and bandwidth requirements of the user's application [2].

Multiple specifications are defined in the PHY layer, each of which is appropriate for a particular frequency band and for a particular application. Information, control data, and statistics are exchanged between PHY layer and MAC CPS through PHY SAP. The details of PHY layer protocols of IEEE 802.16 will be discussed in the WiMAX sections of this chapter, and in AeroMACS chapters.

IEEE 802.16 devices (BS, SS, MS, and RS) may be part of a greater network; therefore, they would require to be interfaced with entities for management and control. The block designated as "Network Control and Management System (NCMS)," placed in the standard protocol reference model of Figure 5.2, contains these entities. The NCMS enables PHY and MAC layers, specified in IEEE 802.16 standard, to be independent from network architecture, transport network, and the protocols used at the backend. NCMS exists at both terminals: On the BS side it is designated as NCMS(BS), and on the user's side, terms NCMS(SS), NCMS(MS), and NCMS(RS) are used for SS, MS, and RS, respectively. All necessary inter-BS coordination is handled by NCMS(BS). There are two points of interface, as shown in Figure 5.2, designated as C-SAP (Control-Service Access Point) and M-SAP (Management-Service Access Point), whose purpose is to unveil control plane and management plane to higher layer protocols [2].

5.3 WiMAX: an IEEE 802.16-Based Technology

The IEEE 802.16 is a large collection of standards; therefore, for any particular driven technology, the scope of the standard needs to be reduced to smaller sets of design choices. A case in point is the WiMAX network, which is an IEEE 802.16 standard-based broadband cellular wireless solution. Consequently, WiMAX uses a subset of the IEEE 802.16 standard's mandatory and optional specifications consisting of selected PHY and MAC layer protocols. WiMAX is the first cellular system that applies OFDMA access technology. This is in contrast with multiple access technologies such as FDMA, TDMA, and CDMA that are dominant in 1G, 2G, 2.5G, and 3G cellular networks. WiMAX enables low-cost mobile access to the Internet and provides integrated wireless fixed and mobile services using single air interface and network architecture.

5.3.1 Basic Features of WiMAX Systems

The WiMAX network architecture provides for IP compatibility and an open environment for innovative applications and services with high QoS.[5] WiMAX technology, although initially perceived as "last mile" connector to the Internet, supports a variety of wireless access in regard to mobility and the location of the subscriber station and the mobile station – ranging from fixed to fully mobile access. The requirements of different types of access are met in two versions of WiMAX that are based on IEEE 802.16-2004 and IEEE 802.16e amendment to the standard. The following list outlines various forms of access supported by WiMAX technology.

1) *Fixed Access*: Entails the wireless access on the part of a single, outdoor and/or indoor, stationary station.
2) *Nomadic Access*: Supports indoor wireless access from multiple stationary locations for subscriber stations.
3) *Portable Access*: Supports low-speed multiple access for subscriber stations with hard handover provision.
4) *Simple Mobility*: Supports low-speed, up to 60 km/h, mobile access with hard handover.
5) *Full Mobility*: Supports high-speed, up to 120 km/h, mobile access with soft handover [9].

---

5 High QoS here does not refer to superior performance, which is normally associated with high data rate, low BER, low latency, and so on. In packet switched networks, high QoS is related to the network ability to prioritization and resource reservation for various services.

Originally, the fixed and nomadic access were featured by WiMAX based on IEEE 802.16 2004, and portability and mobility management are optimized by mobile WiMAX, which was based on IEEE 802.16e. Both of these features are now consolidated in IEEE 802.16-2009 standard.

A key PHY layer feature of WiMAX networks, inherited from IEEE 802.16 standards, is their access technology OFDMA, employed for both DL and UL. OFDMA enhances performance against frequency selective fading effect and enables bandwidth scalability. Hence, WiMAX essentially applies "Scalable OFDMA (SOFDMA)," which allows the standard to expand over several bandwidths and several spectral ranges, as such WiMAX channel bandwidth may be scaled to up to 20 MHz. Thus, *scalability* is an important character of mobile WiMAX.

WiMAX was originally designed to support technologies that make *triple-play* service offerings possible. As a result, it is possible for WiMAX technologies not only to support high-speed broadband Internet access but also IP voice and IP video services (VoIP and IPTV). This enables WiMAX to become a replacement for DSL, Cable TV, and telephony services simultaneously, that is, WiMAX was designed to become what used to be called the "information superhighway" in the 1990s.

With regard to duplexing formats, WiMAX predominantly supports TDD; however, FDD and simplex FDD are also included in WiMAX protocols. An important aspect of TDD architecture lies in the fact that TDD supports the exchange of asymmetric traffic.

AMC is another key feature of PHY layer of WiMAX networks. WiMAX supports a variety of modulation and coding schemes and selects a proper combination thereof, referred to as *burst profile*. These burst profiles are selected in an adaptive fashion, depending upon channel conditions, and for the purpose of network throughput optimization.

Two levels of error control are provided in WiMAX standards. Primarily, WiMAX standards define an adaptive modulation coding scheme at the physical layer. Second, a multilayer (link layer and transport layer) ARQ and HARQ error controls are included in the WiMAX protocols.

One of the fundamental promises of the IEEE 802.16 MAC architecture is quality of service (QoS). The WiMAX solution is also friendly to IP-based end-to-end network architecture. The IEEE 802.16e/mobile WiMAX MAC layer is so structured that enables an overall end-to-end IP-based QoS [10].

WiMAX PHY layer supports the implementation of advanced antenna techniques. As such beamforming, spatial multiplexing, MIMO with space-time coding, and so on are available to enhance the overall spectral efficiency and system capacity without requiring additional bandwidth.

WiMAX offers several first class *security* features. It provides extensible authentication protocol (EAP) for authentication, AES (Advanced

Encryption Standard) that is based on CCM (Counter with Cipher-block chaining Message authentication code) authenticated encryption, as well as CMAC (Cipher-based Message Authentication Code) and HMAC (Hash Message Authentication Code) schemes for message protection. Regarding user's credentials confirmation several methods are supported that includes SIM (Subscriber Identify Module) and USIM (Universal SIM) [10].

For *mobility* management, optimized handover schemes are supported by Mobile WiMAX. Latency levels less than 50 ms ensure real-time applications such as VoIP perform without service degradation [10].

OFDMA offers a form of *frequency diversity* for WiMAX systems. A promising technique that produces another form of diversity, called *spatial diversity*, is produced for WiMAX and AeroMACS networks by the application of the multiple antennae at either the transmitting side or the receiving side or at both ends of the communication link. Accordingly, each antenna pair creates an independent channel in space and therefore several uncorrelated copies of the same signal is delivered to the receiver. This technique is collectively called multiple input multiple output (MIMO). MIMO may be implemented in WiMAX and AeroMACS systems without requiring additional band-width. MIMO is a relatively new innovation, yet, it plays a central role in delivering high-speed and reliable wireless broadband services over an extended coverage area. MIMO is gradually becoming an integral part of modern wireless networks and in fact is considered to be one of the key technology components of the emerging 5G cellular networks.

WiMAX multiple access format, OFDMA, groups thousands of orthogonal subcarriers to form user's subchannels. When neighboring WiMAX cells use the same subchannels for signal transmission, *co-channel (or co-subchannel) interference* occurs. There are several methods available to mitigate this form of interference. WiMAX applies a widely accepted method known as fractional frequency reuse (FFR). In this technique frequency reuse factor (FRF), or equivalently cluster size of the cellular layout, is not constant but rather adaptive. In the simplest form of fractional frequency reuse format, the cells are divided into "inner" and "outer" regions. Inner regions are closer to the base stations and less susceptible to co-channel or co-subchannel interference. For these regions, FRF is one, in other words, all subchannels are accessible to all users in this region. However, for the outer cell regions, the subchannels are divided in the same way as they are in traditional cellular networks, that is, disjoint sets of subchannels are allocated to adjacent cell edges. This is a static form of adaptive FRF. Many new algorithms for dynamic adaptive FRF has been devised recently [11].

In order to form any driven technology from the IEEE 802.16 standard, the scope of the standard must be reduced, in large part to ensure interoperability. The selection of a subset of IEEE 802.16 standards for creation of WiMAX, or any other IEEE 802.16 compliant technology for that matter, is performed by composing *profiles*. For WiMAX, the *WiMAX Forum* is charged with the task of defining two types of profiles: *system profiles* and *certification profiles*. A system profile is a subset of mandatory and optional PHY and MAC layer protocols selected from IEEE 802.16 standards, which suits some application services. A certification profile is established when the operating frequency, channel bandwidth, FFT size, and duplexing technique are defined for a system profile. Each system profile may instigate several certification profiles, as different parts of the globe may have to employ different bands of frequencies for the same application. In other words, systems and devices that comply with the same system profile are not necessarily interoperable, as they have been designed to operate over different spectral ranges. This requires that multiple certification profiles be created following the definition of a single system profile. For instance, the first WiMAX system profile, developed by the WiMAX Forum, was based on IEEE 802.16-2004 for stationary nodes. Subsequent to this system profile establishment, at least five certification profiles were developed.

The need for system and certification profiling stems from the practical requirement that WiMAX devices that are produced by different vendors should be in compliance with the base technology, and the systems that conform to a given system profile and a particular certification profile simultaneously are interoperable. In order for a WiMAX product to obtain certification, two tests need to be passed successfully.

1) *Compliance Test* ensures that the product conforms to the specifications defined by the system profile.

2) *Interoperability Test* confirms that the product is interoperable with other products that are compliant with the same system and certification profiles. This is a more stringent test that might take longer time to complete, as the vendors may have to make changes in their design of the product.

It is noteworthy to mention that the same procedure is essentially applied for the certification of AeroMACS products.

The Mobile WiMAX protocol reference model is pretty much the same as that of IEEE 802.16-2009 standard, as illustrated in Figure 5.2.

Since WiMAX is indeed a full-fledged cellular standard, it is in direct competition with the existing cellular networks such as CDMA 2000 and 3GPP (Third-Generation Partnership Project) LTE networks. Although

WiMAX is still considered an "unproven" technology, several clear advantages of this technology should be pointed out. Primarily, the PHY layer of WiMAX network, consisting of OFDM, OFDMA, and MIMO, is more suitable for transmission of high-speed signal. Second, WiMAX is an IP-based network, so there is no need for development and deployment of the costly backbone network. Moreover, WiMAX standard is much simpler than competing technologies. In the next section, a number of these features are described in some details.

5.3.2 WiMAX Physical Layer Characterization

The WiMAX network is set up for a fixed MAC layer that supports flexible PHY layers, that is, a number of completely different PHY layers are available that can be exploited to match the available bandwidth as well as local regulations. Our interest is in the AeroMACS network, which is powered by mobile WiMAX standard using OFDMA over 2–11 GHz band. Mobile WiMAX is the first standard-based broadband wireless-access solution that allows the convergence of mobile and fixed broadband networks through a common wide-area radio-access technology and flexible network architecture. It has been predicted that with the availability of sufficient bandwidth, and with appropriate integration of MIMO technology, WiMAX solution will be capable of supporting data rates in excess of 1 Gb/s over the air, for a wide range of high-quality IP-based applications and services [12].

WiMAX Forum has published several WiMAX system profiles designated by "Release" numbers. These system profiles define selected design options and interoperability specifications that are adopted from the parent standard. The original WiMAX system profile, "Release 1" published in 2009, is, by and large, based on IEEE 802.16e-2005, although some enhancements are also accepted from IEEE 802.16-2009 [13]. Subsequently, Release 1.5 was published, which is conformant to IEEE 802.16-2009 standards, and is composed of a common part, a TDD part, and a FDD part [14–16]. More recently, in 2012, 2013, and 2014, Release 2.0, Release 2.1, and Release 2.2, have been issued by WiMAX Forum, which are based on IEEE 802.16m standard [17–19]. PHY and MAC layers specifications in each of these system profiles are the modified and evolved version of the previous ones. Specifically, PHY layer parameters such as bandwidth, supported DL and UP data rates, MIMO configurations, and so on are different in these Releases. In particular, Release 2 versions are of interest in AeroMACS networks, as they support mobility and the type of high vehicular speeds that are encountered on the airport surface. In what follows, the main PHY layer features of mobile WiMAX networks are presented.

5.3.2.1 OFDMA and SOFDMA for WiMAX

As we have mentioned previously, OFDM is a multiplexing scheme with a number of advantages to be exploited in WiMAX and AeroMACS networks. Chief among these advantages are relative immunity against frequency selective distortion when broadband signals are transmitted through multipath channels and computational efficiency due to applicability of IFFT and FFT algorithms for implementation of some important portions of the OFDM system. Furthermore, as the delay spread exceeds the value for which the system is designed, the system adaptively changes the modulation scheme to a more conservative one and reduces the channel code rate, in order to cope with the channel condition. In other words, AMC scheme is a natural match for OFDM. Robustness against deep fades, which ordinarily create burst errors, is another advantage of OFDM. Subchannelization randomizes the effect of fading over various subcarriers that combined with interleaving and channel coding can remove, or at least mitigate, the effect of deep fades. When narrowband interference afflicts the signal, it would affect only a small number of subcarriers; in this manner, the OFDM provides protection against this type of interference. Alongside with all these advantages, OFDM suffers from a few shortcomings, chief among them is high PAPR (peak-to-average power ratio) leading to power inefficiencies and susceptibility to phase noise and frequency dispersion [20].

Scalability is a key feature of mobile WiMAX. Accordingly, as the available bandwidth increases, the FFT size is scaled from 128 to 2048 such that the subcarrier spacing always remains fixed. A constant tone spacing keeps many OFDM parameters fixed. These parameters include OFDM symbol duration, OFDM frame time duration, cyclic prefix, and so on. Consequently, no readjustment of bandwidth-related quantities is necessary. The cyclic prefix selected for WiMAX in Release 1 and Release 1.5. is 1/8. In Releases 2.0, 2.1, and 2.2, the CP could assume values of 1/4, 1/8, and 1/16.

WiMAX standards are developed to operate over 2–11 GHz and 10–66 GHz, otherwise no frequency range of operation is specified. A *band class* is a particular spectrum that is supported by the WiMAX BS. The band class may consist of several subbands. In Mobile WiMAX, a band class is characterized by its width, associated WiMAX channel bandwidth, the applied duplexing technique, and the list of center frequencies.

Oversampling is a common practice in WiMAX networks through which a multitude of benefits are realized. Primary reason for oversampling is to allow adjacent secondary channels to overlap spectrally without interference. Oversampling also reduces PAPR and assists in achieving the spectral emission masks. The sampling frequency f_s;

Table 5.1 WiMAX PHY layer parameters.

BW	FFT size	Sampling Frequency	Tone spacing	Useful symbol time	OFDMA symbol time	CP	No. of data subs. DL PUSC	No. of pilot subs. DL PUSC	No. of active subs.
1.25 MHz	128	1.4 MHz	10.94 kHz	91.43 µs	102.94 µs	1/8	72	12	84
5 MHz	512	5.6 MHz	10.94 kHz	91.43 µs	102.94 µs	1/8	360	60	420
10 MHz	1024	11.2 MHz	10.94 kHz	91.43 µs	102.94 µs	1/8	720	120	840
20 MHz	2048	22.2 MHz	10.94 kHz	91.43 µs	102.94 µs	1/8	1440	240	1680

subcarrier (tone) spacing f_{ts}, oversampling factor ζ, and WiMAX channel bandwidth BW are related in the manner shown by Equations 5.1–5.3.

$$\zeta \triangleq \frac{f_s}{BW} \tag{5.1}$$

$$f_s = \left\lfloor \zeta \frac{BW}{8000} \right\rfloor 8000 \tag{5.2}$$

$$f_{ts} = \left\lfloor \zeta \frac{BW}{8000} \right\rfloor \times \frac{8000}{\|FFT\|} \tag{5.3}$$

where $\lfloor x \rfloor$ is the floor function of x (the greatest integer less than or equal to x) and $\|FFT\|$ is FFT size, that is, the number of FFT points. As Equation 5.2 indicates, the sampling frequency comes out to be a multiple of 8000. For bandwidths that are multiples of 1.25 MHz, the oversampling ratio is selected as $28/25 = 1.12$. With these values, the tone spacing is calculated to be 10.94 kHz. This tone spacing strikes a balance between two conflicting requirements regarding delay spread and Doppler spread for operating over combined fixed and mobile environments. Table 5.1 summarizes the PHY layer parameters of Mobile WiMAX networks for various bandwidths. We note that bandwidth 2.5 MHz corresponding to FFT size 256 is not included, as this mode is dedicated to fixed WiMAX OFDM networks.

5.3.2.2 Comparison of Duplexing Technologies: TDD versus FDD

WiMAX predominantly supports time division duplexing, however, frequency division duplexing is also supported in the PHY layer.[6] Within

---

6 Mobile WiMAX Release 1.0 is based on TDD exclusively; however, the subsequent Releases, that is, Releases 1.5, 1.6, 2.0, 2.1, and 2.2, support both TDD and FDD.

FDD, simplex (half duplex) mode is also defined for supporting less complex links in which one radio channel is time-shared between the transmitter and the receiver. A brief comparison between the two major duplexing formats, vis-à-vis their suitability in WiMAX networks, follows.

A major advantage of time division duplexing technology lies in the fact that TDD enables adjustment of the downlink–uplink ratio and therefore can efficiently support asymmetric DL/UL traffic. This is in contrast with FDD in which DL and UL channels always have fixed and generally equal bandwidths. The important point to note here is that unlike voice services that require symmetric use of UL and DL, data and video services, by and large, need to utilize the system capacities in an asymmetric fashion; therefore, TDD is more appropriate for these applications, which form the bulk of WiMAX load, particularly as it pertains to the Internet access.

TDD "time shares" a single channel for downlink and uplink, while FDD requires a pair of radio channels to accommodate the links. This implies that, in general, TDD renders a higher level of network spectral efficiency. And since TDD possesses channel reciprocity, it can naturally support link adaptation for applications such as MIMO and other closed-loop advanced antenna technologies. Moreover, transceiver designs for TDD implementations are less complex and therefore less expensive than those of FDD [10].

Unlike FDD schemes, in order to limit the effect of interference, TDD requires a network level synchronization. This is one of the down sides of using TDD; nevertheless, TDD is the preferred duplexing mode in WiMAX and AeroMACS networks for the reasons listed earlier.

5.3.2.3 Subchannelization for Mobile WiMAX

As was discussed in Chapter 4, subchannelization is the clustering of a number of subcarriers and designating them as subchannels. The subchannels are then allocated for data transmission and reception. Subchannelization, therefore, is a mechanism through which OFDM multiple access system, OFDMA, is created. The exact number of subcarrier and how they are selected to form subchannels is defined in what is known as *subcarrier permutation*. As chapter 4 explains, there are two general classes of subcarrier permutation methods: *adjacent subcarrier permutation* and *distributed subcarrier permutation*. The selection of a subcarrier permutation mode is based on optimization of system performance, given the particular usage mode, channel model, specific multiple antenna configuration applied, and so on. In distributed subcarrier permutation, the subcarriers selected for the subchannel, perhaps in quasi-random fashion, spread across the channel frequency spectrum. This permutation provides a form of frequency diversity that is particularly advantageous in mobile applications. In adjacent subcarrier permutation, selected subcarriers for the subchannel are spectrally contiguous. This permutation mode is more suitable for adaptive modulation and coding and facilitates channel estimation and instantaneous feedback from the transmitter to the receiver more readily. Consequently, this mode is better suited for fixed, portable, nomadic, and low-speed mobile cases.

The distributed subcarrier permutation is further classified into partial usage subcarrier (PUSC) and full usage subcarrier (FUSC). These permutations are different in the number of pilot subcarriers per subchannel or the number of symbols per slot. A *slot* is the smallest time–frequency subchannelization resource unit that can be allocated by a WiMAX system to a link. The exact definition of a slot depends upon the subchannelization (subcarrier permutation) and whether it is used for DL or UL. Each slot spans over time and frequency domains and consists of one subchannel crossed over one, two, or three OFDM symbols, in accordance with the selected subchannelization scheme [20]. Alternative methods of distributed permutations include optional PUSC (OPUSC), optional FUSC (OFUSC), tile usage subchannel 1 (TUSC1), and tile usage subchannel 2 (TUSC2). On the adjacent permutation side, the only mode is adaptive modulation coding (AMC). Some of these permutations are used for DL exclusively, whereas some are used for both DL and UL. PUSC is the only mandatory subcarrier permutation, all others are optional. For more detailed information on subcarrier permutation and subchannelization, the reader is referred to Refs [2] and [21]. The only standard mandatory permutations, PUSC and AMC, are briefly discussed in the following paragraphs.

For WiMAX networks in which reuse factor is one, all subchannels are used in all sectors; that can lead to intersector interference in both DL and UP. However, with reuse factor of three and three sectors in the cell, intersector interference can be mitigated by utilizing PUSC. In PUSC, the basic idea is to divide all subcarriers into several groups and allocate each group to a sector. Hence, only a fraction of subchannels are assigned to a sector of the WiMAX cell, therefore the name "partial usage subchannels." The remaining subchannels are assigned to the other sectors. One slot of PUSC DL consists of two OFDM symbols by one subchannel, where each subchannel comprises 24 data subcarriers (for FFT 2048, BW = 20 MHz). The 2048 subcarriers are distributed as follows: 1440 data subcarriers, 367 guard subcarriers, 240 pilot subcarriers, and one DC subcarrier. Similar subcarrier distribution figures are specified in the standard for other bandwidths.

With DL PUSC, the active subcarriers (data and pilot) available in 1 slot are grouped into clusters containing 14 contiguous subcarriers per symbol period. The data and pilot subcarriers for even and odd symbols are, then, arranged as shown in Figure 5.3 [10].

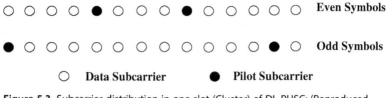

Figure 5.3 Subcarrier distribution in one slot (Cluster) of DL PUSC; (Reproduced from Ref. [2].)

The formation of subchannels for DL PUSC is outlined below.

1) Using a quasi-random numbering scheme (defined by the IEEE 802.16e standard), the clusters formed as shown in Figure 5.3 are renumbered. Clusters prior to renumbering are contiguous in the frequency domain and are called *physical clusters*, clusters after renumbering are referred to as *logical clusters*.
2) Logical clusters are then divided into 6 groups for FFT sizes of 1024 and 2048 and into 3 groups for FFT sizes of 128 and 512.
3) The first one-sixth (or one-third) of the logical clusters are placed in group1, the next one-sixth (or one-third) in group 2, and so on. Step 1 essentially ensures that the clusters included in each group are selected pseudo-randomly from across the entire subcarrier space.
4) A subchannel is formed by selecting two clusters from the same group.

Therefore, a subchannel consists of 48 data subcarriers and 8 pilot subcarriers. The number of clusters, and thus the number of subchannels, varies with channel bandwidth and FFT size. For instance, with FFT size 512 (BW 5 MHz), as Table 5.1 shows, there are 420 active subcarriers; hence, 30 clusters can be formed. There are three groups in this case. PUSC may allocate various groups or a subset of them to the transmitters of different sectors [22].

In UL PUSC, the time–frequency plane is divided into 3×4 rectangular sections called *tiles*. A tile is composed of 4 contiguous subcarriers from 3 OFDM symbols, consequently a tile is composed of 12 subcarriers, 4 of which are pilot subcarriers and the rest are data subcarriers. Tiles for UL PUSC are analogous to clusters in DL PUSK. In order to assemble UL PUSC slot, the active subcarriers are first divided into tiles. The tiles formed originally take up subcarriers in ascending frequency order. These tiles are called *physical tiles*. Physical tiles are converted to logical tiles using a pseudo-random renumbering algorithm, in more or less the same fashion that physical clusters are transformed to logical clusters. Then six logical tiles are selected (effectively from across the entire spectrum) and bunched together. Thus, a UL PUSC slot contains 72 subcarriers, 48 of which are data subcarriers. Figure 5.4 depicts the structure of a UL PUSC slot with its constituent logical tiles.

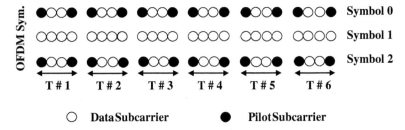

Figure 5.4 The structure of an UL PUSC slot consisting of six basic tiles – a subchannel.

A UL PUSC subchannel consists of six logical tiles. In fact, Figure 5.5 represents a UL PUSC subchannel.

PUSC is a distributed subcarrier permutation, distinguished by its data and pilot subcarriers distribution methods, with interesting features when

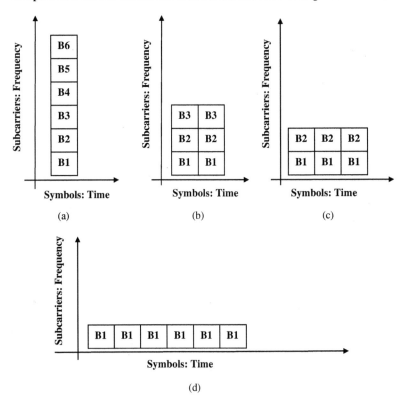

Figure 5.5 Possible modes of subchannelization for AMC: (a) six bins over one OFDM symbol (default configuration), (b) three bins spread over two OFDM symbols, (c) two bins over three OFDM symbols, and (d) one bin repeated over six OFDM symbols.

applied to WiMAX or any other cellular network. PUSC may be used in a *segmented*[7] *mode*, the so-called "PUSK 1/3," in which the reuse factor is 3 and groups are divided between sectors. PUSC may also be used in no-segmentation mode in which all groups may be allocated to a sector of WiMAX BS or all groups are assigned to all sectors, the so-called "PUSC with all subchannels".

The only adjacent subcarrier permutation technique applied in mobile WiMAX is AMC, which is particularly suitable when advanced antenna systems are employed. In AMC, a *bin* is constructed by bunching nine contiguous physical subcarriers together, where the middle subcarrier is a pilot tone and the rest are data tones. In this permutation, DL AMC and UL AMC bins have the same structure: a group of four consecutive rows of bins from what is known as a *physical band* or *AMC band*. An AMC slot, or subchannel, is defined as a collection of six bins and is expressed by a time–frequency matrix of the form $N_{bin} \times N_{sym} = 6$, where N_{bin} is the number of contiguous bins and N_{sym} is the number of contiguous OFDM symbols. In other words, an AMC subchannel is formed by six contiguous bins over one OFDM symbol (the default type), or two sets of three contiguous bins spread over two OFDM symbols, or three sets of two contiguous bins spread over three OFDM symbols, or one bin across six OFDM symbols, as shown in Figure 5.5 [21].

Figure 5.5 demonstrates that in order to form an AMC subchannel, six bins are required, but flexibility is provided with four possible configurations that are available to match various circumstances. Thus, the AMC subchannels are composed of bins that are in the same "band", either in frequency or time.

The AMC permutation has superior channel estimation capabilities, and because AMC subchannels are composed of contiguous subcarriers, it is well suited for adaptive modulation coding. The SS/MS may switch from PUSC (or other distributed permutations) to AMC when the mode changes from regular antenna system to adaptive antenna system or to MIMO [21].

5.3.2.4 WiMAX TDD Frame Structure

The frame composition defined in the original IEEE 802.16-2009 with time division duplexing implementation consists of downlink and uplink subframes [24]. Figure 5.6 illustrates the frame structure of IEEE 802.16e TDD WiMAX networks, which is based on Release 1.0 of WiMAX Forum system profile.

---

7 In the context of WiMAX systems, a segment is a subdivision in the set of available subchannels used for deployment of one MAC instance [23].

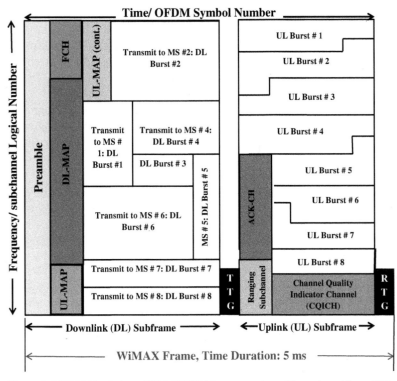

Figure 5.6 IEEE 802.16e-based TDD WiMAX frame structure. (Reproduced from IEEE 802.16-2009 Standard [2].)

Mobile WiMAX profile Release 1.0 is based on TDD; however, later Releases support both TDD and FDD. The IEEE 802.16-2009, as well as earlier mobile versions of the Standard, allow for different values of frame time duration. The selection of the frame time duration involves a trade-off between network latency performance and overhead efficiency. In other words, the shorter frame time duration results in a better latency performance; however, it would demand that the time and frequency-consuming overhead parts of the frame be required more often. As illustrated in Figure 5.6, the frame time duration for TDD mobile WiMAX is selected to be 5 ms.

The vertical axis in this frame structure is frequency (subcarriers/subchannels logical number [2]) and the horizontal axis is time (OFDM symbol number). The network time–frequency resource is divided into frames. Each frame consists of downlink and uplink subframes. The preamble is mainly used for time synchronization.

The downlink MAP (DL-MAP) and uplink MAP (UL-MAP) define the burst[8] start time and burst end time, modulation type, and error control for each MS/SS. The MAP's lengths and usable subcarriers are defined by the Frame Control Header (FCH). Since, due to the nature of wireless media, the channel state condition changes over time, AMC is employed by WiMAX to facilitate the adaptive selection of modulation-coding combination that matches the channel conditions. As we have mentioned earlier, the AMC subcarrier permutation is most suitable for implementation of adaptive modulation coding. Both the MS and BS can conduct channel estimation based on what the BS decides for selection of the most efficient modulation and coding scheme. A Channel Quality Indicator (CQI) is used to convey the channel state condition information. In the downlink direction, the IEEE 802.16e standard requires that all DL data bursts be rectangular.

In order to avoid interference between downlink and uplink subframes of the same WiMAX frame, a small guard time known as Transmit Time Gap (TTG) is placed between the subframes. Similarly, to allow transition from the uplink subframe of a WiMAX frame to the downlink subframe of the next frame, a guard time called Receive Time Gap (RTG) is left between them, as shown in Figure 5.6.

The frame duration of 5 ms allows for 48 OFDM symbols (each has time duration of around 100 μs). In what follows, detailed information regarding the overhead and control parts at the beginning of the WiMAX TDD frame (used to ensure optimum network operation) are provided.

- *Preamble:* This is the first OFDM-DL symbol of the frame and is used by the MS for BS identification, time synchronizing between the MS and the BS, and channel estimation. It should be noted that the uplink subframe, generally, does not contain preamble.
- *Frame Control Head (FCH):* This is a special management burst that follows the preamble. It provides the frame configuration information such as MAP (Media Access Protocol) message length and coding scheme, as well as usable (active) subchannels.
- *DL-MAP (Down Link-Media Access Protocol):* Defines for each MS the location of their corresponding burst, that is, their OFDM subchannel allocation as well as other control information in the DL subframe, addressed to each MS.

---

8 The term *burst* refers to a contiguous portion of the data stream that uses the same modulation and coding schemes, as well as the same preamble. These physical characteristics are called *burst profile*. Each burst profile is identified by a code: either a DL Interval Usage Code (DIUC) or an UL Interval Usage Code (UISC) [23].

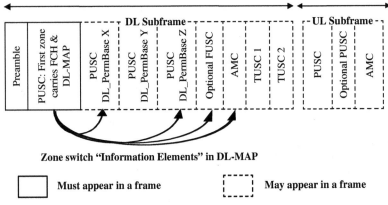

Zone switch "Information Elements" in DL-MAP

☐ Must appear in a frame   ┌┄┄┐
                                 ┆   ┆ May appear in a frame
                                 └┄┄┘

Figure 5.7 OFDMA frame structure with multiple zones available in WiMAX. (Reproduced from IEEE 802.16 standard [2].)

- *UL-MAP:* This is a burst that defines for each MS the location, that is, subchannel allocation, in the upcoming UL subframe, as well as other control information.
- *UL Ranging[9]:* The UL ranging subchannel is allocated for MSs to perform closed-loop time, frequency, and power adjustment as well as bandwidth requests.
- *UL CQICH (Channel Quality Indicator Channel):* The UL CQICH subchannel is allocated for the MS to feedback channel state information.
- *UL ACK:* The UL ACK is allocated for the MS to feedback DL HARQ acknowledgement.

The WiMAX frame includes multiple zones for various subcarrier permutations such as PUSC, FUSC, AMC, and the rest. A permutation zone is a set of OFDMA symbols that use the same permutation type. The transition between zones is shown in DL-MAP. Figure 5.7 shows a WiMAX frame with multiple zones, and possible switching from mandatory to optional choices.

The largest fundamental time–frequency resource unit in WiMAX and AeroMACS networks is a frame that encompasses all subcarriers. Figure 5.7 reiterates the fact that a frame time duration, in TDD scheme, is divided into two subframes that provide time slots for downlink and uplink communications. Each subframe is further divided into permutation zones within which one particular subcarrier permutation is used [22].

---

9 Ranging is a process, defined in IEEE 802.16 standards, through which the SS/MS carries out three tasks: (i) acquires the correct network timing offset for the purpose of aligning with the frame to be received from the BS, (ii) requests power adjustment, and (iii) downloads burst profile change [23].

5.3.2.5 Adaptive (Advanced) Modulation and Coding (AMC)

WiMAX PHY layer supports several digital modulation schemes, as well as a number of forward error correction coding formats; combined and applied in conjunction with OFDMA. OFDMA performs efficiently for LOS communication links; however, under NLOS conditions, additional performance- enhancing techniques are needed. One such technique supported by IEEE 802.16 standard is AMC. The AMC algorithm interfaces the user with the highest possible order modulation and greatest possible coding rate that can be supported by the level of SNIR at the link's receiver in an adaptive fashion; the combination is called a burst profile. In other words, when the channel is clear, the system throughput is as high as possible, and when the channel is poor, conservative modulation and coding schemes are used, resulting in lower system throughput. In this manner, each user is provided with the highest possible data rate that can be supported in their respective links [20].

The WiMAX physical layer supports 4-QAM (QPSK), 16-QAM, and 64-QAM digital modulation schemes for DL and UL paths on a mandatory and optional basis. Figure 5.8 illustrates the constellation of 16-QAM scheme alongside with Gray coded bit sequence assignment to each waveform. Similar constellation diagrams may be composed for QPSK and 64-QAM (see Chapter 7).

In WiMAX standard, significant portion of error control is carried out in connection with AMC; however, the MAC layer also contributes to

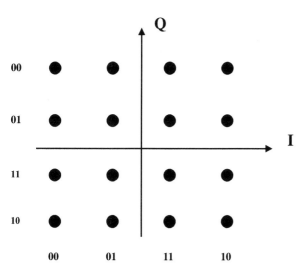

Figure 5.8 16-QAM constellation diagram with gray coded bit sequence assignment.

Table 5.2 FEC schemes supported by WiMAX.

Mode/path	Downlink	Uplink
Mandatory	TBCC Rates: 1/2, 2/3, 3/4, 5/6	TBCC Rates: 1/2, 2/3, 3/4, 5/6
Optional	1. CTC Rates: 1/2, 2/3, 3/4, 5/6 2. BTC Rates: 1/2, 3/4 3. Repetition codes Rates:1/2, 1/3, 1/6 4. LDPC codes Rates: 1/2, 2/3, 3/4, 5/6 5. RS codes Rates are not specified	1. CTC Rates 1/2, 2/3, 3/4, 5/6 2. BTC Rates: 1/2, 3/4 3. Repetition codes Rates: 1/2, 1/3, 1/6 4. LDPC codes Rates: 1/2, 2/3, 3/4, 5/6

overall error control strategy through the application of ARQ and HARQ schemes. In the physical layer, a combination of randomization, interleaving, and forward error correction (FEC) is supported. A variety of FEC schemes, such as tail-biting convolution codes (TBCC)[10] [25], convolutional Turbo codes (CTC), block Turbo codes (BTC), repetition codes, LDPC, and Reed-Solomon (RS) codes that are constructed over $GF(256)$, are supported by IEEE 802.16 standards. Table 5.2 lists various coding scheme selections supported by WiMAX standard as either mandatory or optional choices.

As Table 5.2 indicates, TBCC is the only mandatory FEC; however, WiMAX System Profiles require that all compliant systems support CTC as well [13]. In general, a coding scheme is selected for a particular case based on required signal reliability, that is, BER or FER (frame error rate), channel characteristics, throughput/goodput requirements, latency concerns, hardware/software complexity, and so on. If the channel is infested with frequent deep fades (which is the case in certain parts of the airport surface channel), coding schemes with burst error-correcting capabilities are most suitable. Table 5.3 illustrates some possible modulation-coding combinations, that is, burst profiles supported by WiMAX adaptive modulation-coding scheme.

---

10 In tail-biting convolutional coding, the encoder memory is initialized with the last block of data sequence, ensuring that the initial and final states of the encoder are the same [25].

Table 5.3 Modulation coding combinations in WiMAX AMC.

Link	DL	UL
Coding/Modulation	QPSK, 16-QAM, 64-QAM	QPSK, 16-QAM, 64-QAM
TBCC: Rates	1/2, 2/3, 3/4, 5/6	1/2, 2/3, 5/6
CTC: Rates	1/2, 2/3, 3/4, 5/6	1/2, 2/3, 5/6
BTC: Rates	1/2, 3/4	1/2, ¾
Repetition codes: Rates	1/2, 1/4, 1/6	1/2, 1/4, 1/6

A key objective of AMC protocols is to optimize the utilization of the air link. A modulation scheme is selected for each user based on the quality of the corresponding radio channel. As Table 5.3 illustrates, over the UL 4QAM (QPSK), 16QAM, and 64QAM, modulation schemes are supported, where 64QAM is optional and the others are mandatory. When it becomes necessary to adapt a new burst profile for the DL path, the BS applies one of the two available power adjustment rubrics corresponding to either maintaining constant constellation peak power or maintaining constant constellation mean power. In order to sustain constellation peak power, the constellation corner waveforms are transmitted at equal power levels regardless of modulation type. In the latter case, the signal is transmitted at equal mean power levels regardless of modulation type. In-phase (I) and Quadrature (Q) baseband signals are pulse-shaped by root-raised cosine filters prior to modulation [26].

AMC algorithms and adaptation of various burst profiles in WiMAX standards operate based on channel measurement conducted by the network (the matching of modulation and coding formats and other protocols with channel conditions is also referred to as *link adaptation*). However, except for a suggested simple channel quality table lookup method, the detail of the algorithms to be used for various scenarios are not specified. The implications are that different applications, such as AeroMACS, might need to develop their own adaptation algorithm to be applied in their particular circumstances for optimization with respect to critical system criteria such as QoS, throughput, latency, and so on. There are 52 possible burst profiles defined for WiMAX networks; however, most practical WiMAX systems furnish only a fraction of all available burst profile supported by the standard [20].

Owing to the fact that the physical layer of mobile WiMAX is quite flexible with large number of possible modulation and coding combinations, physical layer data rates take on different values depending on selected PHY layer attributes and protocols. Major parameters that directly influence the physical layer data rate are channel bandwidth, modulation format, and the error control scheme. Table 5.4 shows some physical layer data rates for DL and UP paths for channel bandwidths 5 MHz and 10 MHz. Subchannelization permutation is PUSC and the frame duration is 5 ms. There are 48 OFDM symbols in each frame of which 44 frames are available for data transmission [10].

The bandwidth for AeroMACS channels has been selected to be 5 MHz. As Table 5.4 indicates, for 5 MHz channel bandwidth, the DL data rate can increase by about 15-fold, as the channel conditions change from highly impaired to clear. Thus, the AMC is a mechanism that promises data reliability, while optimizes signal throughput and goodput. The application of some modern radio technologies, such as MIMO, can increase the data rates beyond what is shown in Table 5.4.

Table 5.4 Physical data rates for mobile WiMAX with PUSC permutation.

Channel bandwidth			5 MHz		10 MHz	
Modulation	Code, rate	Repetition code rate	DL rate (Mbits/s)	UL rate (Mbits/s)	DL rate (Mbits/s)	UL rate (Mbits/s)
QPSK	CTC, 1/2	1/6	0.53	0.38	1.06	0.78
QPSK	CTC, 1/2	1/4	0.79	0.57	1.58	1.18
QPSK	CTC, 1/2	1/2	1.58	1.14	3.17	2.35
QPSK	CTC, 1/2	1	3.17	2.28	6.34	4.70
QPSK	CTC, 3/4	1	4.75	3.43	9.50	7.06
16-QAM	CTC, 1/2	1	6.34	4.57	12.67	9.41
16-QAM	CTC, 3/4	1	9.50	6.85	19.01	14.11
64-QAM	CTC, 1/2	1	9.50	6.85	19.01	14.11
64-QAM	CTC, 2/3	1	12.67	9.14	25.34	18.82
64-QAM	CTC, 3/4	1	14.26	10.28	28.51	21.17
64-QAM	CTC, 4/5	1	15.84	11.42	31.68	23.52

5.3.2.6 ARQ and Hybrid ARQ: Multilayer Error Control Schemes

Automatic Repeat Request (ARQ) scheme is a general error control strategy in which an error detecting code, such as a parity check code or CRC (Cyclic Redundancy Check) code, is used to determine whether or not a received code word is erroneous. If the received code word is error free, the receiver sends a positive acknowledgment (ACK) to the transmitter through a back channel. When error is detected in the received sequence, the receiver sends a negative acknowledgment (NACK) and requests for a retransmission of the code word, again via the back channel. The process continues until the code word is successfully received. In HARQ system, a combination of FEC and ARQ is applied. The function of the FEC portion is to detect and correct frequently occurring errors. When a rare error pattern, containing more errors that is beyond the error correcting capability of the FEC, occurs, the ARQ portion is invoked and the receiver asks for a retransmission of the sequence. A proper combination of FEC and ARQ would provide greater data reliability and higher system throughput than when FEC or ARQ strategies are applied alone [25].

WiMAX supports *type I chase combining* HARQ scheme in the PHY layer. In this format, a retransmitted code word is combined with the previously received code word and fed to the input of FEC decoder. WiMAX also optionally supports *type II incremental redundancy* HARQ format. In this technique, each retransmission uses a different encoding technique that has higher error control capability. In effect, the number of parity check bits are increased (the code rate is decreased) in each retransmission [20].

In WiMAX networks, the ARQ function is performed in both PHY layer and MAC layer, as such it is a multilayer procedure. In the PHY layer, as was alluded to earlier, HARQ is applied to mitigate the effect of wireless channel impairments, to enhance the transmitted signal robustness, and to increase system throughput. In the MAC layer, the main function of the supported ARQ scheme is to provide feedback on successfully received and missing data blocks. Physical layer HARQ is faster and more reliable than MAC layer ARQ, owing to the fact that HARQ takes advantage of faster responding physical layer ACK and NACK packets, transmitted from the receiver. HARQ is supported by both DL and UL channels [27].

5.3.2.7 Multiple Antenna Techniques, MIMO, and Space-Time Coding

The ever-increasing demand for capacity and robustness, that is, higher speed and greater quality (reliability), communications through wireless networks, requires provision of continually higher data rates. In particular, Internet access and multimedia communications require data throughputs that are several orders of magnitude greater than that of ordinary voice and text transmissions. Conventional methods to increase network capacity and overall system throughput, include increasing channel bandwidth and introduction of higher level modulations and efficient coding schemes. More recently, several new signal processing procedures and radio technologies have been employed in the PHY layer of WiMAX networks (as well as in other wireless networks), to further increase data rates and reliability without expanding the network bandwidth. Channel condition-sensitive adaptive combination of modulation and coding (AMC), application of multiple smart antennas at both ends of a communications link, and fractional frequency reuse are among these methods. In this section, multiple smart antenna and MIMO techniques are briefly discussed. The use of MIMO techniques and the associated *space-time coding* (STC) has rapidly become new frontier of research and development in wireless communications and is expected to be massively applied in future wireless networks such as 5G. Historically, antenna technologies have played a central role for capacity enhancement even in classical wireless networks. Sectorization, discussed in Chapter 2, was an early antenna technology that increased cellular network capacities by several folds. The improvements gained from multiple smart antenna applications is closely related to the concept of diversity.

Diversity techniques, in general, are defined as transmission of the same information through multiple uncorrelated (or weakly correlated) physical or virtual channels. In wireless fading channels, diversity techniques are applied to enhance performance-related parameters such as data rate and reliability, capacity, and spectral efficiency, as well as for extending the network's radio outreach. The diversity idea for wireless systems is based on the fact that individual channels experience different levels of fading, and therefore, with multiple versions of the same signal available at the receiver, it is possible to select the best received signal, or use some diversity combination technique to achieve the best possible result. There are different methods of diversity that may be employed to enhance the performance of wireless communication systems. *Space or Antenna Diversity*, or *spatial diversity*, which is classified as either receive diversity or transmit diversity, is the use of two or more antennas at the receiver or at the transmitter with sufficient spacing to create uncorrelated wireless channels. Signal reception through several antennas, in the case of *receiver diversity*, enhances the probability that if one signal is undergoing deep fade, the other signals are not. Receive diversity was the original form of space diversity in cellular networks that enabled improvement in terms of link budget and combating co-channel interference. Later on, it was understood that the same gain may be achieved by using multiple antennas at the transmitter, that is, through the application of *transmit diversity*, which is the communication of the same signal using multiple transmit antennas. In *time diversity*, multiple versions of the same signal are transmitted through the same channel at different times. Time diversity is implemented either by actually transmitting the same signal several times or by employing some form of error correction coding, where redundancies are added to the information bearing signal. Frequency diversity involves the transmission of the same signal through more than one radio channel. Frequency diversity maybe realized either by actually transmitting the signal modulating more than one RF carrier or by spread spectrum techniques. OFDM transmission technique may also be viewed as a form of frequency diversity scheme. Another form of diversity is *polarization diversity* in which independent communication links can be created based on the fact that vertically and horizontally polarized waveforms are orthogonal, and therefore independent.

For cellular networks such as LTE-based 4G and 5G, WiMAX, and AeroMACS, receive diversity may be applied for the uplink, while transmit diversity is implemented in the downlink, both at the base station. The vast interest in such schemes is due to the fact that significant performance improvement is possible without adding extra bandwidth, hardware/software system complexity, and power consumption to the user's system. The IEEE 802.16 standards and WiMAX technologies support a number of advanced signal processing techniques involving multiple element antenna systems to implement various forms of diversity. Beamforming, transmit and receive diversity, spatial multiplexing, and the general MIMO antenna schemes are among them [13–17].

The term *smart antenna* refers to a directional antenna with a built-in intelligent signal processing algorithm that enables the identification of the direction of the arrival (DOA) of the signal. *Beamforming*, also referred to as *smart antenna system* (SAS) and *adaptive antenna system*, is a technique in which an adaptive smart antenna array gain is steered to a track that allows power concentration of the antenna array toward a desired direction. In WiMAX networks, beamforming is supported for both UL and DL paths. In the UL, in effect, beamforming coherently combines the signals received by all elements of the antenna array. The signal processing involved in the DL is very similar to that of UL. There are two methods of beamforming. Phased array systems, or *switched beamforming*, provide a finite number of predefined antenna radiation patterns. An algorithm estimates the DOA of the signal, based on which the antenna array switches to one of the predefined patterns that best matches the signal DOA. *Adaptive array system* (AAS), or *adaptive beamforming* on the other hand forms a radiation pattern that is exactly adjusted to the scenario at hand, that is, directed toward the target, in the real time. The complexity and cost of AAS is much higher than those of switched beamforming system.

Spatial multiplexing is intended to increase data rate but not necessarily signal robustness. In this multiple antenna technique, multiple independent data streams are transmitted through the same wireless channel simultaneously, each stream using its own transmit and receive antennae. The signal transmission capacity of WiMAX network applying such scheme can be increased linearly with the number of pairs of receive/transmit antenna. For instance, using six antennas (three on each side) can virtually triple the system data rate. Spatial multiplexing is feasible under the condition that relatively high levels of SINR are available [20].

MIMO antenna technology takes advantage of several diversity techniques, in particular space diversity and time and/or frequency diversity. The MIMO idea relates to the simultaneous transmission of multiple uncorrelated or independent signals through the wireless media using multiple antennas at the transmitter side and multiple antennas at the receiver end. The signals may be separated at the receiver as uncorrelated versions of the same signal by appropriate signal-processing techniques, provided that multipath channels defined by various links from a transmit antenna to the corresponding receive antenna are sufficiently decorrelated. Multiple antenna schemes such as beamforming and spatial multiplexing, discussed earlier, are considered as special cases of MIMO. We now concentrate on brief characterization of the general MIMO idea.

MIMO techniques may be classified in various ways. Single-user MIMO (SU-MIMO) serves a lone user via multiple antennas, whereas a multiple user MIMO (MU-MIMO) system transmits data to a number of users simultaneously using multiple antennas. In open-loop MIMO, channel state information (CSI) is not required. This is in contrast with closed-loop MIMO, in which the knowledge of CSI is needed by the transmitter. Since closed-loop MIMO requires CSI information, it does not perform well in high-mobility environment such as an airport surface in AeroMACS applications. Cooperative MIMO (CO-MIMO), also known as distributed MIMO (Network MIMO or Virtual MIMO), groups multiple communication devices into a virtual antenna array. In this way, MIMO performance improvements can be achieved exploiting spatial domain of wireless channel and without requiring multiple antennas in any of the devices. This is an interesting MIMO configuration, particularly when the wireless network is fortified with multihop relays.

Figure 5.9 illustrates a general configuration of an open-loop SU-MIMO – consisting of N transmit antennas, M receive antennas, with STC incorporated in the system. The illustrated MIMO system is open loop; however, in most practical cases, it is assumed that full or partial

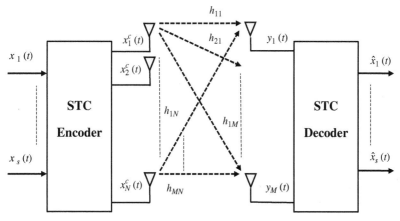

Figure 5.9 A general $N \times M$ MIMO system diagram with incorporated space-time coding. There are s transmitted signals.

CSI is available at the receiver side. The beamforming may also be incorporated alongside space-time coding. The MIMO performance enhancements are achieved either in the form of diversity gain to improve data reliability or in the mode of space multiplexing to transmit multiple data streams simultaneously; thereby increasing the effective data rate or a combination thereof. In the system shown in Figure 5.9, the maximum possible diversity gain, relative to single-input single-output (SISO) case, is $N \times M$, and the maximum multiplexing rate is $Min(N, M)$.

In fact, MIMO offers a trade-off between multiplexing rate and diversity gain to the network designer [27–28]. Hence, various combinations of data rate increase and diversity gain are at the disposal of the system designer. In general, as the number of antennas increases in a MIMO system, the diversity gain is reduced relative to that of the maximum theoretical gain. An exception is the CO-MIMO configuration where gains close to maximum level is attainable, owing to the fact that "multiple antennas" are well spread apart.

Referring to Figure 5.9 and under the assumptions of narrowband signal and frequency flat fading channels, h_{ij} is a complex number representing the gain of the wireless channel between receive antenna i and transmit antenna j. These channels are assumed to be Rayleigh, Rician, or in some cases even Gaussian. Terms designated as $x_k^c(t)$ denote the space-time coded version of the input data. The received signals are represented by $y_j(t)$ and the space-time coding decoder reproduces the estimation of the original signal, which are denoted by $\hat{x}_i(t)$. In general, the $N \times M$ MIMO channels of the sort shown in Figure 5.9 can be characterized by the transmission matrix given in Equation 5.4.

$$H = \begin{bmatrix} h_{11} & h_{12} & \cdots & h_{1N} \\ h_{21} & h_{22} & \cdots & h_{2N} \\ \vdots & \vdots & \cdots & \vdots \\ h_{M1} & h_{M2} & \cdots & h_{MN} \end{bmatrix} \tag{5.4}$$

The MIMO channel output is therefore computed by matrix Equation 5.5.

$$Y = H \times X^c + E \tag{5.5}$$

where $Y = \begin{bmatrix} y_1 & y_2 \dots y_M \end{bmatrix}^T$ is channel output vector, $X^c = \begin{bmatrix} x_1^c & x_2^c \dots x_N^c \end{bmatrix}^T$ is channel input vector, and E is the noise vector given by $E = [e_1 \quad e_2 \dots e_M]^T$. The noise is assumed to be AWG, and the adverse effect of cellular mobile and fading channels are accounted for in the transmission matrix H.

Channel coding idea, in general, refers to mapping of data blocks into symbol sequences containing redundancies that are exploited to detect and correct transmission errors at the receiver. In MIMO system, the redundancies are created by STC techniques; however, the redundancies generated by STC are not meant to be directly used for detection and correction of errors. STC was first introduced by Tarokh, et al., as a novel method of transmit diversity for performance enhancement of wireless communication systems [29–31]. Originally, two main categories of STCs were introduced: space-time block codes (STBC) and space-time trellis codes (STTC). It should be noted that unlike standard block or convolutional (Trellis) codes applied to signals for transmission over SISO or AWGN channels, STCs do not generally provide coding gain but rather offer transmit and receive diversity, that is, diversity gain. If it is desired to generate coding gain, the STCs may be concatenated with a conventional error correction coding scheme.

A simple, but rather elegant and ingenious, method of space-time coding was proposed by Alamouti in 1998 [32]. This STBC scheme does not require feedback from the receiver to the transmitter and provides diversity by using two transmit antennas, while the received signal is processed over a single antenna. Figure 5.10 illustrates Alamouti's space-time coding scheme. According to Alamouti's method, during a given symbol period two signals are simultaneously transmitted from two transmit antennas (antenna one shown by T1, and antenna 2 shown by T2). The signals are designated as s_1 and s_2, as shown in Figure 5.10. Each of these signals may carry multiple data bits using a complex modulation constellation such as MPSK, QAM, or even an OFDM symbol. In the next symbol period, antennas T1 and T2 simultaneously transmit signals $-s_2^*$ and s_1^*, respectively. The channel can be modeled as a

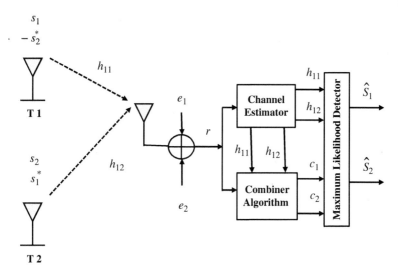

Figure 5.10 Alamouti's space-time coding scheme providing a simple transmit diversity [32].

complex multiplicative term consisting of a magnitude response as well as a phase response [32]. Assuming fading is stationary within time duration of two consecutive symbols, the channels between transmit antennas T1 and T2 and the receive antenna may be modeled by Equations 5.6 and 5.7.

$$h_1(t) = h_1(t + T_s) = \alpha_1 e^{j\theta_1} \triangleq h_{11} \tag{5.6}$$

$$h_2(t) = h_2(t + T_s) = \alpha_2 e^{j\theta_2} \triangleq h_{12} \tag{5.7}$$

where T_s is the time duration of each symbol. The received signals can, therefore, be calculated as follows.

$$r_1(t) \triangleq r_1 = h_{11}s_1 + h_{12}s_2 + e_1 \tag{5.8}$$

$$r_2(t) \triangleq r_2 = -h_{11}s_2^* + h_{12}s_1^* + e_2 \tag{5.9}$$

where r_1 and r_2 are received signals at time t and $t + T_s$, e_1 and e_2 are complex random variables representing received noise and interference.

Using the outcome of channel estimator (estimation of h_{11} and h_{12}), the combiner algorithm synthesizes the following two functions to be used in the maximum likelihood (ML) detector for the estimation of the transmitted signals s_1 and s_2.

$$c_1 = h_{11}^* r_1 + h_{12} r_2^* = (\alpha_1^2 + \alpha_2^2)s_1 + h_{11}^* e_1 + h_{12} e_2^* \tag{5.10}$$

$$c_2 = h_{12}^* r_1 - h_{11} r_2^* = (\alpha_1^2 + \alpha_2^2)s_2 - h_{11} e_2^* + h_{12}^* e_2^* \tag{5.11}$$

The right sides of Equations 5.10 and 5.11 result when Equations 5.8 and 5.9 are substituted into the left sides of these equations, respectively. The ML detector receives c_1, c_2, h_{11}, and h_{12} from the previous blocks, as illustrated in Figure 5.10. Conceptually speaking, a ML detector measures the Euclidean distance (or Hamming distance in the case of ML decoding of some algebraic codes) between a received noisy signal(s), which has undergone some predetection processing, and all possible transmitted waveforms (or pair of waveforms), and selects the signal (or signals) that bears the minimum distance value as its estimation of the transmitted signal(s). The Euclidean distance between two N-dimensional vectors $X = (x_1 \quad x_2 \ldots x_N)$ and $Y = (y_1 \quad y_2 \ldots y_N)$ is defined as in Equation 5.12.

$$
\begin{aligned}
d(X, Y) &\triangleq \sqrt{(x_1 - y_1)^2 + (x_2 - y_2)^2 + \ldots (x_N - y_N)^2} \\
\to d^2(X, Y) &\triangleq (x_1 - y_1)^2 + (x_2 - y_2)^2 + \ldots (x_N - y_N)^2 \\
\to d^2(X, Y) &= (X - Y)(X^* - Y^*)
\end{aligned}
\tag{5.12}
$$

In this case, the predetection processed noisy signals are c_1 and c_2 expressed in Equations 5.10 and 5.11. The ML detection procedure, therefore, corresponds to making a decision in favor of the pair of signals, s_i and s_j, that minimize the Euclidean distance between the pair c_1 and c_2 on the one hand, and all possible transmitted signal pairs on the other. It turns out that the ML detection for the two transmitted signals can be decoupled, that is, ML detection rule can be separately applied for estimation of each of the signals. The ML detection algorithm is reduced to the following rule for estimation of the each of the signals [32]:

Select s_i if and only if

$$
(\alpha_1^2 + \alpha_2^2 - 1)|s_i|^2 + d^2(c_j, s_i) \leq (\alpha_1^2 + \alpha_2^2 - 1)|s_k|^2 + d^2(c_j, s_k) \quad \forall i \neq k, \quad j = 1, 2
\tag{5.13}
$$

When the modulation scheme is MPSK, and the transmitted signals are of constant envelope type, the signals carry equal energy of $E_s = |s_i|^2$ and the ML detection rule is simplified to that of expression (5.14).

Select s_i if and only if

$$
d^2(c_j, s_i) \leq d^2(c_j, s_k) \quad \forall i \neq k \quad j = 1, 2
\tag{5.14}
$$

In summary, the Alamouti scheme presents a clever method of space-time coding that does not require channel information at the transmitter, and for which no bandwidth expansion is needed. Alamouti further extended this technique to the case of two antennas at the transmitter side and m antennas at receiver end.

A wealth of literature is available on MIMO (see Refs [33,34] for overview treatments) and in subsequent development of STC. For a thorough treatment of STC-related topics, the reader is referred to Refs [35,36].

5.3.2.8 Fractional Frequency Reuse Techniques for Combating Intercell Interference and to Boost Spectral Efficiency

In Chapter 2, the cellular architecture and the relationship between cluster structure and co-channel interference was discussed. Table 2.1 illustrates how signal to co-channel interference ratio (SIR) varies against changes in cluster size for a fixed value of path loss exponent. In fact, Table 2.1 demonstrates that for a fixed cell area, as cluster size increases, so does the SIR; however, the network capacity decreases. The available duplex radio channels are divided between the N (cluster size) cells within a cluster. A parameter that determines what fraction of all duplex radio channels are accessible to each cell is $1/N$, the FRF. For instance, when FRF is unity in a cell, it implies that all radio channels are available for use in the cell, whereas with cluster size 7 (FRF equal to $1/7$) only one out of every seven radio channels are accessible to the user in the cell.

As we discussed earlier, OFDMA groups large number of orthogonal subcarriers to form user's subchannels in WiMAX networks. The intra-cell channels, by and large, are assumed to be orthogonal and therefore, theoretically, pose no interference. However, when neighboring WiMAX cells use the same subchannels for signal transmission *co-subchannel interference*, or *intercell interference* occurs. A number of viable techniques are available for what is known as intercell interference coordination (ICIC), which is a strategy to minimize the effect of intercell interference in the network by intelligently allocating resources to each cell. Among ICIC schemes are power control, opportunistic spectrum access,[11] interference cancellation, multiple antenna and MIMO techniques, FFR, and so on [37].

WiMAX applies, among other techniques, FFR, which is particularly well suited for OFDMA cellular-based networks. In this technique, within WiMAX cell boundaries, FRF is not a constant but rather adaptive. The central idea of FFR is splitting the available bandwidth into subbands and allocating them to different parts of WiMAX cellular grid, such that the MSs that are moving along the boundary of adjacent cells do not pose interference to each other, and interference received (or created) by MSs travelling in the interior region of the cell is reduced [38]. In the simplest form of FFR format, the cells are divided into "inner" and "outer" regions

---

11 Opportunistic spectrum access refers to a procedure in which radio stations identify unused portions of the spectrum that is not used by the primary user of the spectrum and utilize that spectrum without adverse impact on the primary user. This scheme is widely applied in cognitive radio.

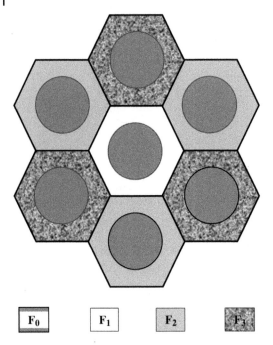

Figure 5.11 A static (strict) FFR configuration over a seven-cell hexagonal grid. Subband F_0 is shared by the inner region of all WiMAX cells.

F_0	F_1	F_2	F_3

and the available bandwidth is partitioned into four subbands of F_0, F_1, F_2, and F_3. Inner regions are closer to the base stations and by reducing the transmitted power, they become less susceptible to co-channel or co-subchannel interference. Subband F_0 is allocated to these regions within which FRF is one, that is, all radio channels (or subchannels) are accessible to all users in the interior parts of all WiMAX cells. However, for the outer cell regions, the subbands F_1, F_2, and F_3 are divided between the regions of various cells with FRF of $1/3$ and with a higher transmitted power than that of inner regions. This configuration is classified as a static, or "strict," FFR scheme [38]. Figure 5.11 illustrates a seven-cell hexagonal cellular grid with static fractional frequency reuse composition. The blocks underneath the cellular layout define how subbands are used to cover the inner regions and outer regions of the cellular grid. As Figure 5.11 demonstrates, the subband that is allocated to the outer region of one cell is different from the subbands that are assigned to the outer regions of the adjacent cells. In this manner, intercell interference is reduced and spectral efficiency is boosted. A direct comparison between this static FFR scheme and the classical cellular network spectrum division method (discussed in Chapter 2) reveals the sources of potential spectral efficiency enhancement in the FFR is that at least part of the allocated spectrum is used in all cells throughout the network.

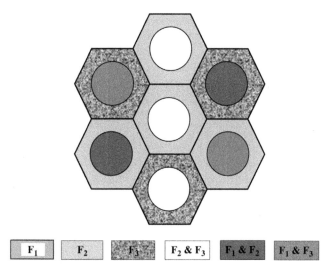

| F_1 | F_2 | F_3 | F_2 & F_3 | F_1 & F_2 | F_1 & F_3 |

Figure 5.12 An alternative method of static FFR configuration, known as soft FFR. The boxes underneath the hexagonal seven-cell grid define the spectrum distribution.

In the past couple of decades or so, several static, dynamic, and adaptive FFR schemes have been developed to support ICIC and spectral efficiency in various wireless networks. Figure 5.12 illustrates another FFR scheme that is sometimes referred to as soft FFR [38], applied to a hexagonal cellular grid. In this FFR approach, the available bandwidth is divided into three subbands of $F_1, F_2,$ and F_3. The subbands are then allocated to inner and outer regions of a cell as indicated in Figure 5.12.

In this configuration, each cell edge and outer region is supported by a single subband and higher transmitted power, whereas the inner region of the same cell is allocated with the remaining subbands and uses lower transmit power. Adjacent cell edges are assigned with different subbands. Other cells use similar reuse pattern. Clearly, the overall frequency reuse factor is one-third in this scheme.

Reference [39] examines several static and dynamic FFR techniques, while Ref. [38] presents a thorough analytical evaluation of the two FFR schemes that were described above, assuming a Poison point random process for modeling of base stations location. The focus of this analysis is on cellular downlink path and the results point to a trade-off between superior interference reduction of strict FFR and the better resource efficiency on the part of soft FFR, while both schemes outperform classical cellular network reuse strategy in spectral efficiency.

5.3.2.9 Power Control and Saving Modes in WiMAX Networks

The definition of the term power control in wireless mobile networks may depend upon the context in which it is being deliberated. However, in a broad sense, power control for cellular networks may be described as the proper selection of the transmitted power, for both UL and DL traffic, that satisfies certain criteria or optimizes specific parameter(s) in the network. Accordingly, *power control algorithms* have been developed for various wireless communications scenarios.

In WiMAX networks, power control procedures are supported for both UL and DL. The serving BS controls the power level per subframe and determines the MS's transmitter output power. This is carried out by WiMAX system requirement that MSs report on two power quantities: maximum available power and current normalized transmitted power. The BS also uses this information to determine the current burst profile to be used by the MS.

The support of power control procedure is mandatory in the uplink. The power level assigned for UL is calculated to compensate for long-term attenuation, large-scale fading, and short-term fading, to mitigate the effect of intercell and intracell interference, as well as making up for implementation losses [27]. The uplink power control algorithm is based on information provided by the open-loop and closed-loop power control mechanisms. The open-loop power control mechanism enables the MS to directly determine the transmit power based on transmission parameters provided by the serving BS, downlink CSI, uplink channel transmission quality, and interference information obtained from the DL. Clearly, the required information is available without requiring frequency interaction with the serving BS. The open-loop control is normally used to combat the effects of channel variations and for compensation of implementation losses. However, the dynamic channel fluctuations are countered by closed-loop power control mechanism, where the serving BS plays an active role in issuing direct power control commands to the MS. The basis of the serving BS command is the measurement it conducts on UL CSI and interference information using UL data, as well as information provided by control channels. The MS, upon the reception of power control command from the serving BS, adjusts its transmitter output power accordingly [27].

The power consumption and efficiency are critical issues for mobile applications as they are directly related to the lifetime of the battery used in the mobile unit. Power management feature of Mobile WiMAX and IEEE 802.16e standards support two energy saving modes for power efficiency: Sleep Mode and Idle Mode. In Sleep Mode, the MS takes prenegotiated leave of absence from air interface engagement with the serving BS. In other words, the MS becomes simply unavailable for either UL or DL traffic. Sleep Mode is intended to minimize MS power usage and to reduce the usage of the serving BS air interface resources [10]. The Sleep Mode may be enacted when the MS is in the active mode. In the IEEE 802.16m standard, the MS is provided with several listening and sleep windows. The "listening window" refers to time intervals in which the MS is available for reception and transmission of data and control signals [27]. The Idle Mode is a WiMAX power conserving feature that allows the MS to become periodically available for DL broadcast traffic messaging, such as paging, without registration at any specific BS, as the MS roams around an environment covered by multiple BSs. Clearly, the Idle Mode removes the requirement for handoff and other normal operation from inactive mobile terminals, while still providing the MS with a timely method of alerting her about pending DL traffic. In the meanwhile, the network and the BSs are released from air interference engagement for handoff and other routine operations [10], which would in turn conserve power for both the network and the MS.

5.3.3 WiMAX MAC Layer Description

In this section, a brief overview of WiMAX MAC layer protocols is presented. The MAC, according to OSI (Open System Interconnection) model, is actually a sublayer of Data Link Layer. The IEEE 802.16 standards were essentially developed for delivery of broadband services. Figure 5.2 illustrates the protocol structure proposed by the IEEE 802.16 standards, which is identical to that of WiMAX networks. Consequently, WiMAX network MAC layer consists of the same three sublayers with exactly the same functionalities that were outlined in Section 5.4, that is, the service-specific CS, CPS, and the security sublayer (SS). The convergence sublayer can interface with a variety of higher layer protocols such as IP, ATM, Ethernet, and so on. The IEEE 802.16 standards' (as well as WiMAX networks') MAC layer is based on a proven international telecommunication standard – originally developed for cable TV, known as Data Over Cable Service Interface Specification (DOCSIS). This standard permits the addition of broadband data transfer to an existing cable TV system, and as such it can support bursty traffic with a high peak rate.

From the networking point of view, the primary function of the MAC layer is the provisioning of interface between physical layer and higher level layers. Packets of data that are sent to MAC layer, perhaps from the Network Layer or most probably from IP Layer, are called MAC service data units, or MAC SDUs. The MAC layer, subsequently, reformats and converts the SDUs to MAC protocol data units, or MAC PDUs, and delivers them to the PHY layer for transmission over the air. At the receiver side, the MAC layer converts PDUs received from the PHY layer to SDUs and delivers them to the Network layer or IP Layer.

5.3.3.1 WiMAX MAC CS; Connections and Service Flows

The CS interfaces the WiMAX MAC layer to the network layer; as such it is responsible for transformation or mapping of SDUs received through CS SAP (see Figure 5.2) to MAC PDUs. This functionality includes classification and mapping of SDUs into appropriate MAC connections over the air. A *connection* is a MAC level linkage between BS/ABS on the one hand and SS/MS/AMS/RS on the other, in either UL path or DL path, for the purpose of transportation of service flow's traffic [4]. *Service flow* is a unidirectional logical connection between two MAC peers. A connection is for a single type of service exclusively, in other words one connection cannot carry audio and date simultaneously, for instance. A MAC connection is identified by a 16-bit long CID. A service flow (SF) is actually a MAC connection with a set of QoS parameters associated with it. A service flow is characterized by the following parameters:

1) A 32-bit long SFID
2) A 16-bit long CID.
3) Provisioned QoS parameters: the recommended QoS parameters.
4) Admitted QoS parameters, which are the actual QoS assigned to the service flow.
5) Active QoS parameters – the QoS parameters that are being provided at a given time.
6) The authorization module.

A MAC SF with a set of security parameters is represented by a security association (SA). It should be noted that all IEEE 802.16 traffic is carried through connections.

The second task for MAC CS is the header compression, which includes packet header suppression and robust header compression for applications such as Voice over IP (VoIP). The packets that are delivered from the network layer to the MAC layer may have very large headers. For instance, some IPv6 (Internet Protocols Version 6) packet headers may be as long as 120 bytes. Very often the headers contain redundant information and therefore there is no need to transmit the entire header through the radio channel, which is a scarce communication resource. The process of removing the header redundancy and compacting the header is known as header compression [4].

Protocol data unit (PDU) formatting is also carried out in the MAC CS.

5.3.3.2 The MAC CPS Functionalities

The major responsibilities of the CPS MAC are mobility control, resource management, support for signal transmission over the air, and provision of control and support for the physical channel that is characterized by

PHY layer. In addition, CPS has an essential role in QoS support over the air [27]. The functionalities of CPS include selection of burst profiles, that is, modulation-coding combination, bandwidth management (including bandwidth request processing and bandwidth allocation), fragmentation and packing,[12] and feedback of CSI. In addition, CPS is responsible for access control, call admission control (CAC), ARQ support, multicast and broadcast services, and MAC PDU processing.

The maximum size of a PDU is 2048 bytes, which is divided into three fields. The first 6 bytes of the MAC PDU define the "generic MAC header" (GMH), and the last 4 bytes are used for the provision of an optional CRC-related redundancies. The remaining bytes are dedicated to a variable-length payload information [27].

QoS parameters of a service over the air interface such as throughput, packet latency, jitter, packet loss, and so on are controlled and managed by MAC management, scheduler, and ARQ modules of CPS. MAC CPS arranges for and maintains QoS for each and every MAC connection [27].

5.3.3.3 WiMAX Security Sublayer

The security sublayer is responsible for data integrity control, data protection against theft, user and device authentication, and securing key exchange. Security sublayer consists of two protocol suites of encapsulation for encryption of data packets and privacy key management (PKM). The main issues to be addressed by security sublayer are data encryption and authentication. All known security attacks and risks should be considered when security system algorithms are developed.

The IEEE 802.16 standards provide protocols in support of cryptography and authentication that are inherited by WiMAX technology. An encapsulation protocol is used for data encryption for the purpose of securing data packets across the WiMAX network. This protocol combines encryption and authentication algorithms by defining a set of supported cryptographic suites that are applied to MAC payloads according to some defined rules. The PKM, in their original form, provide the secure distribution of keying data from BS to MS/SS, where BS and MS synchronize keying data. This is a one-way authentication. The IEEE 802.16-2009 standard, however, has made several modifications and enhancements on the original (IEEE 802.16-2004-based) PKM protocols, now referred to as PKMv2. These new features include three-party protocol (supplicant, authenticator, and authentication deriver), data

---

12 When the received SDU packets are too large to be placed in a single PDU, the CPS divides them into multiple segments. This process is called *fragmentation*. On the other hand, when SDU packets are too short, CPS packed them together in order to save overhead. This is referred to as *packing*.

integrity and authentication and over-the-air encryption, user and device mutual authentication, application of EAP to support mutual credential types, and so on [2].

5.3.3.4 WiMAX MAC Frame and MAC Header Format

The IEEE 802.16 standard defines a global network address for subscriber stations (SS) and mobile stations (MS). Accordingly, each WiMAX SS/MS is provisioned with a unique 48-bit address, which is referred to as IEEE 802.16 MAC address, by the manufacturer that is originally supplied by the IEEE Registration Authority. In IEEE 802.16 standard management messages, MAC address is used to address the SS/MS prior to CID allocation, and it is transferred as part of context during handover process [40]. The CID is a unique 16-bit identifier for a given BS–MS pair that brands the logical connection between the peer MAC layer of the BS and MS. CID changes as MS moves on and is handed over to another BS. The CID is used to identify all information exchanged between the BS and MS/SS after the initial registration and authentication. This reduces overhead associated with using the MAC address, which is 48 bits long. In addition, the MS is assigned with a logical identifier consisting of two parts. First part is an MS identifier that uniquely labels the MS during entry[13] and/or reentry within a cell. Second part is a flow identifier that uniquely identifies the management connections to the MS. The IEEE 802.16 Standards assigns a 48-bit long Base Station Identification code, abbreviated as BSID, to each BS.

A MAC PDU is called a MAC frame, which starts with a MAC header. A MAC frame may be followed by payload information. The payload information is of variable length which, in turn, makes the MAC frame length variable as well. As was mentioned earlier, a MAC frame consists of three parts as depicted in Figure 5.13.

MAC Header 6-byte Long	Optional Payload of Variable Length	CRC (Optional) 4-byte Long

Figure 5.13 The makeup of a MAC frame (MAC PDU) with maximum length of 2048 bytes.

---

13 Network entry is the process through which a MS detects a cellular network and establishes connection with it. The key steps taken in entry procedures are synchronization with the serving BS, acquiring necessary system information, initial ranging and basic capability negotiation, authentication, authorization, and registration [27].

Two MAC header formats are defined in the standard. First is GMH, which is the header of MAC frames that carry either MAC management messages or CS data. The CS data could be either user data or higher layer management message. The frames with GMH are the only ones that are used for downlink. Second mode of MAC header is the "header without payload," in which MAC header is not followed by payload data or any CRC redundancy bits. This type of MAC header was introduced by IEEE 802.16e standard. MAC frames without payload are used in various applications. For instance, bandwidth request uses a special form of MAC frame without payload [4].

5.3.3.5 Quality of Service (QoS), Scheduling, and Bandwidth Allocation

The concept of quality of service hardly lends itself to a concise definition, although some, in a very broad sense and for packet switched networks such as WiMAX, have informally defined QoS as the probability of a packet successfully traveling between two points in the network. However, an end-to-end WiMAX network consists of access service network (ASN), connectivity service network (backhaul network) (CSN), and IP (core network), which are independent networks with their own individual architectures and protocol stacks, and therefore estimation of probability of successful passage of a packet between two points in this compound network system may not be feasible. Nonetheless, QoS is a key descriptive element of service delivery by WiMAX networks or any other telecommunication network for that matter.

A more practical definition for QoS may be expressed as *the ability of a network to provide a desired level of service quality for a specified type of traffic.* In other words, QoS in wireless networks, or in any telecommunications network, means, at least conceptually, the probability that the network fulfills the promise of a given traffic agreement. Thus, QoS manifests itself in the ability of the network constituent elements (such as BSs, relays, routers, applications, hosts, and so forth) to have some level of assurance that its traffic and service requirements would be satisfied. Therefore, QoS is a measure of how reliable and consistent a network operates. The QoS within a packet network is traditionally characterized based on three parameters that play major role in the network performance.

1) *Latency:* Latency is a measure of time delay experienced in the WiMAX network that corresponds to the time that it takes from the initiation of sending data until it arrives at its destination. The latency in WiMAX networks has four components. Time that it takes for the user to access the BS in the WiMAX ASN, time required for signal passage through the backhaul network, the time delay over the IP network, and the time duration from the destination BS to the end

user. Typically, the dominant component is the time delay that takes place over the IP network.

2) *Packet Loss:* Packet loss presents a challenge in all packet-switching networks. It is the loss of data packets during transmission between the source and the destination. Packet loss may occur for a variety of reasons. The major reasons for packet loss are link congestion, high network latency, overloading or failure of switching device (router, switch, firewall, etc.), software issues in some part of the network, and sometimes even faulty hardware and cabling. Normally the dominant cause is high network latency or overloading of switches that are unable to process or route all the incoming data packets.

3) *Jitter:* Jitter is a measure of time variability of the packet latency across a network. Ideally, a network with constant latency has no jitter. However, as the nature of data changes from text to voice and to wideband signals, it takes a variable amount of time for a packet to arrive at its destination. Quantitatively, packet jitter is expressed as expected value of the deviation from the network mean latency.

The QoS concept is, in fact, applied to cope with the effect of latency and jitter. The packet loss problem has to be dealt with in conjunction with network design.

The key network concepts related to QoS are *service flows* and *bandwidth grant provisions*. A *provisioned service flow* possesses SFID but it may not carry traffic, rather it might be waiting to be activated for usage. An *admitted service flow* goes through the process of activation. An *active service flow* is an actuated service flow with all required resources allocated to it. Normally, upon an external request for a service flow, the BS/MS will check for required resources to satisfy the requested QoS. The service flows are the main mechanism for providing a requested QoS [41].

In order to support a broad range of applications, WiMAX defines five categories of QoS through scheduling schemes that are supported by the BS MAC scheduler for information transmission over any given connection. These services support different applications and provide various QoS, therefore require different bandwidths that are defined by MAC scheduler (some literature have called them WiMAX QoS). We briefly define these five classes of QoS-related scheduling for the general WiMAX networks and provide more detailed information in Chapter 7 in connection with the AeroMACS network case.

The Unsolicited Grant Services (UGS) is a scheduling method originally designed to support real-time constant bit rate traffic such as voice and VoIP. Hence, this service is appropriate for fixed-rate transmission, with no need for polling. Real-Time Polling Services (rtPS) is designed to support variable rate data packet traffic such as MPEG streaming. In this service, the ABS/BS polls the AMS-MS/SS/RS periodically to provide request opportunities. Non-Real-Time Polling Services (nrtPS) is appropriate for delay-insensitive applications. This service allows the SS/MS/RS to use contention request and unicast request for bandwidth bid. Unicast request opportunities are offered regularly in order to ensure that the AMS/MA/SS/RS has chance to request bandwidth in congested network environment. Best Effort Services (BE) service is meant for low-priority services, as such BE does not specify any service requirement. Similar to nrtPS, BE offers contention and unicast request opportunities but not on a regular basis. This WiMAX service is normally used for Internet services such as email and browsing. Data packets are carried as space becomes available. Delays may be incurred and jitter does not pose a problem. Extended Real-Time Polling Services (ertPS) is a scheduling algorithm that combines the efficiencies of UGA and rtPS. The purpose of this service is to support real-time service flows that generate variable size data packets, therefore requires a dynamic bandwidth allocation, on a periodic basis. One typical example that uses this scheduling scheme is Skype.

WiMAX provides broadband access for various services that require different QoS. The MAC layer schedules traffic flows and allocates bandwidth such that the QoS requirement for each service is met. The QoS can include one or more measures that are known as "QoS parameters." In what follows, some common WiMAX QoS parameters are explained.

- *Maximum Sustained Traffic Rate (MSTR)*: This represents the maximum allowable data transmission rate for the service expressed in bits per second. MSTR defines the maximum bound, not a guaranteed rate that is always available.
- *Minimum Reserved Traffic Rate (MRTR)*: This parameter specifies the minimum information rate reserved for a service flow and it is expressed in bits per second. This is the minimum average rate.
- *Maximum Latency Tolerance (MLT)*: This parameter specifies the maximum latency between the reception of a frame by the ABS/BS or AMS/MS/RS/SS on its network interface and the forwarding of the packet to its RF interface. MLT is normally expressed in milliseconds.
- *Jitter Tolerance:* Packets arrive at the destination with different latencies this creates packet delay variation (PDV) or as it is commonly known delay jitter. Jitter tolerance defines the maximum delay variation that is tolerated for the connection and it is expressed in milliseconds.
- *Request/transmission Policy*: This QoS parameter allows for specifying certain attributes of the service flow. For instance, it allows the selection of an option for PDU formation.

- *Traffic Priority*: Traffic priority specifies the priority assigned to a service flow. To define priority a value between 0 and 7 is assigned to each service flow. Given two service flows that have identical QoS parameters and different priorities, the higher priority service flow will be given lower delay and higher buffering performance [41].
- *Maximum Burst Rate (MBR)*: Maximum number of OFDM bursts transmitted per second.
- *ARQ Type:* This entails the selection of an ARQ technique.
- *SDU Type and Size:* This is related to the size and type of service data units to be used.

WiMAX Scheduling Methods include but not limited to the following techniques:

- Earliest Deadline First (EDF)
- Weighted Fair Queue (WFQ)
- Round-Robin (RR)
- Deficit Round-Robin (DRR)
- First-In-First-Out (FIFO)

In so far as WiMAX bandwidth request mechanisms are concerned, the IEEE 802.16 standards do not provide any information as to how bandwidth should be allocated to different applications. A mechanism is needed for bandwidth request that naturally has direct bearing on QoS.[14]

One has to distinguish between *QoS parameters* and *QoS metrics*, which normally include *throughput and goodput, latency, jitter, packet loss*, and so forth.

WiMAX defines a set of QoS parameters that include the key factors of *maximum tolerable latency* and *delay variation (jitter)* over the air interface connections. Combined with QoS mechanism over the backbone wireline or wireless networks, WiMAX QoS can provide predictable end-to-end QoS. When the network supports relays, the QoS analysis

---

14 The IEEE 802.16 standards do not mention how to allocate bandwidth to different applications and how to improve the performance of the whole network. Whenever an AMS/MS/SS desires to use the uplink, it requests for transmission opportunities (TO) in the units of time slots, through special purpose information elements referred to as *BW_Requests*. The BS accepts these requests over a period of time and compiles a time slot allocation map describing the channel allocation, UL_MAP, to inform all subordinate SSs of the available transmitting time slots. This MAP message is broadcast to all subordinated SSs in the next time frame. SS will provide service to part of applications connected to it with the granted bandwidth from BS. Since most of the research work about WiMAX bandwidth requesting and allocating modeling are circuit switched based, little work had been done with the TDD mode, which is an important aspect of the WiMAX network.

must also include two additional independent wireless links: AMS/MS-RS and RS-ABS/BS links.

5.3.4 WiMAX Forum and WiMAX Profiles

As was pointed out earlier, the IEEE 802.16 standards were developed to create wireless alternatives to wireline metropolitan area networks. These standards specify PHY and MAC layer protocols; however, the standards do not guarantee interoperability between standard-compliant devices. There are several reasons as to why the standards cannot ensure interoperability.

1) The IEEE 802.16 standard does not cover higher layer protocols (such as network layer) beyond PHY and MAC layer.
2) The standard does not define an end-to-end network architecture.
3) The IEEE 802.16 standards do not provide sufficient details (unlike 3GPP LTE protocols) to allow full interoperability in PHY and MAC layers.
4) The standard contains large number of PHY and MAC layer options, to the extent that designing cost-effective devices that can cover all possible options is impossible [22].

To address the key issue of interoperability, and to promote the application and deployment of WiMAX networks as a new IEEE 802.16 standard-based broadband access technology and for "the delivery of the last mile wireless broadband access" to the Internet or other networks, a nonprofit organization called "The WiMAX Forum" was formed in 2001. The main activities of WiMAX Forum may be classified in six different categories.

1) Development and publishing of "WiMAX System Profiles" by initiation of several long-term projects to ensure interoperability and to complement IEEE 802.16 standardization activities.
2) Establishing specifications of protocol and radio conformance testing for the purpose of certification of BS/ABSs, SSs, and MS/AMSs.
3) Development of "WiMAX certification Profiles."
4) Forming certification laboratories across the globe to manage conformance and interoperability testing to ensure that all WiMAX-certified products work flawlessly with one another.
5) Defining end-to-end network protocol specifications and architectures consistent with IEEE 802.16 standards and WiMAX Forum system profiles.
6) Developing an ecosystem and fostering an industry alliance, involving a large number of companies that are engaged with various aspects of

WiMAX technology, such as circuit/device/subsystem development, management systems, software applications, system integration, and so on. [27].

5.3.4.1 WiMAX System Profiles and Certification Profiles

The IEEE 802.16-2009 and IEEE 802.16j amendments consist of large number of protocol suites, specifications, and functionalities. Consequently, any technology that is constructed based on these standards selects a subset of available protocols, parameters, and features to be used in typical implementation cases. WiMAX is said to be an interoperable implementation of IEEE 802.16 standards. To address the issue of selection of proper subset of IEEE 802.16 standard mandatory and optional protocols for WiMAX, the WiMAX Forum has formed the Technical Working Group (TWG). The TWG is in charge of developing the technical conformance specifications, system profiles, and certification test schemes for air interface to ensure interoperability and compliance of MS/AMS,[15] SS, and BS/ABS[16] with IEEE 802.16 standards. For complete listing and description of the charges of various WiMAX Forum Working Groups, the reader is referred to Ref [27].

The TWG of WiMAX Forum defines two types of profiles: system profiles and certification profiles. A *system profile* is a subset of mandatory and optional PHY and MAC layer elements selected from IEEE 802.16 standards. Two categories of system profile have been defined and published by WiMAX Forum. First is *fixed system profile*, which is based on IEEE 802.16-2004 – the standard based on which WiMAX networks with stationary nodes were structured. Second is the *mobile system profile*, which is based on IEEE 802.16e scalable OFDMA standards.

A *certification profile* is established when the operating frequencies, channel bandwidth, FFT size, and duplexing technique are defined for a system profile. Initially, 5 fixed certification profiles and 14 mobile certification profiles were introduced by the WiMAX Forum. Table 5.5 shows the list, designations, and the main parameters of the five fixed system certification profiles published by WiMAX Forum [4]. All these profiles are based on OFDM PHY layer with PMP topology.

---

15 Advanced Mobile Station (AMS) is an IEEE 802.16m compliant mobile unit that is backward compatible with the legacy system (referred to as MS in this text). The AMS is capable of communicating BSs and ABSs.

16 Advanced Base Station (ABS) is the IEEE 802.16m compliant base station, which is backward compatible with legacy base station (referred to as BS in this text). ABSs are capable of communicating with other ABSs, BSs, MSs, AMSs, and RSs.

Table 5.5 Fixed WiMAX certification profiles.

Profile designation	Operating frequency (GHz)	Bandwidth (MHz)	FFT size	Duplexing scheme
3.5T1	3.5	7	512	TDD
3.5T2	3.5	3.5	256	TDD
3.5F1	3.5	3.5	128	FDD
3.5F2	3.5	7	256	FDD
5.8T	3.5	10	1024	TDD

5.3.4.2 WiMAX Mobile System Profiles

Several WiMAX mobile system profiles, which are based on IEEE 802.16e standard, have been published by WiMAX Forum over the past decade or so. These profiles are designated as "WiMAX System Profile Release" with a numerical identifier attached to them. Table 5.6 shows the key parameters of the original WiMAX mobile system certificate known as "WiMAX System Profile Release 1.0." This profile is based on scalable OFDMA PHY layer, which implies that bandwidth and FFT size are scalable, and the topology is PMP [13]. It should be mentioned that the mobile speeds allowed in this profile are well below the speeds that are dealt with on the airport surface.

Table 5.6 WiMAX mobile system certification profile Release 1.0.

Operating band of frequencies (GHz)	Bandwidth (MHz)	FFT size	Duplexing scheme
2.3–2.4	5	512	TDD
	8.75	1024	
	10	1024	
2.305–2.320	3.5	512	TDD
	5	512	
	10	1024	
2.496–2.690	5	512	TDD
	10	1024	
3.3–3.4	5	512	TDD
	7	1024	
	10	1024	
3.4–3.8	5	512	TDD
	7	1024	
	10	1024	

A modified version of Release 1.0, that is, "WiMAX Forum Mobile System Profile Release 1.5," which accommodates both TDD and FDD (full duplex and half duplex) technologies, were made public in 2009. This Release is based on IEEE 802.16-2009 Standard. Three different documents were published by WiMAX Forum on this Release. "WiMAX Forum Mobile System Profile Specification Release 1.5 Common Part" is the main document published by WiMAX Forum that specifies the WiMAX air interface aspects that are common for both TDD and FDD formats [14]. The other two documents are related to the FDD [15] and TDD [16] specifications. The main objectives of this revision are as outlined below.

1) To include FDD (full-duplex and half-duplex; HFDD) technology.
2) To enable advanced services such as "location-based services" and improve broadcast/multicast services.
3) To enhance network capacity and coverage for low-mobility usage by adding closed-loop MIMO.
4) To improve MAC layer efficiency for services such as VoIP, by lowering MAP overhead.

Table 5.7 shows some frequency-related parameters of WiMAX mobile system certification profile of Release 1.5 [14–16,27].

It should be noted that the maximum speed of MS supported in Release 1.5 is 120 km/h. This means that mobile nodes with higher speeds may not be able to maintain connection with the base station. For AeroMACS applications, higher speeds are required, as the landing and takeoff speeds of many commercial aircrafts are well above 120 km/h.

In September 2010, WiMAX Forum approved a new WiMAX system profile designated as "Release 2.0," which is based on IEEE 802.16-2009 standard amended by IEEE 802.16m, and the corresponding "air interface specifications" document was published in May 2012 [17]. Later on two modified versions of this system profile was approved and published under the titles of "Release 2.1" [18] and "Release 2.2" [19].

Release 2.0 and subsequent versions Release 2.1 and Release 2.2 reflect new technological features, as the IEEE 802.16m standard is designed to meet the IMT-Advanced requirements[17] [40]. Services that are provided in Release 2.0 and beyond include support for legacy services (backward compatibility) more thoroughly than is provided by Releases 1.0 and 1.5. The architecture of IEEE 802.16m systems is flexible, in order for the Release 2.0 to support new services required by next generation of cellular

---

17 IMT Advanced requirements envisions a comprehensive and secure end-to-end IP broadband solution for 4G and 5G cellular networks. It is expected that IMT-A provides services such as broadband Internet access, IP telephony, streamed multimedia, and so on.

Table 5.7 WiMAX mobile system certification profile Release 1.5 frequency parameters.

MS transmission bands (GHz)	BS transmission bands (GHz)	MS duplexing mode	BS duplexing mode	Channel bandwidth (MHz)
2.345–2.320	2.305–2.320	HFDD	FDD	2×3.5, 2×5, 2×10
2.496–2.572	2.614–2.690	HFDD	FDD	2×5, 2×10
3.400–3.500	3.500–3.600	HFDD	FDD	2×5, 2×7, 2×10
1.710–1.770	2.110–2.170	HFDD	FDD	2×5, 2×10
1.920–1.980	2.110–2.170	HFDD	FDD	2×5, 2×10, and optional 2×20
0.776–0.787	0.746–0.757	HFDD	FDD	2×5, 2×10
0.788–0.793 0.793–0.798	0.758–0.763 0.763–0.768	HFDD	FDD	2×5
0.788–0.798	0.758–0.768	HDFF	FDD	2×10
0.698–0.862	0.698–0.862	Dual mode TDD/ HFDD	TDD or FDD	TDD: 5, 7 FDD: 2×5, 2×7, 2×10
1.785–1.805 1.880–1.920 1.910–1.930 2.010–2.025	1.785–1.805 1.880–1.920 1.910–1.930 2.010–2.025	TDD	TDD	5, 10

network as listed in Ref. [40]. Furthermore, Release 2.0 and beyond support different QoS for various services, including IMT-Advanced requirement for end-to-end latency, throughput, and error performance [42].

The scalable channel bandwidths for Release 2.0 are 5–40 MHz within the operating frequencies that are listed below.

1) 450–470 MHz
2) 698–960 MHz
3) 1710–2025 MHz
4) 2110–2200 MHz
5) 2300–2400 MHz
6) 2500–2690 MHz
7) 3400–3600 MHz

The bands 450–470 MHz, 1710–2025 MHz, and 2110–2200 MHz are not supported by the previous Releases.

Release 2.0 supports both paired and unpaired frequency allocation with fix duplexing frequency separation. An interesting aspect of FDD mode in

Table 5.8 Peak supported spectral efficiency in Release 2.0 [42].

Link path	MIMO mode	Peak spectral efficiency (bits/(s Hz))
DL	2 × 2	8.0
UL	1 × 2	2.8
DL	4 × 4	15.0
UL	2 × 4	6.75

Release 2.0 is that UL and DL channel bandwidths may be different, for instance, UL channel bandwidth may be 5 MHz, while the DL channel bandwidth is selected to be 20 MHz. In TDD mode, since both UL and DL use the same RF band, UL and DL transmission capacities may be varied by changing time durations allocated for each path. In other words, the DL/UL time ratio is adjustable in Release 2.0. This is in fact an important advantage of TDD technique over FDD format.

Release 2.0, following the provisions of IEEE 802.16m standard, supports SU-MIMO and MU-MIMO, beamforming, and other advanced antenna techniques. For BS a minimum of two transmit and two receive antennas is supported, while for MS a minimum of one transmit and two receive antennas is adopted. This is consistent with 2 × 2 DL MIMO configuration and 1 × 2 UL MIMO arrangement [42].

With regard to spectral efficiency[18] (bandwidth efficiency), IEEE 802.16m and WiMAX System Profile Release 2.0, support the peak spectral efficiencies as specified in Table 5.8 [42].

Rows two and three in the Table 5.8 are corresponding to the minimum DL and UL peak spectral efficiency achievable with simplest MIMO configuration for BS and MS but under ideal conditions. The last two rows of the table provide maximum DL and UL peak spectral efficiencies achievable, when MS and BS are equipped with a higher order MIMO configurations. Clearly, higher order MIMO configurations may be applied for increasing the peak spectral efficiency, which is a clear indication that how effectively MIMO technologies can boost wireless networks' transmission capacity. It is noted that peak spectral efficiencies

---

18 Spectral efficiency, or bandwidth efficiency, is a key characterizing parameter based on which various digital communication systems performance can be compared. It is generally defined as the number of information bits per second that is transmitted through every unit bandwidth of the channel. It may also be referred to as data rate normalized by bandwidth, as such its measurement unit is bits/second/Hz.

presented in Table 5.8 are valid for all possible channel bandwidths and duplexing technologies, assuming 100% of radio resources are available.

IEEE 802.16m and WiMAX Release 2.0 support mobility at higher speeds than the ones supported by Releases 1.0 and 1.5. For stationary stations and low-speed MSs (pedestrian users), mobility is optimized. For medium-speed vehicular MSs, that is, speed range 10–120 km/h, mobility is supported but graceful degradation is expected as the MS speed increases. For vehicles whose speeds ranges from 120 to 350 km/h, BS–MS connectivity is maintained [42]. This speed range is of interest in AeroMACS networks, since the commercial aircraft landing and takeoff speeds are within this range.

5.3.5 WiMAX Network Architecture

The WiMAX Network Reference Model (NRM) is a logical representation of the network architecture. The NRM identifies functional entities and reference points (RPs) over which interoperability is achieved between functional entities [41]. As we have discussed earlier, the IEEE 802.16 standards specify and standardize the network air interface and radio link for various entities such as SSs, MSs, BSs, and RSs. The air interface set of defined rules include PHY and MAC layer protocols. This is one of the high points of WiMAX technology that leaves the higher layers of the protocol stack open to other bodies like the Internet Engineering Task Force (IETF) to fill in and complete the standard for end-to-end communications. This is different from 3G systems where 3GPP (Third Generation Partnership Project) sets the standard for a wide range of interfaces to ensure intervendor and internetwork interoperability. In order to define end-to-end network connection, the WiMAX Forum has formed two working groups to define the architecture. The first is the Network Working Group (NWG), which is charged with the task of developing (or adapting some already existing suite of protocols) end-to-end network specifications, architecture, and protocols for WiMAX networks. The second is the Service Provider Working Group (SPWG), which assists in defining requirements and priorities. In other words, a major duty of NWG and SPWG is to go beyond what IEEE 802.16 standard offers and develop profiles for end-to-end broadband networking architectures. An end-to-end architecture is necessary if WiMAX is to compete for market share with its peer industries. To that end, the WiMAX Forum has developed an end-to-end network reference model architecture that is based on all-IP protocol structure.

The WiMAX network reference model attempts to provide a unified network architecture for supporting various mobility modes implemented in the network: fixed, portable, nomadic, partial and full mobile categories. From the implementation point of view, the objective of NRM is to allow multiple realization options for various functional entities and achieve interoperability among all of these different realizations. Interoperability is a central theme in WiMAX and AeroMACS networks design. "Interoperability is based on the definition of communication protocols and data path treatment between functional entities to achieve an overall end-to-end function, for example, security or mobility management" [41].

5.3.5.1 WiMAX Network Reference Model as Presented by WiMAX Forum
The WiMAX NRM consists of three logical entities and a number of RPs, as shown in Figure 5.14. First among these logical entities is the MS and/ or AMS (advanced mobile station), which is a mobile device. The second is ASN that is composed of a complete set of functionalities required for providing radio access to the WiMAX network for MS/AMS/RS/SSs.

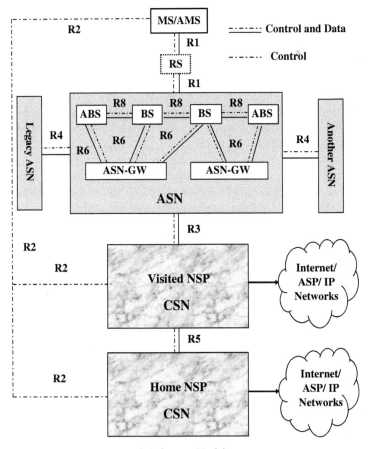

Figure 5.14 WiMAX Network Reference Model.

In other words, ASN is essentially an access network infrastructure to which mobile devices and stationary nodes are connected. Radio-related functions of ASN are the responsibility of BSs and RSs that are logical and physical parts of ASN. Another important component within ASN is the ASN gateway (ASN-GW) that is essentially the WiMAX router. The last logical/physical entities of WiMAX NRM is the CSN, which contains a collection of network protocols and functions that deliver IP connectivity services to WiMAX users. The CSN may consist of several network components such as routers, AAA[19] (Authentication, Authorization, and Accounting) proxy/servers, home agents, user databases, Internetworking gateways, and so on. In short, The CSN provides connectivity to the Internet, ASPs (Application Service Providers), public networks, and corporate networks [40]. There are also three business players involved in the operation of an end-to-end WiMAX network. First is Network Access Provider (NAP) that establishes, operates, and maintains access infrastructure for WiMAX network. An NAP is an operator that may own several ASNs and provide access between various ASNs via RP R4. The NAP is connected to another business enterprise called the Network Service Provider (NSP). The function of the NSP is to deliver IP connectivity and WiMAX services to the subscribers with established Service Level Agreements (SLA). The last of these business actors is ASP that is an enterprise that delivers application functionalities and services across the WiMAX network, and to multitude of users. The NSP is responsible for providing IP connection to WiMAX networks in accordance with agreements made with the WiMAX terminal uses at the service layer. As a result, the NSP can enter an agreement with other ASPs or ISPs to provide specific services. In the roaming mode, the NSP must enter an agreement with other NSPs. When an MS/AMS is located in the area served by an NSP other than the home NSP (HNSP), the NSP that provides WiMAX services to the MS/AMS terminal is called visiting NSP (VNSP). The CSNs are owned and operated by NSPs.

The general philosophy for the design of WiMAX network architecture has been based on several principles. First and foremost, the architecture is based upon functional decomposition concept. The required features are divided into functional entities with no specific implementation assumptions for their physical realization. The architecture specifies well-defined reference points between these functional entities. This is to ensure multivendor interoperability. Second, a modular and flexible

---

19 "AAA refers to a framework, based on IETF protocols (RADIUS or Diameter), that specifies the protocols and procedures for authentication, authorization, and accounting associated with the user, MS, and subscribed services across different access technologies" [41].

approach is adopted to guarantee that a broad range of physical/logical implementation options are acceptable for deployment. An important feature of the architecture is the decoupling of access network and its support technologies from the IP connectivity network. This enables the unbundling of access infrastructure from the IP connectivity protocols. The network supports a variety of business models, as well as the coexistence of fixed, nomadic, portable, and mobile usage models [20,43].

5.3.5.2 Characterization of Major Logical and Physical Components of WiMAX NRM

As it was alluded to earlier, the basic WiMAX network architecture may be logically characterized by the NRM in the form of a block diagram, illustrated in Figure 5.14. This model identifies key functional entities of the network; that is, MS/AMS, ASN, and CSN, as well as reference points over which the network interoperability requirements are described.

The mapping of functions into physical devices within the functional entities is left to the equipment vendor to decide, provided that the implementation meets the functional and interoperability requirements. This leads to different profiles defined by the particular mapping of the functions to network elements. In one network architecture, the functions are mapped onto two network elements; the BS and the ASN-GW. Other functional mappings and network architectures are possible as well.

We now describe the major logical and physical components of WiMAX NRM in more details.

1) Mobile Station (MS), Advanced Mobile Station (AMS), and stationary Subscriber Station (SS) are user equipment sets, providing connectivity for a single host or multiple hosts to WiMAX network.

2) ASN collectively represents a complete set of functionalities required for providing radio access to the WiMAX network for MS/AMS/RS/SSs. Some key functions performed by ASN are as follows:

   i) Layer 2 (Data Link Layer) IEEE 802.16 standards connectivity in correspondence with WiMAX system profiles.

   ii) Network discovery and selection of the user's preferred NSP.

   iii) Transfer of user, device, and service credentials; that is, authentication, authorization, and accounting message to the selected NSP.

   iv) Relay functionality for establishing layer 3 connectivity with the MS/SS, that is, IP address allocation.

   v) Radio resource management (RRM) based on requested QoS from the NSP.

   vi) Mobility management within the ASN, such as handover and location management.

The ASN may be consisting of one or more WiMAX BS/ABS (Advanced Base Station) and single or multiple ASN-GW as shown in Figure 5.14.

3) CSN is a collection of network protocols and functions that deliver IP connectivity services to WiMAX users, that is, it plays the role of the backhaul network. The CSN may consist of several network components such as routers, AAA proxy/servers, home agents, user databases, Internetworking gateways, and so on. In short, The CSN provides connectivity to the Internet, ASPs, public networks, and corporate networks. The major network functions provided by CSN are outlined below.

   i) Managing the IP address allocation.

   ii) Providing authentication, authorization, and accounting services for the users through a server or a proxy server.

   iii) QoS management according to the user's contract. The CSN of the home NSP distributes the user's profile to the NAP, either directly or through the visited NSP.

   iv) Subscriber billing.

   v) Support for roaming between NSPs through inter-CSN tunneling.[20]

   vi) Inter-ASN mobility management and mobile IP home agent functionalities.

   vii) Connectivity to the Internet, IP multimedia services, location-based services, peer-to-peer services, and broadcast and multicast services

4) Legacy Base Station (BS) or Advanced Base Station (ABS) are logical and physical entities that are responsible for providing air interface to MS/AMS/SS/RS and cover the radio-related functions of the ASN. The interface is carried out according to MAC and PHY layer protocols of IEEE 802.16e (or IEEE 802.16j) subject to applicable interpretations and parameters in the WiMAX Forum system profile. Additional functions that may be part of the BS are handoff triggering and tunnel establishment, radio resource management, QoS policy enforcement, and traffic classification, session management, and multicast group management.

5) ASN-GW acts as a layer 2 traffic aggregation point within an ASN. ASN-GW is a basic WiMAX router. Additional functions that may be part of the ASN-GW include intra-ASN location management and paging, radio resource management and admission control, caching of subscriber profiles and encryption keys, AAA client functionality, establishment and management of mobility tunnel with base stations, QoS and policy enforcement, foreign agent functionality for mobile IP, and routing to the selected CSN [27].

---

20 Tunneling (encapsulation) refers to the ability of two-packet networks to exchange packets and data through an intermediate network without revealing the protocol details to the intermediate network [41].

5.3.5.3 Visual Depiction of WiMAX NRM

Figure 5.14 is an illustration of WiMAX NRM. Each of the major components shown in the figure represents a grouping of functionalities. These functionalities may be realized by a single physical entity, or they may be realized in a distributed fashion, that is, the functionality may be distributed over several physical and logical elements that as a whole implement the task of the functionality. Once again, the key objective of the NRM is to allow multiple implementation possibilities for each and every functionality and yet achieve interoperability among all these realizations [41].

A Home Network Service Provider (HNSP) is a business actor that has SLA with WiMAX subscribers to authenticate and authorize subscriber sessions and to provide accounting services such as charging and billing. Generally speaking, the subscribers may receive the WiMAX service from HASP or from a visited NSP (VNSP), with whom HASP has a roaming agreement.

5.3.5.4 The Description of WiMAX Reference Points

WiMAX Network Reference Model, graphically shown in Figure 5.14, contains several interoperability reference points. Reference points are conceptual links that connect two sets of operations that reside in different functional entities of the network main components, that is, ASN, CSN, and MS/AMS. Two protocols associated with an RP shall be able to originate and terminate in different functional entities, that is, a given protocol associated with an RP may not always terminate in the same functional entity. Reference points are not necessarily physical interfaces. They become physical connections when functional entities that they interface reside on different physical devices.

It is noted that reference points R1 through R5 are inter-ASN RPs and the rest are by and large intra-ASN reference points. The reference point R1 is the interface between MS/AMS and the ASN, or in fact a BS. The connection is provided either directly or through an IEEE 802.16j-based (or IEEE 802.16m-based) relay station.

Table 5.9 provides description for eight different RPs that are defined in WiMAX NRM, in accordance with WiMAX Forum "WiMAX System Profile Release 2.2" [41].

5.3.6 Mobility and Handover in WiMAX Networks

Mobility and mobility management are issues of utmost importance in any wireless mobile network. In general, mobility in cellular networks means the ability of a mobile user to move around a service area (home ASN), within a defined speed range and with a given access technology, or

Table 5.9 WiMAX major reference points descriptions [41].

RP	End points	Description
R1	AMS/MS/SS/ RS-ASN	IEEE 802.16e/802.16j air interface protocols implementation; PHY and MAC to support legacy MS and AMS to communicate with legacy BS and ABS respectively/ may include additional management plane protocols
R2	AMS/MS/SS/ RS-CSN	A nondirect-protocol logical interface for authentication, authorization, IP host configuration management, and mobility management
R3	ASN-CSN	Control protocols to support AAA, mobility management capabilities, policy enforcement, tunneling and other data plane methods to transfer IP data between ASN and CSN
R4	ASNGW- ASNGW	A set of control and data path protocols originating/ terminating in an ASN-GW that coordinates AMS mobility between Asns and ASN-GWs. It is the only interoperable RP between ASN-GWs of one or two different ASNs
R5	CSN-CSN	A collection of control and data path protocols enabling Internetworking between CSNs operated by HNSP and VNSP
R6	BS/ABS- ASNGW	A set of control and data path protocols for communication between BS/ABS and ASNGW within a single ASN. The data path consists of intra-ASN and inter-ASN tunnels between the BS/ABS and ASNGW. The control protocols include function for data path establishment, modification, and release control in accordance with MS/AMS mobility
R7		This RP was defined and included in Release 1.0; however, it was deprecated in Release 1.5 and beyond
R8	BS/ABS-BS/ ABS	A collection of intra-ASN control message flows between legacy BS and ABSs and between ABSs to ensure fast and seamless handovers. The data plane includes protocols that allow inter-BS/ABS communication protocols and data transfer. Additionally, protocols that control data transfer BS/ABSs involved in handover of a certain MA/AMS

cross into a different service area (visiting ASN) with the same or different access technology, without losing its connection to the destination application, while the network keeps track of exact whereabouts of the mobile station. The primary areas of interest in broadband cellular network mobility-related issues are location management and handover arrangements. A less important mobility-related matter, but yet critical for AeroMACS networks, is the maximum allowed vehicular speed in the mobile network. Our focus in this section will be on handover (HO) process. The HO process is broadly classified into *horizontal handover* and *vertical handover*. Horizontal handovers are the ones that occur between the cells of a homogeneous network, while vertical handovers take place in heterogeneous cases and between cells of different networks [44]. A short overview of horizontal handovers is presented below.

The simplest mobility management scenario is related to an MS that is in motion within the same service area; however, for the sake of service continuation, the MS needs to be "handed over" from one BS to another. *Handover* or *handoff* (HO) process is the ability of an MS, SS, or even a mobile RS to terminate communications with one BS and recommence it with a different BS without interruption in the ongoing session. Handover is a required and necessary feature in any cellular mobile network. The HO process takes place due to a variety of reasons. In majority of cases, handover occurs when an MS moves beyond the radio range of one BS and enters the coverage area of another, that is, when the signal strength from a neighboring cell exceeds the signal strength from the current cell. Under these circumstances, the MS is interfaced with, normally, a new pair of radio channels assigned by the network through the associated BS. Handover can also be initiated due to changes in the radio channel conditions that substantially weaken the signal received by the MS. In other cases, when a cell's traffic capacity is reached, the network originates a HO that connects the MS to a different BS whose traffic volume is not saturated.

The original IEEE 802.16 standard was designed for fix nodes and had no provisions for mobility and handover. The IEEE 802.16e and subsequent amendments stipulated enhancements in the PHY and MAC layers to support mobility at pedestrian and vehicular speeds. The addition of handover protocols in different versions of IEEE 802.16e presented a challenge in WiMAX technologies in terms of backward compatibility. In IEEE 802.16-2009, several enhancements for HO process were introduced [2,27]. Handover process presents a number of challenges. Primarily, HO has to be a fast process (on order of 50–150 ms). The other issue is that HO involves not only the PHY and MAC layers but also third layer (network layer), and as such it is dependent on the network architecture [23].

The IEEE 802.16 mobile standards and WiMAX technologies support three basic handover methods. *Hard handover* (HHO) is the simplest handover mode in which the MS disconnects its radio link with the serving BS (SBS); the prehandover BS, before establishing its radio connection with the target BS (TBS); the posthandover BS. The flow of traffic is interrupted while the handover process is in progress; for this reason this type of handover is also referred to as *break before make* technique. HHO can be initiated by the MS by means of a particular MAC management message (MOB_MSHO-REQ), or by the BS using another MAC management message (MOB_BSHO-REQ). The HO process is completed when the TBS receives another MAC message from the MS (MS_HO-IND) [27].

Two other modes of HO, generally classified as *soft handovers* (SHO), are macrodiversity handover (MDHO) and fast base station switching (FBSS); they are also known as *make before break* techniques. In other words, the MS start communication with TBS(s) before terminating its connection with SBS, so that there is no interruption in the ongoing communications. In both of these methods, multiple handover candidate TBSs are registered in a list called diversity set. In MDHO, the MS may communicate with one or more candidate TBSs in the diversity list simultaneously, whereas in FBSS the MS is in communication with a single TBS drawn from the diversity set. A key difference between HHO and SHO techniques is that in SHO methods the MS indicates its TBS selection through a fast feedback channel as opposed to sending a MAC management message. Consequently, the SHO techniques are faster, and the MS might be communicating with both SBS and TBS simultaneously before it completely breaks up with the SBS, therefore the title of soft handover. Although SHO techniques impose lower latencies, they require backhaul channels and more overhead than HHO. Currently, in WiMAX and AeroMACS networks, HHO is mandatory and MDHO and FBSS are optional.

5.3.7 Multicast and Broadcast with WiMAX

Multicast and broadcast services (MCBCS), as opposed to point-to-point unicast features, refer to downlink point-to-multipoint transmission of multimedia packets from WiMAX BSs to multiple receiving terminals in the designated MCBCS area. In Mobile WiMAX, MCBCS is an optional network capability introduced in WiMAX Forum System Profile Release 1.0, and is supported in subsequent Releases [27].

WiMAX end-to-end IP network architecture and associated technologies, such as OFDMA and flexible subchannelization, smart antenna and MIMO system, AMC, and so on, enable the delivery of multimedia content locally, regionally, or nationally. This delivery takes place across a large Internetworking system whose network components may not share the same cell sizes, mobility protocols, frequency reuse setups, and so on.

The broadcast services are normally offered to all users that are authorized to access the WiMAX network as an "added value"; however, the multicast services are typically provided on the basis of user subscription [27].

The basic operations that provides MCBCS over WiMAX networks are outlined below.

1) ASPs generate the contents for MCBCS and deliver them to MCBCS controller that is assembled within the NSP domain.
2) The MCBCS controller transmits the contents through "MCBCS channels" to a multicast enabled router that passes on the contents to the core network within the NSP domain (see Figure 5.14).
3) The MCBCS contents, through another sets of multicast enabled routers, are delivered to geographical areas known as MCBCS zones that may include a number of "multicast-broadcast service (MBS) zones."

IEEE 802.16-2009, which is the parent standard of AeroMACS, defines MBS zones that may include several WiMAX base stations [27].

5.4 Summary

Mobile WiMAX standards have enabled a scalable integrated fixed and mobile broadband access technology through a common architecture. The technology is based on IEEE 802.16 standards.

In this chapter, a brief overview of IEEE 802.16 standards and WiMAX technology, as the parent technologies of AeroMACS, is presented. Key PHY layer and MAC layer features of WiMAX that are of significant importance in AeroMACS systems are emphasized.

The mobile WiMAX networks air interface is assembled based on fading resistant OFDMA multiple access technology. The application of smart antenna system and MIMO concept have defined a new frontier in pushing the envelope in research and development of future advanced high-speed, high-capacity, reliable wireless networks. After a brief review of MIMO concept, we have discussed the simplest method STC scheme, known as Alamouti code. More sophisticated space-time coding schemes are discussed in a wealth of references published since the proposal of Alamouti code, such as Ref. [35]. Other advanced smart and multi-antenna system such as intelligent beamforming, spatial multiplexing, and so on, which also form key building blocks of present and future WiMAX networks, are discussed.

A high point of WiMAX technology is the fact that only PHY and MAC layer protocols have been specified, and the higher layer protocols and the core network architecture are left undefined to be filled by other available technologies. To this end, IP network protocols that are well known for their simplicity have been accepted to define an end-to-end architecture for WiMAX networks [45]. In short, OFDMA, MIMO, and IP

architecture form the backbone of WiMAX technology, which is inherited by AeroMACS networks as well. The typical mobile WiMAX cell
radius is in the range of 1–3 miles for the C-band spectrum of 2–6 GHz,
which is well suited for AeroMACS applications.

References

1 IEEE Standard for WirelessMAN-Advanced Air Interface for Broadband
   Wireless Access Systems, September 2012.
2 IEEE 802.16–2009 Standard for Local and Metropolitan Area, Part 16:
   Air Interface for Broadband Wireless Access Systems, May 2009.
3 C. Eklund, R. B. Marks, K. L. Stanwood, and S. Wang, "IEEE Standard
   802.16: A Technical Overview of the WirelessMAN™ Air Interface for
   Broadband Wireless Access," *IEEE Communications Magazine*, June
   2002.
4 L. Nuayami, *WiMAX Technology for Broadband Wireless Access*, John
   Wiley & Sons, Inc., New York, 2007.
5 IEEE 802.16–2001, "IEEE Standard for Local and Metropolitan Area
   Networks – Part 16: Air Interface for Fixed Broadband Wireless Access
   Systems," April 2002.
6 IEEE P802.16a/D3-2001: "Draft Amendment to IEEE Standard for Local
   and Metropolitan Area Networks – Part 16: Air Interface for Fixed
   Broadband Wireless Access Systems – Medium Access Control
   Modifications and Additional Physical Layers Specifications for 2–11
   GHz," March 2002.
7 IEEE 802.16's Relay Task Group, IEEE 802.16j-09, June 2009. Available
   online at http://www.ieee802.org/16/relay/.
8 V. Genc, S. Murphy, Y. Yu, and J. Murphy, "IEEE 802.16j Relay-Based
   Wireless Access Network: An Overview," *IEEE Wireless
   Communications*, 15(5), 56–63, 2008.
9 WiMAX Forum, "Fixed, Nomadic, Portable and Mobile Applications for
   802.16–2004 and 802.16e WiMAX Networks, November 2005.
10 WiMAX Forum, "Mobile WiMAX – Part I: A Technical Overview and
   Performance Evaluation," August 2006.
11 Y-J Choi, C. S. Kim, and S. Bahk "Flexible design of Frequency Reuse
   Factor in OFDMA Cellular Networks," *Proceedings of 2006 IEEE ICC*,
   pp. 1784–1788, June 2006.
12 S. Ahmadi,"An Overview of Next-Generation Mobile WiMAX
   Technology," *IEEE Communications Magazine*, pp. 84–98, June 2009.
13 WiMAX Forum®, "Mobile System Profile," Release 1.0, Version 09,
   2009.

14 WiMAX Forum®, "Mobile System Profile," Release 1.5, Common Part, Version 1, 2009.

15 WiMAX Forum®, "Mobile System Profile," Release 1.5, FDD Specific, Version 1, 2009.

16 WiMAX Forum®, "Mobile System Profile," Release 1.5, TDD Specific, Version 1, 2009.

17 WiMAX Forum®, "Mobile System Profile," Release 2.0, Air Interface Specifications, 2012.

18 WiMAX Forum®, "Mobile System Profile," Release 2.1, Air Interface Specifications, 2013.

19 WiMAX Forum®, "Mobile System Profile," Release 2.2, Air Interface Specifications, 2014.

20 J. G. Andrews, A. Ghosh, and R. Muhamad, *Fundamentals of WiMAX Understanding Broadband Wireless Networking*, Prentice Hall, 2007.

21 M. Maqbool, M. Coupechoux, and P. Godlewski,"Subcarrier Permutation Types in IEEE 802.16e," Department of Information TELECOM Paris Tech, April 12, 2008.

22 A. F. Molisch, *Wireless Communications*, 2nd edn, John Wiley & Sons, Inc., New York, 2011.

23 L. Nuaymi, *WiMAX Technology for Broadband Wireless Access*, John Wiley & Sons, Inc., New York, 2007.

24 B. Kamali, R. A. Bennett, D. C. Cox, "Understanding WiMAX: An IEEE-802.16 Standard-Based Wireless Technology," *IEEE Potentials*, September/October, 2012.

25 S. Lin and D. J. Costello, *Error Control Coding*, 2nd edn, Prentice Hall, 2004.

26 R. Prasad and F. J. Velez, *WiMAX Networks Techno-Economic Vision and Challenges*, Springer, 2010.

27 K. Etemad and M-Y Lai (eds.), *WiMAX Technology and Network Evolution*, Wiley IEEE Press, 2010.

28 L. Zheng and D. N. C. Tse, "Diversity and Multiplexing: A Fundamental Tradeoff in Multiple Antenna Channels," *IEEE Transactions on Information Theory*, 49(5), 1073–1096, 2003.

29 V. Tarokh, N. Seshadri, and A.R. Calderbank, "Space-Time Codes for High Data Rates Wireless Communications: Performance Criterion and Code Construction," *IEEE Transactions on Information Theory*, 44, 744–765, 1998.

30 V. Tarokh, H. Jafarkhani, and A.R. Calderbank, "Space-Time Block Coding from Orthogonal Designs," *IEEE Transactions on Information Theory*, 45, 1456–1467, 1999.

31 V. Tarokh, H. Jafarkhani, and A.R. Calderbank, "Space-Time Block Coding for Wireless Communications: Performance Results," *IEEE Journal on Select Areas in Communications*, 17, 451–460, 1999.

32 S. M. Alamouti, "A Simple Transmit Diversity Technique for Wireless Communications," *IEEE Journal on Select Areas in Communications*, 16(8), 1451–1458, 1998.

33 D. Gesbert, M. Shafi, D. S. Shiu, P. J. Smith, and A. Naguib, "From Theory to Practice: An Overview of MIMO Space-Time Coded Wireless Systems," *IEEE Journal on Select Areas in Communications*, 21(3), 281–302, 2003.

34 S. N. Diggavi, N. Al-Dhahir, A. Stamoulis, and A. R. Calderbank, "Great Expectation: The Value of Spatial Diversity in Wireless Networks," *The Proceedings of the IEEE*, 92(2), 219–270, 2004.

35 H. Jafarkhani, *Space-Time Coding: Theory and Practice*, Cambridge University Press, 2005.

36 E. G. Larsson and P. Stocia, *Space-Time Block Coding for Wireless Communications*, Cambridge University Press, 2003.

37 G. Boudreau, et al., "Interference Coordination and Cancellation for 4G Networks," *IEEE Communications Magazine*, 47(4), 74–81, 2009.

38 T. D. Novlan, R. K. Ganti, A. Gosh, and J. G. Andrews, "Analytical Evaluation of Fractional Frequency Reuse for OFDMA Cellular Networks," *IEEE Transactions on Wireless Communications*, 10(12), 4294–4305, 2011.

39 J. Li, X. Wu, and R. Laroia, *OFDMA Mobile Broadband Communications: A System Approach*, Cambridge University Press, 2013.

40 ITU, "World Mobile Telecommunication Market Forecast," ITU-R Report M.2072, 2005.

41 WiMAX Forum®, "Network Architecture: Architecture Tenets, Reference Model and Reference Points Base Specification," Release 2.2, 2014.

42 M. Cudak,"IEEE 802.16m System Requirements," IEEE Task Group m, submitted January 2010.

43 WiMAX Forum®, "WiMAX End-to-End Network System Architecture. Stage 2: Architecture Tenets, Reference Model, and Reference Points," Release 1.0 V & V Draft, August 8, 2006.

44 S. K. Ray, K. Pawlikowski, and H. Sirisena, "Handover in Mobile WiMAX Networks: The State of Art and Research Issues," *IEEE Communications Surveys & Tutorials*, 12(3), 376–399, 2010.

45 Bo Li et al., "A Survey on Mobile WiMAX", *IEEE Communications Magazine*, December 2007.

6

Introduction to AeroMACS

In the late 1980s, due to rapid growth in both commercial and general sectors of civil aviation, spectrum congestion in aeronautical VHF band became a concern for the aviation community in the United States and in Europe. In the mid-1990s, the Europeans implemented the 8.33 AM scheme within areas of high congestion, as a short-term resolution for the anticipated spectrum depletion problem [1]. Traditional airport surface communication systems consist of, primarily, fiber optic and copper cable loops buried under the airport surface. In occasions where wireless safety-related information exchange with mobile airport assets is required, VHF 25 kHz voice and data channels are used, that is, the same radio channels that are used for A/G and G/A communications (see Chapter 1).

In addition to general concerns about inability of the legacy system to safely manage future levels of air traffic, the modernization of air transportation system (envisioned under the JPDO *Nextgen* plan in the United States, and SESAR program in Europe) for the first quarter of the twenty-first century, demanded transition from analog system to high-speed digital network-centric technologies. The EUROCONTROL-FAA Future Communication Study (FCS) has determined that no single communication technology can satisfy all physical, operational, and functional requirements of various aeronautical transmission domains such as enroute, airport surface, ramp area, and oceanic [2].

6.1 The Origins of the AeroMACS Concept

In order to accommodate the large volume of data that is planned by *Nextgen* for information exchange between different mobile units, such as aircraft, and other entities, including airport air traffic management (ATM) and terminals, airline operational control (AOC), national airspace system (NAS), and so on, wideband wireless communication

AeroMACS: An IEEE 802.16 Standard-Based Technology for the Next Generation of Air Transportation Systems, First Edition. Behnam Kamali.

technologies are required. To this end, the 2007 ITU World Radio-communication Conference (WRC-2007) took an action that allowed the development of a new international standard of aeronautical mobile route services (AM(R)S) to support airport surface communications in the MLS (microwave landing system) extension band. The authorized C-band is 5091–5150 MHz and an additional band 5000–5030 MHz may become available in the future. In order to minimize the network development cost, and to take advantage of continual improvements, a standard-based communications technology was preferred over costlier alternatives of proprietary systems, which normally reach obsolescence before the system is fully deployed [2]. It was then recommended that this new aviation specific communication technology should be based on the IEEE 802.16e standard [3]. The proposed standard supports fixed and mobile ground to ground data communications applications and services.

The 802.16e-based system and its interoperable version, WiMAX technology, as we explored them in the previous chapter, are multinode wireless networks that support both fixed and mobile stations. Furthermore, the network is scalable in the sense that it can accommodate small, medium, and large airports by merely incorporating additional cells and/or relays, as stipulated by IEEE 802.16 standards and WiMAX technology. This is all possible since WiMAX technology is not only based on cellular architecture, but also can be fortified with multihop relays as specified by IEEE 802.16j amendment (see Chapter 8). Consequently, all areas of an airport surface and, if necessary, areas beyond airport property in which some airport assets (lighting, navigational aids, weather sensors, wake vortex sensor, etc.) are located, can be covered. This network-centric wireless system is capable of accommodating all mobile communications requirements for aircraft and various ground vehicles, as well as enabling links to fixed airport assets. Aircraft that are parked at a gate or taxiing would also be able to access the system, as long as they are not in flight [2]. As we have mentioned in earlier chapters, this WiMAX-based network is referred to as the aeronautical mobile airport communications system (AeroMACS).

For existing airports, AeroMACS may be initially interconnected with the airport cable loop system to create a hybrid network. For new airports with no cable infrastructure, AeroMACS can be installed readily and at far lower cost than the cable alternative.

6.1.1 WiMAX Salient Features and the Genealogy of AeroMACS

The IEEE 802.16e-based WiMAX standard was selected for AeroMACS on account of several major features and capabilities that it affords. The IEEE 802.16-2009-based WiMAX technology supports vehicular speed up to 120 km/h, which is sufficient to accommodate fixed airport nodes, airport surface vehicles, parked and taxing aircrafts, but may not be high enough to match aircraft landing and takeoff speeds. However, as it was discussed in Chapter 5, the new versions of IEEE 802.16e standard, such as IEEE 802.16m amendment, allow speeds of up to 350 km/h for the mobile unit without loss of connectivity. This is a direct benefit of selecting a standard-based technology for AeroMACS, as the parent standard evolves so do the descendant technologies.

A major desirable feature of WiMAX standard is its access technology, OFDMA, which is a robust technique against multipath propagation in airport radio channel, and works well in NLOS environments such as airport surface area, with potentially many large aircraft in motion simultaneously. WiMAX is capable of transmitting up to 50 Mbits/s of data over cells of radius 1–3 miles over the C-band range of 2–6 GHz, which covers the allocated AeroMACS spectrum. With regard to new frontier technologies for performance and throughput enhancements in wireless networks, that is, the smart antenna systems (covered in Chapter 5), WiMAX supports a full range of MIMO and related technologies such as spatial multiplexing, beamforming, and space-time codes. Adaptive modulation and coding (AMC), applying modulation schemes QPSK, 16-QAM, and 64-QAM, and error control coding techniques such as convolutional and convolutional turbo codes with variable code rates, LDPC, and Reed–Solomon codes, as well as decoding algorithms such as Viterbi decoding, iterative decoding, Berlekamp–Massey decoding, are all supported by WiMAX technology. Variable code rates, in convolutional type coding formats, allows for adaptation to link conditions, and therefore enables maintaining of an acceptable level of QoS. Hard and soft handover, including macrodiversity handover (MDHO) and fast base station switching (FBSS), are provided as mandatory and optional choices, respectively. Necessary data and network security is provided through device/user authentication, flexible key management protocol, traffic encryption, control and management plane message protection, and security protocol optimizations for fast handovers [2]. Other salient features of WiMAX include the possibility of deployment of an end-to-end IP networking. Insofar as security procedures are concerned, IEEE 802.16-2009 standard and corresponding WiMAX technology support mechanisms for authentication, authorization, encryption, and fast handover. Quality of Service in WiMAX that is passed on to AeroMACS (see Chapter 7) is defined based on throughput rate, time delay and jitter, scheduling, packet error rate deletion, and resource management [4]. Last but not the least, AeroMACS networks are potentially capable of deployment of multihop relays as a cost-effective method of extending the radio outreach of a WiMAX cell, performance improvement, or both (see Chapter 8).

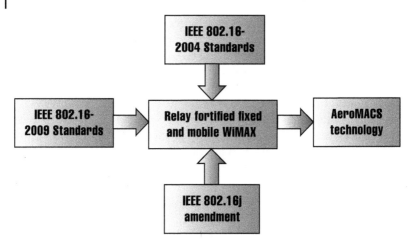

Figure 6.1 Genealogy of AeroMACS technology.

The idea of standard-based telecommunications technology development was conceived in the last quarter of the twentieth century. This chapter attempts to present a complete picture of the overall process of developing a new technology based on an already established standard; in this case IEEE 802.16 standards. As we have discussed earlier, the parent technology of AeroMACS is WiMAX that is composed of a selected subset of IEEE 802.16 standard (specifically IEEE 802.16-2009) design options. Throughout this text our assumption is that the IEEE 802.16j-based multihop relay systems (or some other multihop relay system) will be incorporated into AeroMACS networks. This is justified by a number of reasons that are discussed in Chapter 8. Thus, the IEEE 802.16j amendment is also viewed as part of AeroMACS Standard genealogy. Since AeroMACS systems accommodate fixed nodes such as sensors and airport security cameras, the IEEE 802.16-2004, the standard for fixed broadband access and the parent standard of "fixed WiMAX," has a part to play in development of AeroMACS networks, although IEEE 802. 16-2009 supports both fixed and mobile nodes and in fact has superseded IEEE 802.16-2004. The block diagram in Figure 6.1 portrays the creation of AeroMACS networks from its parent standards and technologies.

6.2 Defining Documents in the Making of AeroMACS Technology

The creation of AeroMACS (sometimes referred to as airport surface data link) systems, from a concern about traffic overflow in large airports and eminent depletion of aeronautical VHF spectrum and a need for modernization, to a concept, to technology identification, to standard development and prototype design, to test and validation, and finally to deployment in several major U.S. airports, was a long process that spanned over almost two decades. In this section and the next, a brief overview of AeroMACS concept development, from a historical perspective, is presented. Subsequently, a short synopsis of few documents that have played important roles in the formation of this modern aviation system is deliberated. Without a doubt, there are many other articles, research accounts, test result reports, and other documents that have made important contributions to the development of AeroMACS systems, unfortunately they cannot all be listed and reviewed in one or two short sections.

Spectrum congestion in aeronautical VHF band for civil aviation became a concern for the aviation community in the early 1990s. It was mentioned in Chapter 1 that the demand for aeronautical radio links surpassed what the existing VHF radio channels could supply, without unacceptable level of interference, particularly in the United States, Europe, and parts of East Asia, in the late 1990. The spectrum depletion was more intense in Europe that eventually lead to the introduction of VHF 8.33 AM air-to-ground radio in the European airspace [1], a standard that was never adopted in the United States. Despite the establishment of this new channelization, it was predicted that the VHF spectrum would be depleted again within a few years due to estimated air traffic growth.

The need to investigate new communication technologies operating outside of VHF aeronautical band, in order to increase capacity and to support new applications, functionalities, and services, was recognized at the 11th ICAO Air Navigation Conference (ANC-11) held in Montréal, Canada, in 2003. ANC-11 essentially defined the future path for modernization of aeronautical communications infrastructure through making a number of recommendations [5].

1) *Recommendation 7/3:* Evolutionary approach for global interoperability of A/G communications. This recommendation promotes the continuing use of already implemented systems, the optimization of the available spectrum utilization, and the consideration of transition aspects.

2) *Recommendation 7/4:* Investigation of future technology alternatives for A/G communications. This recommendation addresses the need for investigations to identify the technology candidates to support the future aeronautical communications.

3) *Recommendation 7/5:* Standardization of aeronautical communication systems. Finally, this recommendation emphasizes the need for standardization activities for technically proven technologies which provide proven operational benefits [5,6].

These recommendations established the goals for what later became known as FCS.

At ANC-11 a strong request was expressed by the airlines industry for international cooperation and harmonization in order to achieve the stated objectives in the recommendations, particularly insofar as A/G communications is concerned. These recommendations call for investigation of new terrestrial and satellite-based technologies, on the basis of their potential for ICAO standardization for aeronautical mobile communications use, taking into account the safety-critical standards of aviation and the associated cost issues [5]. Furthermore, an assessment for needs for additional aeronautical spectrum to meet requirements for increased communications capacity and new applications, and assist various countries in securing appropriate additional allocations by the ITU, was deemed necessary. ANC-11 was also successful in requesting that the ITU WRC 2003 (WRC-3) add an agenda item regarding additional radio band allocation at WRC 2007 (WRC-07).

Subsequent to ANC-11 recommendations, EUROCONTROL and FAA (Federal Aviation Administration) established a dedicated working arrangement in accordance with what is called "Action Plan 17 of the EUROCONTROL-FAA Memorandum of Cooperation" (AP17) to move forward in the direction of implementation of this project in a consistent manner in United States and in Europe. ICAO ACP (Aeronautical Communications Panel) was closely coordinating AP17 to achieve worldwide consensus and global harmonization [6]. Under the U.S. AP17 activities, NASA Glenn Research Center (GRC) was asked by the FAA to lead the U.S. technology investigations. In Europe; France, Germany, Spain, Sweden, Austria, and the United Kingdom were actively supporting and contributing to the European investigations. The work undertaken under AP17 was supported by MITRE and ITT (as a contractor for NASA Glenn) Corporations in the United States and by QinetiQ in Europe. The AP17 activities were coordinated with the multiagency Joint Planning and Development Office NextGen initiative. In Europe, the work was coordinated with the EUROCONTROL/European SESAR Program [6].

The actual AeroMACS technology development process commenced through the cooperative Eurocotrol–FAA research project of "Future Communications Study." This was a technology assessment project focused on description and identification of possible future communications technologies for aeronautics. NASA GRC actively participated in this project on behalf of the FAA. The initial work began in 2004 when NASA GRC and ITT, in close cooperation with Eurocontrol and QinetiQ, commenced the process of FCS technology assessment, evaluating the technical suitability and operational capabilities of over 60 different technologies. One defining principle that was observed throughout the

FCS technology assessment and selection was the maximization of the use of existing technologies and standards. The aviation Future Communication Infrastructure (FCI) was divided into the following three flight domains (as opposed to the conventional eight aeronautical domains described in Chapter 1) [7].

1) *Continental Domain:* Consisting of enroute domain that is within the LOS of terrestrial ATC (air traffic control) communications facilities.
2) *Oceanic and Remote Domain:* Enroute domains outside of LOS of terrestrial ACT communications facilities such as oceanic and polar domains.
3) *Airport Domain:* Comprised of conventional "landing and takeoff," "surface and taxiing," and "parked" domains.

The final results of FCS were published in 2007 under "Action Plan 17 Future Communications Study: Final Conclusions and Recommendations" [8]. This report is considered to be one of the foundational documents in the development of AeroMACS systems. An important finding of this study is the affirmation of the fact that no single communications technology can adequately provide connection between aircraft, airports, and air traffic management system in all aeronautical domains. This is due to operational modes and physical channel characteristics in different flight domains. Therefore, a combination of several wireless technologies is needed for accommodation of communications requirements of different flight domains. Among major achievements of this study, that directly support the objectives of NextGen and SEASAR, are the establishment of future aeronautical communications operating concept and requirements (COCR), the identification of enabling technologies for various flight domains, and the development of a roadmap for the FCI that covers the transitions from 2007 through 2030 [6,8].

The document makes a paradigm shifting recommendation in regard to utilization of aeronautical C-band for the airport domain. It recommends an IEEE 802.16e Standard-based system as the solution for the provision of dedicated aeronautical communication services on the airport surface. The IEEE 802.16e-based technologies, such as WiMAX, are designed for short range, high data rate, multimedia communications, which matches the requirements of airport surface communications in terms of capacity and performance. Specifically, the C-band recommendation calls for identification of portions of the IEEE 802.16e standard mandatory and optional protocols that best suit a wireless mobile communications system for airport surface. The report also requests that performance evaluation and validation for this new aviation-specific standard should be conducted through trials and testbed development. Finally, the report asks for identification of a channelization methodology setup for allocation of safety and regularity of flight services to accommodate a range of airport classes, configurations, and operational requirements [8].

In the same year, the World Radiocommunications Conference (WRC-07) approved the allocation of the 5091–5150 MHz band for the Aeronautical Mobile (Route) Service (AM(R)S), setting into motion the process of creation of aeronautical safety communication on the airport surface. Subsequently, ICAO was tasked to develop the standards and recommended practices (SARPS) for AeroMACS as a broadband wireless communication service operating in the protected aeronautical band of 5091–5150 MHz to allow ground safety operation and ATM on the airport surface.

There are a number of reasons for selection of IEEE 802.16-2009-based WiMAX technology for airport surface domain, some of which have been previously articulated in this chapter. Table 6.1, whose idea is adopted from Refs [4] and [7], highlights key features of the IEEE 802.16 standard that are particularly attractive for airport communications application.

A document was published by EUROCONTROL in 2009, analyzing WiMAX Forum's System Profile Release 1 (Version 1.9.0). This system profile is based on IEEE 802.16-2009. The document stipulates suggestions for possible modifications, as well as some additions to individual mandatory or optional protocol items, to suit airport surface communication requirements. In addition, this document comments on other radio characteristics that are addressed neither in the system profile nor in the certification profile. The final conclusion of this investigation states that EUROCONTROL believes that no, or minimum, modifications and additions are needed to the existing certification profile. However, since AeroMACS will be operating over a different band of frequencies, specifically over 5091–5150 MHz, a new system certification profile needs to be developed [9].

In February of 2010, EOROCONTROL published a subsequent document titled "AeroMACS System Requirement Document" [10], which was revised in July 2010. In this document EUROCONTROL looks again into WiMAX Forum Mobile System Profile Release 1, that is, Mobile WiMAX based on IEEE 802.16-2009 standard as a communication technology for the stipulation of dedicated services required by the airport surface link. EUROCONTROL has only analyzed Revision 1.4.0 of Release 1 for the reason that TDD format is envisioned in this revision. This document, once again, finds no technical obstacle that would make it impossible to apply this technology for AeroMACS application. The document also reports that not all aspects of this profile, although acceptable, are optimum for AeroMACS application, particularly insofar as system throughput is concerned. The document, however, states that these concerns may be addressed when WiMAX Forum System Profile Release 2 is made available.

Table 6.1 Suitable features of IEEE 802.16-2009 Standards for AeroMACS application.

IEEE 802.16-2009 feature	Notes on suitability and deficiency
Mobility	Supports speeds up to 120 km/h; sufficient for aircraft taxiing and emergency vehicles, but below takeoff and landing speed of large jets
Range of coverage	LOS coverage of up to 10 km, sufficient to cover small, medium, and most of large airports, relay-fortified (IEEE 802.16j) network can extend the radio outreach to provide coverage to long and very large airports
Quality of service (QoS)	QoS based on throughput, packet error rate deletion, scheduling, time delay, time jitter, and so on are facilitated
Scalability	Scalability in the sense that S-OFDMA is adaptable to various system bandwidths, and scalability in the sense of adaptability of the underlying cellular structure to airports of various sizes
Open sourced	Only PHY and MAC layer protocols are specified, the higher level protocols are open to adopt, end-to-end IP networking is feasible
Security	AAA mechanisms are provided, encryption protocols are available, digital certification and fast handovers are supported
Obstruction tolerance	Multipath propagation is taken advantage of, facilitating NLOS communications
Cost efficiency	A global commercial standard, taking advantage of *market scale factor*, by using "COTS" products that are manufactured for a much larger consumer market than just that of aviation. Enables airlines and airports to have access to multiple vendors for competitive pricing
Ease of standardization	The base standard is established by the IEEE, one of the world's most qualified standardization institutions; rapid standardization process and timely deployment of the system is expected
Certification	Certification agencies, such as WiMAX Forum, are already in place to define procedure that can facilitate interoperability

6.3 AeroMACS Standardization

A multitude of institutions and agencies are involved in AeroMACS standardization process. FAA, EUROCONTROL, and ICAO (through ACP WGS-Surface) are leading the international effort in AeroMACS standardization and harmonization. In Europe, the Working Group 82 (WG-82) of EUROCAE (European Organization for Civil Aviation

Equipment), a regulatory agency for certifying aviation equipment, is one of the agencies that is contributing to AeroMACS standardization. The Systems Architecture and Interfaces SC (Special Committee) of ARINC AEEC (Airlines Electronic Engineering Committee) has been formed for the purpose of assisting in development of AeroMACS standards. In the United States, RTCA[1] (Radio Technical Commission for Aeronautics) established a special committee SC-223 (Internet Protocol Suite (IPS) and AeroMACS) on August 7, 2009, charged with standard development for airport surface wireless communication systems (AeroMACS) based on WiMAX Forum commercial profiles for WiMAX systems. NASA GRC and ITT were closely cooperating with RTCA SC-223 in their standardization efforts. The standards developed by RTCA and EUROCAE are required to support data communication developments for collaborative decision-making (CDM), surveillance broadcast system (SBS), and system wide information management (SWIM), as well as weather and flight information systems (FIS) activities on the airport surface.

The standardization process leads to development of AeroMACS system certification profile that enables aviation industry, as well as other stakeholders, such as test equipment vendors, integrated circuit manufactures, and system developers to support the development of AeroMACS system, but more importantly, ensures interoperability in the global deployments of AeroMACS networks. The AeroMACS profile is also used as a basis for development of Minimum Operational Performance Standards (MOPS) for AeroMACS avionics

6.3.1 AeroMACS Standards and Recommended Practices (SARPS)

Based on decisions made in WRC-07, an ICAO subgroup of AeroMACS Working Group, Working Group S (WGS), at their first meeting provided a document draft and invited AeroMACS Working Group to consider using the document as a basis for the standards and recommended practices (SARPS) for AeroMACS as a broadband wireless communications service for airport surface [11]. WGS, in its second meeting, composed a refined version of the previous document and submitted it as the consensus of the group "to be an operational concept to be used as a tool for the identification of possible requirements to be reflected in

---

1 "RTCA Incorporated is a not-for-profit corporation formed to advance the art and science of aviation and aviation electronic systems for the benefit of the public. The organization functions as a Federal Advisory committee and develops consensus-based recommendations on contemporary aviation issues." From RTCA website.

the AeroMACS SARPS" [12]. The process of modifying and refining the AeroMACS SARPS continued by ICAO WGS and related documents were published subsequently.

A recent version of AeroMACS SARPS draft published by ICAO WGS is titled *Proposed Aeronautical Mobile Airport Communications System (AeroMACS) SARPS* [13]. In this draft, ICAO provides a detailed account of AeroMACS system as outlined in the following lines.

- *General Requirements:* It includes items such as "AeroMACS shall be the standard that operates exclusively when the mobile node is on the surface of an aerodrome", and "messages are processed according to some defined associated priority".
- *Radio Frequency Characteristics:* TDD is the duplexing mode for AeroMACS, the channel bandwidth for AeroMACS is 5 MHz, operating frequency band shall be 5030–5150 MHz.
- *Performance Requirements:* It includes issues such as outage probability, connection resilience, performance requirements for MSs and BSs, security and data integrity, and so forth.
- *System Interfaces:* AeroMACS shall support notification of the communications status and the loss of communications.
- *Application Requirement:* AeroMACS shall support multiple classes of services for provision of appropriate services levels, and in case of resource contention, higher priority services will prevail over lower priority services.
- *Power Restrictions and Minimum Receiver Sensitivity (MRS) Example:* The MS antenna EIRP[2] is upper bounded by 30 dBm ($\leq$ 30 dBm) and BS antenna EIRP is restricted to $\leq$ 39.4 dBm. The minimum receiver sensitivity for MS and BS for various modulation coding combinations are shown in Table 6.2 [4,13].

Table 6.2 provides an example of AeroMACS receiver sensitivity values, assuming various combination of modulation schemes and convolutional coding (CC) of rates shown in the table, for AWGN channel and defined for BER equal to 10^{-6} [4].

Standards and Recommended Practices are intended to assist various countries in managing aviation safety risks in coordination with their service providers. In view of the fact that global air transportation system grows more complex as time progresses, the safety management

---

2 EIRP: Effective isotropic radiated power is a measure of antenna radiation directivity. EIRP is equal to the amount of power an isotropic antenna needs to expend in order to radiate the same power density in all directions as the directive antenna radiates in the direction of its maximum gain.

Table 6.2 AeroMACS minimum receiver sensitivity (MRS) [4,13].

Modulation scheme	Convolutional code rate	Repetition factor DL	MS MRS (dBm)	BS MRS (dBm)
64-QAM	3/4	1	−74.4	−74.5
64-QAM	2/3	1	−76.4	−76.5
16-QAM	3/4	1	−80.4	−80.5
16-QAM	1/2	1	−83.9	−84.0
QPSK	3/4	1	−86.4	−86.5
QPSK	1/2	1	−89.5	−89.5
QPSK	1/2	2	−92.4	−92.5

provisions should support a continual evolutionary and proactive strategy. Based on recommendations in a draft provided by ACP WGS [14], ICAO approved the final version of AeroMACS SARPS on March 2, 2016, it became effective on July 11, 2016, and the applicable date will be November 7, 2019. Countries are expected to begin working on implementation process from the effective date, in anticipation of achieving completion of the new provisions by the applicability date [15,16].

6.3.2 Harmonization Document

A document that is relevant to international technical cooperation, particularly in regard to achieving the necessary level of interoperability between the NextGen and the SESAR visions, was agreed upon between the United States and the European Union, and was published under the title of "State of Harmonization Document" in late 2014 [17]. A key objective of this collaboration is to harmonize and secure air traffic management modernization efforts in support of the ICAO's Global Air Navigation Plan (GANP) with the Aviation System Block Upgrade[3] (ASBU) program. "Both NextGen and SESAR recognize the need to integrate the air and ground parts of their air traffic management systems by addressing efficiency needs of flight trajectories planning and execution and the seamless sharing of accurate information. The US–EU joint harmonization work will assure that modernization and advances in air

---

3 ASBU is the framework through which ICAO GANP conducts harmonization of avionics capabilities and the required air traffic management ground infrastructure.

navigation systems worldwide can be made in a way that supports cooperation, clear communication, seamless operations, and optimally safe practices. The collaborative harmonization work between the United States and the European Union has taken place under the Memorandum of Cooperation (MOC) between the United States of America and the European Union on Civil Aviation Research and Development that was signed in March 2011, and specifically under Annex I, which covers 'SESAR-NextGen Cooperation for Global Interoperability'" [17].

This harmonization agreement directly supports ICAO standardization efforts for AeroMACS. Under this umbrella cooperation pact, the United States and the European Union are also supporting and engaging in initiatives fostered by other international standardization bodies, such as RTCA and EUROCAE. The major areas of collaboration under this agreement are transversal activities; information management; trajectory management; communication, navigation, and surveillance (CNS); and airborne interoperability [15].

6.3.3 Overview of Most Recent AeroMACS Profile

The WiMAX Forum®, as was mentioned in Chapter 5, is an industry consortium charged with development of technical specifications, in the form of system and certification profiles, for underlying of the Forum's certified products. The IEEE 802.16-2009 standard provides the minimum requirements for supporting global interoperability. The AeroMACS profile should consist of a selected subset of IEEE 802.16-2009 (i.e., WiMAX) standard mandatory and optional protocols and features, suitable for airport surface communications application. In August 2010, RTCA SC-223 and the WiMAX Forum formed a joint committee to facilitate development of an AeroMACS profile. It was agreed that AeroMACS profile should closely follow the format and substance of WiMAX Forum commercial and industrial profiles. Concurrently, RTCA SC-223 and EUROCAE WG-82 have been closely cooperating to support the composition of AeroMACS Profile. This joint U.S.–E.U. effort is coordinated by ICAO, in part, to ensure that the recommendations of ANC-11 regarding interoperability are implemented, and in part to assist expediting ICAO's approval of international AeroMACS standard [7].

Given that the IEEE 802.16-2009 is the parent standard of AeroMACS networks, the AeroMACS profile is required to categorize protocols and features of the IEEE standard into mandatory, optional, and not applicable. As such the profile specifies requirements for the unique adaptations to the current IEEE 802.16-2009 standard to provide wireless data communications to mobile platforms on an airport surface [18,19].

WiMAX Forum Aviation Working Group (AWG)[4] is tasked with adapting AeroMACS requirements into the WiMAX profile documents. Within the WiMAX Forum enterprise, a Technical Working Group (TWG) and a Certification Working Group (CWG) are established for composition of AeroMACS profile based on WiMAX profile documents. This is to ensure that the aviation industry and all other AeroMACS stakeholders are active participants in support of AeroMACS technology development. It should be noted that AeroMACS profile is an evolving document in the sense that, as time progresses, it incorporates more features of the IEEE 802.16-2009 (WiMAX) to improve performance or to add more services, that is, more complex part of the standard will be integrated into the profile.

The AeroMACS Profile, designated as "RTCA document DO-345" and "EUROCAE ED-222", was finally published in November 2013 in the European Union, and in December 2013 in the United States. This document, created through the joint effort of RTCA SC-223 and EURO-CAE WG-82 specifies "the unique adaptations required to enable the current IEEE 802.16-2009 standard to specify technical requirements for wireless data communications to mobile stations on an airport surface" [17–18]. The technical section of the profile describes the selection of parameters essential to meet the aviation communication requirements. The requirements section of the profile concentrates on different spectrum ranges, focusing on efficient use of channel assignments and interoperability requirements prescribed by WiMAX Forum.

The current AeroMACS profile is developed based on WiMAX Forum™ Mobile System Profile Release 1.0 Version 0.9 [20]. This is despite the fact that maximum speed supported by this version of WiMAX system profile is 50 nautical miles, which is well below the landing and takeoff speed of large commercial jet aircrafts. The speed limitation issue, as mentioned in Chapter 5, is addressed in the WiMAX Forum system profile version 2.0 and beyond.

The AeroMACS profile document is organized into three sections. In the first section, designated as "Purpose and Scope", the AeroMACS technology is introduced and the Terms of References (ToR) agreement for the profile development is described, and the required cooperation

---

4 The WiMAX Forum AWG acts as a focal point for global aviation industry interest in AeroMACS, as a technology for worldwide aviation applications by defining technology profiles and air interface specifications and certification for AeroMACS networks and products, to ensure the development of a healthy ecosystem of interoperable aviation-centric product solutions. The AWG is the single source for coordinated recommendations and requirements that drive AeroMACS specifications. The AWG activities are reviewed and coordinated by ICAO, FAA, EUROCONTROL, EUROCAE, and RTCA.

between RTCA, EUROCAE, and WiMAX Forum is emphasized. A brief description of the other two sections follows.

6.3.3.1 The AeroMACS Profile Background and Concept of Operations

The joint FAA–EUROCONTROL FCS recommends the application of commercially available IEEE 802.16 2009 standard, over the C-band 5091–5250 MHz, as a solution to support the future aviation requirements on the airport surface [21]. This new communications standard not only addresses the problem of traffic congestion over the legacy VHF band, but also provides a capability to support data intensive applications projected under the FAA's NextGen and the European SESAR program.

The profile document presents material on the concept of operations and potential applications to be supported by AeroMACS. The system is designed to support data and voice communication services on the airport surface for enhancing the safety and regularity of flight. The transmission requirements for these services range from the distribution of fixed sensor data to the transmission of extensive graphical weather information; accordingly, the transmission speed starts from about fraction of kilobits per second to multiples of 100 megabits per second. The AeroMACS profile includes some unique features that are beyond the selection of settings from WiMAX system profile Release 1.9 for implementation of AeroMACS networks. These features include the frequency band of operation, that is, 5000–5150 MHz, which is outside of the band supported by Release 1, network synchronization settings, and spectrum masks [18].

In order to develop IEEE 802.16-2009-based AeroMACS system for airport surface, the following steps are taken.

- Identification of portions of parent standard, best suited for AeroMACS application.
- Identification of any required missing functionalities in the parent standard, and development of replacement aviation-specific standard, to be submitted to appropriate standardization agencies.
- Evaluation and validation of aviation-specific standard performance for supporting wireless mobile networks operating in the given C-band experimentally (trial and testbed).
- Determination of channelization methodology for allocation of safety and regularity of flight services for a range of airport sizes and classes, configurations, and operational requirements.
- Removal of prior limitation in the allocated C-band (MLS Extension band) for support of AeroMACS.
- Enabling the ICAO to rely on the protected aviation band for development of airport surface mobile communications standard with short range coverage and high-data throughput [18].

Airport authorities, airlines, and civil aviation authorities (NAS and Global Airspace System) are the principal aviation agencies that deliver various services and are in need of transportation of application information over AeroMACS. The user applications for transport over Aero-MACS are classified into the following five different functional categories.

- ATM/ATC
- Aeronautical information services (AIS) and meteorological data (MET)
- Aircraft operator
- Airport authority
- Airport infrastructure

These applications have different information contents, and they may require different performance characteristics, security needs, QoS, and throughputs [18]. Examples of potential applications belonging to the five categories that may be transported through AeroMACS include Digital Notice to Airmen (D-NOTAM) – an application created and managed by the government agencies to alert pilots of hazards in the NAS – , pre-departure clearance (PDC), digital automatic terminal information system (D-ATIS), where PDC and D-ATIS data are currently transported using ACARS, and 4D trajectory data link (4DTRAD) in future advanced avionics to manage the end-to-end aircraft trajectory. Some applications related to airport infrastructure such as Airport Surface Detection Equipment (ASDE-X) has long been identified as an application that could benefit from a wireless communications technology. As modernization of NAS (see Chapter 1) continues, new operational applications requiring access to AeroMACS network will continue to appear, therefore, in order to optimize the utilization of the available finite bandwidth, it is necessary to provide careful evaluation of existing and identified future applications.

The operational applications may further be classified into mobile applications and stationary mobile applications. ATC communications is a mobile application that carries a significant portion of airport surface data transportation between aircrafts, moving vehicle in runways and taxiways, and supports Tower Data Link System (TDLS) for flight clearances. AOC, advisory, and non-ATC voice and data transportation are other examples of mobile application that is expected to be carried through AeroMACS. SWIM may be routed to airport surface mobile units through AeroMACS. Examples of stationary mobile applications include airport infrastructure services (pending on local rules) such as navigation aid monitoring and maintenance (NAMM), airport surveillance, and voice remote radio.

6.3.3.2 AeroMACS Profile Technical Aspects

An important incentive for projecting commercial WiMAX technology over to available aeronautical C-band (5000–5150 MHz) is to maximize the use of "commercial off-the-shelf, COTS" technology in order to minimize the system cost, and to minimize the additional hardware development required for the implementation of AeroMACS various parts. The major objectives of the technical profile working group is outlined further.

- Ensure international interoperability and obtain agreement on WiMAX BS and SS/MS equipment settings, as referred to in the profile.
- Coordinate with WiMAX technical agencies (such as WiMAX Forum) to adapt the existing WiMAX profiles for provision of wireless data communications on the airport surface that meets the aeronautical C-band spectral requirements.
- WiMAX Forum engagement ensures that the AeroMACS stakeholders that are needed to support the development are engaged with the commercial development of the AeroMACS technology.

The technical section of the AeroMACS profile details a design flow that is based on the WiMAX profile Release 1.9. This is followed by detailed requirements that are agreed upon by RTCA SC-223 subcommittee, which includes requirements related to network synchronization and the spectrum mask. The next phase of the AeroMACS document outlines a subset of WiMAX Forum features taken from the WiMAX profile document that enable interoperability. Finally, a table and justification list is included to ensure international operability of the WiMAX standard. WiMAX Forum has already published several versions of Release 2 of WiMAX system profiles, therefore, it is anticipated that the technical section of the AeroMACS profile undergoes revision, accordingly.

6.3.3.3 Profile's Key Assumptions for AeroMACS System Design

The AeroMACS profile makes a number of assumptions based on which global cooperation, design compatibility, and system interoperability are fostered. Further is a list of some key assumptions formulated for AeroMACS networks.

1) AeroMACS total radiated power shall stay within the interference thresholds established by the ITU-R for shared spectrum users.[5] System settings may be adjusted to accommodate increased usage as needed. The threshold is defined by 2% noise temperature increase at the satellite receiver.

---

5 AeroMACS Extended MLS C-band (5091–5150 MHz) is shared with Globalstar, a nongeostationary mobile satellite service. See Chapter 7 for more detail.

2) The United States and the European Union agree to use WiMAX Forum Network Reference Model (NRM) and Architecture to define higher level protocols and network functions when developing end-to-end network specifications.

3) The end-to-end network architecture for AeroMACS shall be the same as WiMAX-IP end-to-end system.

4) All future AeroMACS avionics and various infrastructure components will be required to be backward compatible with previously installed AeroMACS equipment.

5) AeroMACS profiles will be maintained under configuration control and will not be changed without undergoing proper review and approvals by all concerned parties.

6) AeroMACS supports roaming between NSPs.

7) AeroMACS will be designed to take full advantage of technical capabilities of IEEE-802.16-2009-based WiMAX technology, while efficiently utilizing the allocated spectrum.

8) In order to provide equitable access and cost-effective services to all users, the government will authorize the use of AeroMACS resources. All non-TSO[6] (Technical Standard Orders) hardware used in the AeroMACS service should have passed WiMAX Forum certification testing for the AeroMACS profile.

9) In the United States, FAA, FCC, and NTIA collaborate to formulate the regulatory framework for AeroMACS service. A similar process is in place in the European Union.

The complete list of AeroMACS profile assumptions can be found in the profile document [18].

6.3.3.4 AeroMACS Radio Profile Requirements and Restrictions

Implementation of AeroMACS systems requires specific settings in regard to radio arrangements. The band of frequencies within which AeroMACS may operate is 5000–5150 MHz, of which the band 5091–5150 MHz is currently allocated by ITU and is globally available for AeroMACS application. AeroMACS unique profile settings and modifications include network synchronization and the transmitter output spectral content and spectral mask. Like any other application, AeroMACS signal requires spectral masking in order to limit its inference

---

6 A TSO is a minimum performance standard, defined by the FAA, used to evaluate an article. An article can be a material, part, component, process, or appliance. Each TSO covers a certain type of article. A "TSO authorization" means receiving authorization for manufacturing an article to a TSO standard. A TSO authorization is both a design and a production approval.

Table 6.3 AeroMACS radio requirements and restriction.

Operating frequency band	Channel BW	FFT size	Center frequency assignment	Spectral mask	Modulation schemes	MAX Doppler velocity
Operation range: 5.000–5.150 GHz	5 MHz	512	Reference frequency: 5145 MHz	Emission mask: M	QPSK 16-QAM 64-QAM*	50 K-Miles/h
Current allocated band: 5.091–5.150 GHz			Range: 5000–5150 MHz $f_C = 5145 - 5n$ MHz $0 \leq n \leq 28$	Adopted from Title 47 CFR, part 90.210	*Optional for UL	92.6 km/h

into adjacent radio channels. Table 6.3 summarizes AeroMACS's various radio requirements and restrictions.

As Table 6.3 indicates, the entire operating band of 5000–5150 MHz is taken into consideration as potential available spectrum for the AeroMACS network implementation. The reference frequency is 5145 MHz, and the frequency band has been channelized for 5 MHz. All preferred channel center frequencies are referenced downward from 5145 MHz in 5 MHz incremental frequencies, thus the center frequencies can be expressed as $f_C = 5145 - 5n$. Accordingly, the next center frequency is at 5140 MHz, the next one at 5135 MHz, and so on. Thus, the reference frequency is used to identify all channels whose center frequencies are included in the list of channel center frequencies that are to be tuned to by AeroMACS.

The AeroMACS spectrum mask requirement is based upon emission mask "M" for all power levels authorized for the AeroMACS service. Emission mask "M" is adopted from Title 47 CFR (Code of Federal Regulation) part 90.210, which lists FCC-defined schemes of signal emission masking and attenuation around and outside of the allocated band. Part 90 lists emission masks for private land mobile radio services [22]. The detail of numerical attenuation requirements for the mask "M" can be found in Refs [18] and [22].

Modulation schemes applied in AeroMACS networks are directly inherited from IEEE 802.16-2009 standard and WiMAX technology, that is, QPSK, 16-QAM, and 64-QAM. However, 64-QAM is an optional modulation scheme in AeroMACS uplink. Details of AeroMACS modulation formats are provided in Chapter 7.

6.3.3.5 AeroMACS Profile Common Part and TDD Format

AeroMACS applies TDD technology and FDD is not considered as an option. A major reason for selection of TDD over FDD is the possibility of dynamic allocation of DL and UL resources to support asymmetric

DL/UL traffic in TDD format. A second reason for this selection is spectral efficiency by virtue of the fact that both DL and UL transmission paths use a single radio channel. However, another advantage of TDD over FDD is its relative simplicity of implementation. The only disadvantage of TDD is its stringent requirement for exact timing and synchronization.

Common Part of AeroMACS profile is derived from that of WiMAX [20]. The WiMAX profile Common Part contains a series of tables that set the operational parameters for the MS/SS and the BS. Many parts of these tables are carried over to AeroMACS and are exploited for quick reference to items for use in AeroMACS systems. Thus, the meaning of the items used for AeroMACS is exactly the same as those specified in the corresponding items in the WiMAX profile. The AeroMACS profile Common Part consists of five sections.

1) PHY profile: Refers to physical layer functionality.
2) MAC profile: Refers to functionality in medium access control layer.
3) Security: Refers to functionality in security sublayer.
4) Radio profile: Indicates permitted radio frequency configuration.
5) Power class profile: Indicates permitted transmitted power configurations [18].

Table 6.4 provides the list of a limited number of important PHY features and parameters for AeroMACS that are selected from IEEE 802.16-2009 standard [23] and WiMAX profile [20].

The AeroMACS PHY profile contains a table that presents the complete list of all items from WiMAX profile and comments on whether or not they are carried over to AeroMACS profile [18].

In regard to MAC layer, the AeroMACS profile provides a table in which various features of WiMAX profile are listed and comments are made on each one, as to whether or not they are carried over to

Table 6.4 AeroMACS main physical layer features and parameters.

Feature/ element	Bandwidth BW	Multiple access method	Sampling factor	Cyclic prefix	Frame time duration	Permutation
Name/Value	5 MHz	OFDMA	28/25	1/8	5 ms	PUSC
Note	Key system resource	Basic system requirement	OFDM parameter for the chosen BW	Selected for standard multipath environment	Basic system requirement	Basic usage mode mandatory

Table 6.5 Some features supported by AeroMACS MAC profile.

Title	Packet header suppression	ARQ	Fragmentation and reassembly	Variable length SDUs packing in MS	Packing ARQ feedback in payload	MAC support for HARQ
Note	It is generally supported	Essential feature	An essential feature	Necessary for UL	Necessary for ARQ	Necessary

AeroMACS profile. Table 6.5 presents few key features of WiMAX MAC profile that is supported by AeroMACS.

A rather complete inventory of MAC features supported by AeroMACS standard is given in the profile document [18]. Chapter 7 covers AeroMACS MAC layer in some detail. The main sources of information on which AeroMACS profile is developed are Refs [20] and [24–26].

6.3.4 AeroMACS Minimum Operational Performance Standards (MOPS)

The AeroMACS profile is used as a basis for development of what is known as "Minimum Operational Performance Standards" (MOPS) for AeroMACS avionics. Generally speaking, MOPS define equipment specifications useful for a host of AeroMACS stakeholders, including manufacturers, system designers, installers, and users. The term "equipment" in MOPS is inclusive of all components and units necessary for the system to perform its intended tasks. "MOPS provide the information needed to understand the rationale for equipment characteristics and requirements stated, describe typical equipment applications and operational goals, and establish the basis for required performance under the standard" [27]. In other words, MOPS contain information pertaining technical characteristics as well as operational performance for AeroMACS equipment and units. This is complemented by Minimum Aviation System Standards (MASPS) to be discussed next.

The final approved version of AeroMACS MOPS was published in the United States under "RTCA DO-346" in February 2014, and in the European Union under EUROCAE "ED-223" in October 2013. This document contains the minimum operational performance standards for AeroMACS on-board (airborne) system and for the ground-based station (BS). Like for any other standard, it is highly recommended that MOPS are closely observed in the process of AeroMACS equipment production and manufacturing. The objectives are providing one means of ensuring that the equipment will perform its intended function(s) satisfactorily under all conditions normally encountered in routine aeronautical operation, and assuring interoperability [4,28].

6.3.4.1 AeroMACS Capabilities and Operational Applications

As a standard-based technology (in this case IEEE 802.16-2009 standards, which is well suited for applications spanned below 11 GHz, as discussed in Chapter 5), AeroMACS enables the aviation community to draw extensive international standardization cooperation, as well as providing commercially available components, units, and broadband services. As Section 6.3.3 demonstrates, AeroMACS profile closely follows the format and substance of WiMAX System Profile Release 1.9, developed by the WiMAX Forum. Consequently, AeroMACS as a broadband transmission system can potentially support a wide range of voice, data, and video communications among mobile and stationary users on the surface of an airport. Moreover, the airport vital functionalities of CNS that supports ATM and ATC can greatly benefit from a secure wideband wireless communications system. Another instance in which AeroMACS can directly affect traffic management, aircraft movement control, and airport security is the capability of supporting an almost real-time graphic data sharing. With sharing of graphic data, the situational awareness can be significantly improved, airport surface traffic movement can be enhanced leading to reduction in congestion and flight delays, and can assist to prevent runway and taxiway incursions [4].

The principal aviation entities that deliver services and are in need of transport of application information through wireless means on airport surface are airport authorities, airlines, civil aviation authorities, and NAS. User applications requiring AeroMACS transmission services have been classified into five functional domain categories, as outlined below [4].

1) *ATM and ATC:* ATM and ATC form the central part of national and global air transportation system, ensuring safe and secure departure of aircraft from an aerodrome, transit through national and international airspace, and landing at the destination aerodrome.

2) *AIS and MET:* The main objective of AIS is providing information that is required for safety, regularity, and operational efficiency of international flights. Another objective is ensuring uniformity and consistency in the provision of such data across the board. ICAO Annex 15 stipulates that aeronautical information (AI) be made available in the form of an Integrated Aeronautical Information Package (IAIP) that contains several components. These components include AI publication (AIP) and supplement, Notice to Airman (NOTAM), and Pre-flight Information Bulletins (PIB).

3) *Aircraft Owner/Operator:* An "aircraft operator" is anyone who is authorize to use or pilot an aircraft for the purpose of air navigation with or without right of legal control over the aircraft.

4) *Airport Authority:* Airport authority refers to entities that are responsible for monitoring and oversight of operation of a single airport or multiple of closely located airports.

5) *Airport Infrastructure:* Airport physical infrastructure includes terminals, taxiways, runways, aircraft parking spaces, security devices, guiding signs, navigation facilitating tools (NAVAIDs) used for aircraft operation, and so on. The quantity of these items available in an airport is directly related to the volume of air traffic the airport is designed to carry, as well as the size of the airport.

It is noted that applications identified in these five domains may require different performance characteristics related to such items as throughput, security needs, QoS, and so on, which could be vastly different.

6.3.4.2 MOPS Equipment Test Procedures

The MOPS document specifies equipment testing procedure as one means of demonstrating compliance with the minimum performance requirement expressed in the rest of the document. The requirements are mostly drawn from the IEEE 802.16-2009 standards. The stated restriction on equipment design, vis-a-vis testing, is that the equipment shall be designed such that the application of specified testing procedures does not have a detrimental effect on the equipment performance, following the completion of the tests. The document recognizes that methods of testing, other than those cited in the document, might be preferred for the equipment. Under these circumstances, the alternate test procedures may be used, provided that, as a minimum, they deliver equivalent information [4].

There are four testing categories described in the MOPS document. Table 6.6 summarizes the main objective of these test procedures.

As Table 6.6 explains the "installed equipment tests" are conducted when "bench tests" fail to determine performance adequately. Installed tests for on-board equipment is conducted under two different conditions: aircraft on the ground and using operational or simulated system inputs, and aircraft in flight using relevant operational system inputs.

The RTCA MOPS document describes, in detail, the scope and limitations of all of test categories described in Table 6.6 [4].

6.3.4.3 Minimum Performance Standard

This is the core of AeroMACS MOPS that defines minimum performance standard for on-board and ground-based equipment. The general requirements for any AeroMACS equipment are as follows:

Table 6.6 AeroMACS MOPS equipment test categories.

Category	Environmental tests	Bench tests	Installed equipment tests	Operational tests
Function/ objective	Electrical/ mechanical performance evaluation	Laboratory assessment for demonstration of compliance	Conducted when performance cannot be adequately determined by bench testing	Conducted by operating personnel to ensure that the equipment functions properly
Notes	Conducted under various environmental conditions expected in actual operation	The results may be used by manufacturer as design guidance	The result may be used for demonstration of functional performance in the operational environment	The operational tests serve as one means of equipment reliability and compliance

1) The equipment shall perform as defined by manufacturer.
2) The proper use of any equipment shall not create hazards for other users of NAS.
3) The equipment shall comply with relevant regulations set by ITU-R, FCC, as well as other applicable bodies.
4) AeroMACS equipment shall be tunable across the band 5000–5150 MHz.

It was mentioned earlier that the current allocated band for AeroMACS is 5091–5150 MHz, however, additional subbands may be allowed for AeroMACS in the future. Table 6.7 provides some detailed information

Table 6.7 AeroMACS current and potential additional operating frequency bands.

Parameter	Core (primary)	Subband I (secondary)	Subband II (tertiary)	Supported channel bandwidth	Step size
Frequency band (MHz)	5091–5150	5000–5030	5030–5091	5 (FFT size 512)	0.25
Notes	Available for channel assignment	Channels may be assigned when national regulations allow	Channels may be assigned, if the ICAO frequency planning permits	Current supported bandwidth S-OFDMA allows for future up and down scaling	Allows for graceful move-away from in-band civilian and military interference

on AeroMACS present and potential future operating frequency bands, as described in RTCA MOPS document [4].

The MOPS document classifies AeroMACS equipment into airborne (onboard, MS) and ground-based (BS) categories, and defines minimum performance standards for each category separately.

Table 6.8 highlights some important general design requirements for AeroMACS airborne (MS) equipment. Airworthiness defines the conditions based on which an aircraft may be judged suitable or unsuitable for safe flight. The United States Code of Federal Regulation, Title 14, Subchapter F, Part 91.7 states that "No person may operate an aircraft unless it is in an airworthy condition".

Insofar as general design requirements for ground-based equipment (BS) is concerned, the following points are presented.

1) The operating frequency bands and associated bandwidths are as provided in Table 6.7.
2) Controls and Indicators: "The user shall not have access to any control, which if wrongly set might impair the technical characteristics of the equipment" [4].

For the complete list of equipment performance standards and standard condition, the reader is referred to RTCA MOPS document [4].

6.3.5 AeroMACS Minimum Aviation System Performance Standards (MASPS)

Another ongoing activity related to the standardization process is EUROCAE development of AeroMACS Minimum Aviation System Performance Standards (MASPS). MASPS define AeroMACS performance requirements as an avionic communications system, as such MASPS describe specifications useful for various AeroMACS stakeholders, such as manufacturers, designers, service providers, and users, for operational use within a defined aerospace. MASPS also outlines possible implementation options, including end-to-end architecture for AeroMACS networks. "MASPS describe the system (subsystems/functions) and provide information needed to understand the rationale for system characteristics, operational goals, requirements, and typical applications. Definitions and assumptions essential to proper understanding of MASPS are provided as well as minimum system test procedures to verify system performance compliance (e.g., end-to-end performance verification)" [27]. The Minimum Aviation System Performance Standards for AeroMACS was finally issued by EUROCAE in September 2016 [29]. In the meanwhile, the AEEC standardization effort is focused on defining required avionics specification for airborne transceiver, capable of operating within

Table 6.8 General design requirements for AeroMACS airborne (MS) equipment.

Item	Software	Airworthiness	Fire protection	Operation of controls	Design assurance
Description	Software design shall follow RTCA Doc.DO-178B	Equipment installation should not impair the airworthiness of the aircraft	All materials used for equipment shall be self-extinguishing	Controls intended for use during flight cannot be operated to harm aircraft operation or equipment reliability	DAL[a] should be adequate to compensate for contribution of the equipment to overall aircraft failure
Notes	RTCA DO-178: "Software Considerations in Aircraft Systems and Equipment Certification"	Airworthiness is the measure of aircraft fit for safe flight. Certificates of aircraft airworthiness are issued and renewed by national aviation authorities	Small parts that would not contribute to fire propagation are excepted. Examples of compliance rules are contained in FAR[b] Part 25, App. F and FAR 23 App. F	Flight safety and equipment reliability concern	DAL for a given hazard classifcation is not same for aircraft types. Contribution of the equipment to an aircraft level failure may depend on the aircraft type and other installed equipment

a) DAL stands for Design Assurance Levels.
b) FAR stands for Federal Aviation Regulations.

the spectral range of 5000–5150 MHz, avionics architecture, interfaces to airborne peripherals, and so on [30]. The technical section of the AeroMACS profile describes the selection of parameters essential to meet the aviation communication requirements.

6.3.6 AeroMACS Technical Manual

International Civil Aviation Organization has published several draft versions of a technical manual for AeroMACS over the past few years. The most recent draft was published under the title of "Manual on the Aeronautical Mobile Airport Communications System (AeroMACS)" in 2016 [31]. It was expected that the final version would be published in the last quarter of 2016. This most recent draft, with some minor changes and modifications will form the final published manual for aeronautical mobile airport communications system.

The manual is organized in three chapters and a single appendix. Chapter 1 of the draft manual covers an overview, some background information, and key features of AeroMACS. The chapter introduces AeroMACS as a driven technology based on an open standard used for commercial communications (IEEE 802.16e standard), and places emphasis on the underlying ICAO aviation objective of capitalizing on existing technologies and benefiting from past development efforts. The chapter concludes with a summary of constraints and limitations associated with using AeroMACS in the airport surface. Table 6.9 highlights these constraints.

Chapter 2 contains the guidance material for issues typically arising in AeroMACS deployments. Applicable services, medium access configuration, frequency allocation, architecture and interfaces to network layers, dimensioning, and BS siting are all discussed in this chapter. Aircraft operations scenarios, consisting of aircraft landing, aircraft parked, and aircraft departure, are covered in this chapter. Aircraft landing refers to the operations performed by an aircraft from touchdown and when the aircraft establishes data link connectivity with the airport network services through AeroMACS, through taxiing, and finally parking at a terminal gate. In the aircraft parked scenario, the aircraft is parked at a gate, its engine is switched off, and some upkeep operations are being conducted. The aircraft is connected to the airport network (see Chapter 7 for information on airport network), ANSP, and Airline Private Networks (APN), while exchanging data through AeroMACS access network. The aircraft departure scenario consists of all operations from pre-departure to takeoff. The operation begins with aircraft pushback from the gate, then starting the engine, taxiing to the runway, and finally taking off. Throughout this operation, the aircraft remains connected to AeroMACS, and is disconnected from AeroMACS as soon as it becomes airborne [31].

Table 6.9 AeroMACS constraints and limitations.

Constraint	Spectrum	Services	Airborne use exclusion	Velocity	Airport domain
Description	Current band of operation: 5091–5150 MHz. Additional band of 5000–5030 MHz will become available in near future	Supports services that are relevant to the safety and regularity of flight operations.	AeroMACS signal transmission is allowed only when the aircraft is on the airport surface	AeroMACS is intended for MS velocities up to 50 nautical miles per hour (92.6 km/h).	AeroMACS systems can support communications exchanges between moving and fixed assets in various operating areas in the airport environment
Notes	It is, however, recommended that AeroMACS radios are manufactured to cover 5000–5150 MHz band	Also, supports information exchanges with ANSPs[a], Aircraft Operators (AOC), and Airport Authorities	This limitation is to avoid interference to coallocated applications such as Fixed Satellite Service (see Chapter 7)	AeroMACS does not preclude operation at higher velocities, which may be commensurate with takeoff and landing speeds of some aircraft	The airport areas covered by AeroMACS include gates, ramps, taxiways, runways, side roads, and even assets that are located outside of the airport property.

a) ANSP stands for air navigation service provider. ANSP refers to any private or public entity that provides air navigation services to airspace users.

These three operations take place in a sequence for an aircraft, in the form of continuous cycles during the aircraft normal daily operations.

Table 6.10 highlights the potential services that the aircraft may be connected to during various aircraft operation scenarios. Airport terminal information messages mentioned in the rightmost column of Table 6.10 includes key takeoff and flight-safety-related information such as availability of active runways, approaches, weather conditions, and NOTAMS. Table 6.10 presents a partial list of such services that the aircraft may be potentially connected to, through AeroMACS network. The AeroMACS manual [31] provides a more thorough list of these services, offering more elaborate description for each one.

Chapter 2 of the manual provides a rather detailed description of some typical airport surface situations for which AeroMACS communications system can be employed [32]. Some of these scenarios are addressed in Section 6.5. An important part of Chapter 2 is provision of an overview of AeroMACS network architecture and network reference model, which are addressed in Chapter 7 of this book. Other issues addressed in Chapter 2 of the AeroMACS manual are network security, support for multicast and broadcast, subnetwork entry and mobility for AeroMACS, upper layer interfaces, and so on.

Chapter 3 of the manual is exclusively dedicated to description of the technical specifications of AeroMACS networks that is related to several sections in the guidance material of Chapter 2. See Chapters 7 and 8 of this book for extensive coverage of technical specification of AeroMACS networks.

Around the time of preparation of this chapter, a document that is related to AeroMACS deployment and technical characterization is published in the form of a FAA memorandum, in which an installation guide for AeroMACS systems in the U.S. airports is provided. The memorandum mentions that the federal use of AeroMACS and its associated frequencies is approved by the National Telecommunications and Information Administration (NTIA). Based on this approval, FAA/NASA-GRC have performed Federal AeroMACS field trials at the Cleveland-Hopkins International Airport (CLE) and NASA-Glenn campus. The FAA Airport Surface Surveillance Capability (ASSC) program office will be using AeroMACS at different airports throughout the United States as an application [32].

6.4 AeroMACS Services and Applications

AeroMACS is a broadband wireless solution, based on a proven communications technology, enabling the implementation of a wide range of

Table 6.10 Services that an aircraft may be connected to on airport surface.

Operation scenario	Landing	Parked	Takeoff
Potential services	1) D-TAXI (Data Link – Taxi) route plan is uploaded to the aircraft (ATC).	1) Avionics software and database are uploaded.	1) Flight plan
	2) OOOI (Off/Out/On/In) status is provided.	2) Maintenance data from avionics/Engine are downloaded.	2) Airport terminal information (ATIS).
	3) ACM transfer of control happens from runway tower to ground tower.	3) Pilot manuals/charts/maps and so on, may be uploaded.	3) Departure clearance and flight preparation.
	4) A-SMGCS[a] surface services are activated in and surface surveillance information is provided to the aircraft (ATC).	4) Electronic checklist/EFB updates.	4) Runway visual range information.
	5) TAXI clearances are provided (ATC tower).	5) Routine maintenance checks are conducted	5) Hazardous weather and operational terminal information.
	6) Flight data is downloaded to airline operation centers (AOC)		6) TAXI route information
			7) Surface surveillance guidance control.
			8) ACM messages. Any other information related to the regularity and safety of flight.

a) A-SMGCS stands for Advanced Surface Movement Guidance and Control System.

applications and services, most of which could not have been supported by the legacy VHF airport communications system. For instance, with AeroMACS graphical data may be shared in a near real-time video broadcast, leading to significant improvement in situational awareness and substantial enhancement in surface traffic movement, which reduces congestion and delays and help prevent runway incursions [7].

A relay-fortified AeroMACS, in particular, can provide temporary communication capabilities to any hot spot that may be created on the surface of an airport due to incidents, emergency situations, outages, and in the event of airport renovation and construction work (see Chapter 8). AeroMACS enables aircraft access to SWIM services for delivery of time-critical advisory information to the cockpit. Collaborative decision-making is supported by AeroMACS, through the possibility of updating large databases. AeroMACS is capable of loading the flight plans into flight management system (FMS) [16].

The WiMAX Forum classifies the services and applications that are required for safety and regularity of flights into three categories [33]. Figure 6.2 portrays this classification in a structural form.

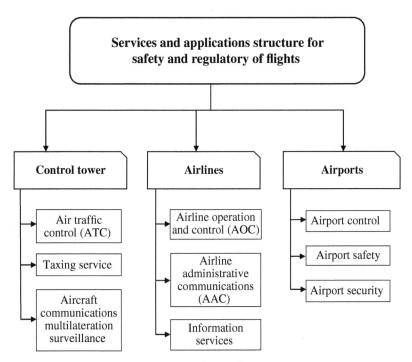

Figure 6.2 Services and applications for flight safety and regularity.

In what follows we explore the historical evolution of classification and identification of applications and services that could potentially be supported by AeroMACS.

In April 2010, a User Applications and Services Survey (USAS) Ad-hoc Working Group (AWG) was organized within the RTCA Special Committee 223 (SC-223), whose purpose was to identify and classify potential user requirements for AeroMACS. It was accepted that the identification should cover all functional areas except passenger communications. Upon the completion of compilation of these requirements, they would be used for providing feedback for validating the needs to be fulfilled by the AeroMACS profile. Originally, the AWG established a set of functional domains to focus their survey research that would lead to various services and applications. Five domains, as it was alluded to earlier, were specified, as shown in the following outline where the approximate number of services to be potentially provided for each category, according to the initial assessment of the AWG, are indicated inside the corresponding parentheses [34].

1) ACT and ATM (33)
2) AIS and MET (96)
3) Airlines and cargo operations, that is, flight operator (90)
4) Airport operations (33)
5) Airport infrastructure (33)

Flight operations refer to activities performed by airline and cargo companies while the aircraft is at a gate, hangar, or repair station. Airport operations, collectively, refer to all activities related to management of the airport physical infrastructure. In addition to physical components (tower, terminals, runways, taxiways, lighting systems, navigational aids, sensors, signs etc.), airport infrastructure includes a data exchange system for transmission of voice, audio, and video. Communication requirements for airport infrastructure are categorized into surveillance, navigational aids, lighting facilities, weather, administrative and communications. In the United States, the airport infrastructure is normally established, managed, and controlled by the FAA.

Following the categorization of applications and services, attempts were made to identify a number of key parameters associated with each application, this is highlighted in Table 6.11.

As the number of services in each category indicates, there are a large number of airport applications and services. Table 6.12 groups these services applications into closely related service clusters for each of the five categories.

AeroMACS can support a wide variety of voice, video, and data communications services for both fixed and mobile scenarios. Nevertheless, it should be noted that not all services and applications within the

Table 6.11 Characterization of airport applications and services.

Item	Description	Comments/examples
Name	The title of the application or service	Notice to the airman (NOTAM)
Owner	The user of the application	FAA, airport authority
Existing or future	Whether the application or service is existing, or it is a planned service for future	Future: D-OTIS[a] Existing: ground clearance
Mobile or fixed	Whether the application involves mobile stations or is provided for stationary nodes	Fixed: surface CNS, weather sensor Mobile: Aeronautical Information Services (AIS)
Category	Is the application classified as ATC service, AOC service, or · others	ATC: Air traffic management Airport operator: Security video
Performance requirements	Different applications may ask for vastly different performance requirements	Performance requirement may be expressed in form of desired rate, latency, availability, and so on.
Security requirements	Different applications may require different security needs	Examples of security requirements: encryption, authentication, authorization
Transport contents	The actual form of signal to be communicated	video, audio, voice, file
End points	Data source and sink	Control tower to cockpit

a) D-OTIS: DL Operational Terminal Information Service.

classes summarized in Table 6.12 may be supported by AeroMACS at the time of initial deployment. However, it is expected that as the technology evolves, increasingly more services will be integrated into AeroMACS delivery package.

The current applications that can potentially be provided by Aero-MACS in support of airport surface services may be classified into three categories.

1) ATC/ATM and infrastructure applications, which includes the safety and critical communications related to aircraft.
2) AOC services, which includes communications between aircraft and airline operational control centers; applications in this category also play an important role in the safety and regularity of flight.
3) Airport authority communication applications that directly affects safety and regularity of flights, involving large number of airport

Table 6.12 Application classes for the five categories of services [34].

Category	Applications classes
ATS/ATM	1) ATS services provided by ACARS 2) Data Communications Safety and Performance Standard; CPDLC, 4-D Trajectory Data Link, and so on. 3) Communications Operating Concept and Requirements (COCR) for the Future Radio System 4) Security
AIS/MET	1) Data Link Aeronautical Update Service (D-AUS), a future AIS service 2) Notice to the Airman (NOTMAN), an existing AIS service 3) Runway configuration information, a future AIS service 4) MET Data Link Weather Planning Decision Service (D-WPDS), a future MET Service
Flight operator	1) Ground operations and services: Provided by airline and cargo companies are primarily ramp activities. Ground support services and associated ramp equipment service aircraft between flights (baggage handling, fueling, etc.) 2) Maintenance information: Information is exchanged while the aircraft is being repaired, inspected, or modified. Wireless connection between aircraft maintenance engineer, pilots, and airline agents is required. 3) Aircraft and company operations: This entails exchange of database and document information. Applications transportable by AeroMACS include Flight Operations Manual (FOM), charts uploads, and EFB.
Airport operation	1) Airport operations and management: Management of airport infrastructure and operations. 2) Airfield management operation: An airfield consists of runways, taxiways, and aprons,[a] also items such as lighting, NAVAID, fuel storage, de-icing, and so on. 3) Law enforcement and airport security. 4) Firefighting, rescue, and medical services.
Airport infrastructure	1) Airport Surveillance Radar (ASR) and Air Traffic Control Beacon Interrogator (ATCBI) are primary air traffic surveillance facilities. 2) The Airport Surface Detection Equipment (ASDE) provides surveillance for the runways and taxiways. 3) The Instrument Landing System (ILS), Distance Measuring Equipment (DME), are among airport navigation facilities.

a) Apron is the area of an airport where aircraft are parked, unloaded or loaded, refueled, or boarded.

entities such as surface vehicles, ground services, runways and taxiways safety management equipment, and so on [31].

These applications and services can be described as either fixed or mobile, depending on mobility mode of the end user. However, because of operational constraints on the international frequency spectrum allocated for AeroMACS, only those services that can directly impact the safety and regularity of flight are candidates for provision by AeroMACS [7]. Table 6.13 summarizes major application in fixed and mobile categories, along with specific examples, as potential candidate to be supported by AeroMACS.

In regard to Table 6.13 and the applications and services that may be provided by AeroMACS, the following observations are made at this juncture.

1) The majority of existing ATS/ATM fixed applications use voice-grade point-to-point links, mostly through cable loops. AeroMACS offers a flexible wireless alternative to wireline media.
2) AIS baseline synchronization service, mentioned in Table 6.13, consists of uploading flight plans to the flight management system, updating terrain, GPS navigational databases, and aerodrome charts to EFB, data delivery to the cockpit (D-AUS), airport/runway configuration information (downlink-operational traffic information system (D-OTIS)), and convective weather information, for example, graphical forecast meteorological information and graphical turbulence guidance (GTG) data and maps [7].
3) As was alluded to earlier, once AeroMACS infrastructure is deployed over an airport for the purpose of enabling migration or augmentation of existing services, the door opens up for integration of many additional potential services into AeroMACS, particularly those that require broadband media.
4) Depending on the situation, an airport user may be connected to one or more categories of applications. For instance, the aircraft or a mobile asset is expected to have connectivity to both ATS and AOC groups of applications. The aircraft, in fact, may also need to be connected to airport authority group of applications, such as for de-icing process. Airport sensor systems, which are generally fixed assets, are expected to connect to an airport authority type of application such as security video. Nomadic and nonaircraft mobile systems may have connections to all the three groups of applications, depending on their needs and authorizations. Determination of the type of user is critical in providing the correct level of service over AeroMACS [31].

References [7 and 31] present a more extensive list of services in all three categories that may potentially be supported by AeroMACS.

Few words are in order regarding multicast services that enable or befit a number of airport applications. Support for the multicast traffic connections is a required feature of the AeroMACS profile. Multicast services can be implemented using the transport traffic connection

Table 6.13 Examples of current candidate services to be potentially supported by AeroMACS.

Application	ATC/ATM	AOC	Airport authority
Category	Mobile: ACT/ATM	Mobile: AIS/MET	Mobile
Examples	1) Selected message services currently carried by ACARS: PDC, OOOI[a] 2) Selected services currently provided by CPDLC: 4D-TRAD,[b] D-SIG,[c] D-LIGHTING	1) AIS baseline synchronization services (Important drivers for AeroMACS design): D-AUS and D-OTIS 2) Convective weather information 3) SWIM	1) In-vehicle and portable mobile cameras for live video, during snow removal, de-icing, security, fire and rescue operations
Category	Fixed: Surface CNS	Fixed	Fixed
Examples	1) Controller-to-pilot voice via RTR[d] 2) NAVAID 3) Surface movement detection and ASR 4) Remote maintenance and monitoring (RMM) 5) AAA	1) Ground operations and services 2) Sharing of maintenance information 3) Flight operations manuals updating	1) Fixed surveillance security video applications, required for safety services 2) Surface detection equipment model X (ASDE-X)

a) Provides time of out, off, on, and in.
b) 4D-TRAD: 4-dimensional trajectory data link.
c) D-SIG: surface information guidance.
d) RTR: remote transmitter receiver.

identification (CID) (see Chapter 7). Under these circumstances mobile stations do not need to be aware of the multicast nature of the service. Many applications would benefit from a multicast service, a partial list of these applications is outlined further [18].

- Traffic information system (TIS, TIS-B)
- Flight information services (FIS): Some FIS functionalities need the option to work in multicast/broadcast manner, some example cases are D-ATIS, D-OTIS, Download-Runway Visual Range (D-RVR), and Digital Significant Meteorological Information (D-SIGMET)[7]
- Graphical weather information (WXGRAPH)
- Airport delay information.
- Notice to airmen (NOTAM)

It should be noted that most of these applications are being considered to be carried by AeroMACS.

6.5 AeroMACS Prototype Network and Testbed

The origins of the idea of establishing a testbed for AeroMACS systems can be traced back to FCS document and even before that. Initially, NASA GRC and Sensis Corporation defined the notion of a testbed to enable exploration and examination of technologies for transformation of the national air transportation system to meet the standards defined by the vision that was later called NextGen. An important objective of establishing a prototype network was to conduct tests on AeroMACS fixed and mobile services performance in an airport setting for the purpose of providing inputs for standard development. The national Testbed was planned to be installed at three Cleveland area airports: Hopkins International, Burke Lakefront, and Lorain County Regional, and at Sensis near Syracuse New York [35]. The Testbed concept in its evolutionary path eventually resulted in what is now known as NASA-CLE CNS Testbed, installed over Cleveland Hopkins International Airport CLE and the nearby NASA GRC campus, with some assets located at Burke Lakefront airport and Lorain County Regional airport. Although our coverage will be on NASA-CLE CNS Testbed, it should be mentioned that AeroMACS technology assessment and field performance evaluation are currently conducted in a number of other locations as well. Harris Corporation trials in Florida over Daytona and Melbourne airports, Sensis Corporation

---

7 SIGMET (Significant Meteorological Information) is a weather and meteorological information for flight safety of all aircraft, which is usually broadcast on the ATIS facilities, or over VOLMET radio stations.

prototype installed testbed at Syracuse Hancock International Airport, and SESAR-Airbus-Indra EUROCONTROL project at Toulouse Airport in France are examples of this sort.

6.5.1 Testbed Configuration

The AeroMACS prototype Testbed, initially, was configured to test wireless interconnection of several Airport Surface Detection Equipment sensor (ASDE-X) systems on the airport surface. And since IEEE 802.16e-based equipment was not available at the time, the sensor data was transmitted to a central processing site using IEEE 802.11-based WiFi system, operating in C-band. The AeroMACS prototype was installed over CLE and parts of NASA Glenn facilities, before the testing of ASDE-X sensor system with IEEE 802.11 was completed. Figure 6.3, adopted from Ref. [35], provides a diagram of the layout of the current AeroMACS Testbed. The prototype Testbed network consists of two AeroMACS (WiMAX) base stations, one installed on CLE property and the second is mounted on a tower adjacent to NASA GRC Flight Research Building hangar. The existence of two BSs in the network provides coverage redundancy and at least two opportunities for a MS to link up with a BS. Each BS contains multiple base transceiver station (BTS) sectors that are implemented for the purpose of increasing coverage area, link sensitivity, and throughput. The BS located in NASA Glenn's property has two BTS sectors that are directed 55° and 200° azimuth from "true north." The BS

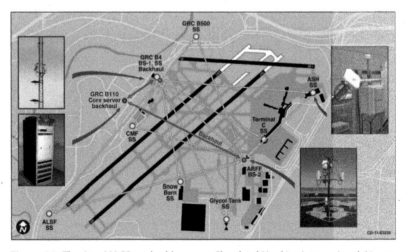

Figure 6.3 The AeroMACS testbed layout at Cleveland Hopkins International Airport and NASA Glenn campus. (Adopted from Reference [37] with permission.)

installed over the CLE area has three BTS coverage sectors directed 45°, 185°, and 295° from true north. The coverage area of each sector is 90° in azimuth determined by the −3 dB pattern roll-off of the BTS sector antenna. These sector-coverage assignments provide coverage redundancy across the desired areas of the runways, most of the taxiways, and much of the ramp zones [36]. The Testbed network also includes ASN–GW CSN functions to provide QoS control, user authentication and authorization, and handoff between BSs and adjacent sectors. It should be noted that the Testbed does not interact with the live airport operations at Hopkins airport, and it is solely configured for experimental use and not as an operational system [7].

As Figure 6.3 shows, there are a number of fixed SSs, microwave backhaul channels connecting the two AeroMACS base stations, and a core sever, in the prototype network. Subscriber stations are placed over ASDE-X sensors, enabling the testing of the sensor system, using IEEE 802.16e network as opposed to IEEE 802.11 WiFi system. The AeroMACS Testbed also includes a mobile element in the form of a moving truck that is called Aeronautical Research Vehicle (ARV). The ARV is a ground vehicle fully equipped with SS equipment, test measurement systems, and video cameras. The AeroMACS SS unit and two antennas are mounted on the roof of ARV. This setup enables the testing of many mobile and fixed aspects of the AeroMACS network under various conditions expected for the airport surface environment. This is accomplished in a cost-efficient and simplified manner by avoiding the expenses and difficulties of using an aircraft for such tests [37], however, later an actual Boing 737 jet aircraft was used for more accurate testing on AeroMACS mobile applications.

6.5.2 Early Testing Procedures and Results

As it has been mentioned earlier, the combination of mobility scenarios and propagation environments on the surface of an airport can be vastly different from one spot to another, all within a relatively small area. This ranges from terminal areas with large number of low vehicular speed or parked MSs and SSs inside a highly multipath environment of LOS and NLOS propagation, to open areas of takeoff and landing runways with low multipath and high-speed aircraft movement.

Originally, the Testbed served to demonstrate the feasibility of using wireless communications for airport surface surveillance, through interconnection of several ASDE-X units and transmitting the sensor data to a central unit, using WiFi network operating in C-band. In 2009, the Testbed was used to examine AeroMACS equipment operating over MLS extension band and utilizing IEEE 802.16e-based BS and MS equipment. During the time frame of 2010 through 2011, AeroMACS prototype testing was conducted on the network consisting of fixed subscriber stations, using the testbed configuration shown in Figure 6.3, under the supervision of NASA GRC and ITT groups [38], with FAA and ICAO guidance [36].

In 2012, two successful AeroMACS demonstrations were conducted over the Testbed. The purpose of the first demonstration, which utilized a Boeing 737 jet aircraft equipped with AeroMACS antenna and subscriber station, was to evaluate AeroMACS performance in bringing safety-critical flight information to the cockpit. The second demonstration tested the suitability of AeroMACS network to transmit and distribute live safety-critical data to the stationary nodes on airport surface. This experimentation provided the first demonstration of a broadband transmission of live safety-critical data over an airport surface, and established the potential of AeroMACS for satisfying many airport surface communication requirements vis-à-vis fixed assets.

6.5.2.1 Mobile Application Testing with ARV

Conducting tests on the mobility requirements of AeroMACS are more challenging and more critical for evaluation of overall AeroMACS network performance. Initially, the ARV was extensively used for mobility-related tests and measurements. In order to prevent intervention with airport operation, the ARV often operated around the perimeter of CLE, along with some critical mobility testing on taxiways and even runways when airport operation was not proceeding. The variety of testing locations enabled measurements for both LOS and NLOS scenarios between the BSs and MS. The AeroMACS testing consisted of AeroMACS mobility testing, testing for MIMO configurations, and testing of vehicle's speed effects [34].

Antenna configurations for the MS consisted of single antenna, double antennas separated by one wavelength, and double antennas separated by ten wavelengths. The results of antenna tests verified the performance superiority of MIMO over SISO configuration. The double antenna with ten wavelength separation outperforms the one wavelength separation case. An interesting result is the verification that MIMO configuration enables a reduction in BS output power, which is an important issue vis-à-vis AeroMACS interference to coallocated applications (see Chapters 7 and 8).

The mobility tests results show robust performance that supports throughput rates of 0.5–4.5 Mbps, depending on MS–BS distance and LOS/NLOS conditions. Insofar as the effects of MS velocity on traffic throughput is concerned, the tests demonstrate essentially no degradation on the throughput levels, up to about 40 KTS[8] (74 km/h). Higher vehicular speeds are addressed when tests are conducted with actual aircraft [36,38–39].

6.5.2.2 The Results of AeroMACS Mobile Tests with Boeing 737–700

A Boeing 737–700 is equipped with an AeroMACS antenna and an AeroMACS MS to conduct mobility tests over the Testbed using an actual aircraft for the first time. The mobility tests conducted with the aircraft were different from the earlier tests performed with ARV in two accounts; the AeroMACS antenna height was significantly greater and the vehicular velocity had a larger range. The AeroMACS MS onboard the aircraft was configured to associate with certain sectors of BS-2 (Figure 6.3). A Sensor Systems Model WLAN S65-5366-720 antenna was selected for the tests. This is a dual-band blade antenna designed for IEEE 802.16 standard-based WiMAX applications.

Essentially, the same series of tests were conducted with aircraft configuration as were run with ARV. Primarily, a set of mobility tests were run to evaluate the effectiveness of various antenna configurations. Tests were conducted to evaluate DL traffic throughput rate, where each BTS operated in 2 × 2 MIMO furnished with space-time coding. The MS antenna configuration was tested in three configurations: SISO, 2 × 1 MIMO (two receive antennas and one transmit antenna) with one wavelength spacing, and 2 × 1 MIMO with ten wavelength antenna separation [37]. The tests were performed using similar taxiway and runway routes at velocities, this time, up to 80 KTS. The initial activity involved the characterization of AeroMACS performance onboard of Boeing 737. This ensued the collection of AeroMACS performance data onboard of an actual aircraft and enabled a comparison with the previously collected ARV tests data, which is considered as "baseline tests." The comparison indicated that performance characteristics, for the most part, are very similar in both cases. An interesting noticeable performance effect is observed in the data related to average traffic throughput rate that shows a mild dependency on vehicular velocity. The AeroMACS MSs installed in the aircraft and ARV performed with close values of average traffic throughput at speed 40 KTS. However, as speed is increased, the AeroMACS MS installed in the Boeing 737 shows a gradual reduction in throughput performance. This is shown in Table 6.14, which provides the

---

8 KTS stands for knots, where one international knot is equal to one nautical mile per hour. Each knot (KT) is equal to 1.151 mile/h or 1.852 km/h. Hence, 40 KTS is approximately 74 km/h.

Table 6.14 Average traffic throughput versus speed.

Vehicle	Speed KTS (km/h)	Average throughput rate Mbps
ARV	40 (74)	4.7
Boeing 737	40 (74)	4.8
Boeing 737	60 (111)	4.2
Boeing 737	80 (148)	3.9

average AeroMACS traffic throughput as a function of vehicular velocity for both ARV and Boeing 737 [36,38–39].

The Testbed measurements, also, show the effects of MS–BS proximity, where the traffic throughput is visibly reduced as the aircraft moves away from serving BS. In summary, the test results indicate that AeroMACS can achieve good performance up to the speed of 50 KTS (92.6 km/h) as required by the current AeroMACS Profile [36].

6.5.2.3 AeroMACS Performance Validation

In addition to characterization of AeroMACS performance onboard of an actual aircraft, ITT Exelis conducted a series of tests over the Testbed as an initial aeronautical mobile validation of AeroMACS performance by successful demonstration of delivery of a safety-critical aeronautical mobile application to the cockpit through AeroMACS. The application chosen for this validation was a weather application known as "InFlight,"[9] capable of presenting the cockpit with critical enroute meteorological data such as AIRMETs (Airmen's Meteorological information), SIGMETs (Significant Meteorological Information), and terminal conditions through Meteorological Aerodrome Reports (METARS). Traditionally, the InFlight system consists of a SIRIUS satellite radio receiver that utilizes a 100BaseT Ethernet connection to interface the unit to an electronic flight bag (EFB) or multifunction display (MFD). The demonstration was about interfacing the satellite radio receiver unit into the AeroMACS testbed and utilizing the wireless data link to provide "last mile" services to the cockpit, which was eventually carried out through AeroMACS connection to WSI via internet [36]. Due to a satellite receiver failure, the configuration was further altered to utilize AeroMACS to connect to WSI demonstration system connected to the internet. This initial testing for AeroMACS performance validation demonstrated that a two-way data link providing weather information that emulates services such as

---

9 InFlight is a WSI Corporation electronic flight bag (EFB) weather application [37].

D-ATIS (Automatic Terminal Information Service), for example, VOLMET,[10] and D-SIGMET messages could be reliably provided over the AeroMACS Testbed [38].

6.6 Summary

In this chapter, preliminary information related to creation, standardization, testing and evaluation (through Testbed), and potential airport surface services that may be conveyed by AeroMACS are explored. AeroMACS is based on an interoperable version of IEEE 802.16-2009 standards, that is, WiMAX (WiMAX Forum System Profile Release 1.09) technology. This was in contrast to assembling a proprietary dedicated technology. Using an established technology standard, such as IEEE 802.16-2009, has many advantages, some of them are listed below.

1) Rapid and cost-effective standardization process and consequently timely deployment of the system by avoiding lengthy technical and political discussions.
2) Relying on the best wireless expertise from different wireless communications manufactures at the global level.
3) The standards are established by one of the world's most qualified standardization institutions, in this case, the IEEE.
4) Certification agencies, such as WiMAX Forum, define procedure that can facilitate interoperability.
5) Taking advantage of *market scale factor*, by using "COTS" (commercial off the shelf) products that are manufactured for a much larger consumer market than just that of aviation. Thus, airlines and airports have access to multiple industrial vendors that enable competitive pricing [40].

It should be noted that some modifications to COTS products might be needed, as they will be applied for MLS Extended band. These products must also be compliant to aviation safety standards as stated in DO-178B and DO 254.

The standardization process starts with technology selection, for AeroMACS the IEEE 802.16-2009-based WiMAX was adopted. For any driven technology, the IEEE 802.16 standard brings with itself a large number of PHY layer and MAC sublayer optional and mandatory protocols to select from. The WiMAX Forum System Profile Version 1.09 was selected as the parent technology based on which AeroMACS profile

---

10 VOLMET, French acronym of VOL (flight) and METEO (meaning weather), is a global short-wave radio network that broadcasts meteorological information for aircraft in flight.

was developed by RTCA. SARPS was developed almost simultaneously with the AeroMACS Profile, developed by RTCA and EUROCAE. Minimum Operation Performance Standards (MOPS) and MASPS were the last pieces of the standardization process for AeroMACS. Finally, the AeroMACS standardization documents became a source for developments of an AeroMACS technical manual and an installation guide document.

References

1 A. Malvern, "Improvement to VHF Air-to-Ground Communications," *IEE Colloquium on Air-to-Ground Communications*, 1997.
2 R. J. Kerczewski, J. M. Budinger, D. E. Brooks, R. P. Dimond, S. DeHart, and M. Borden "Progress on the Development of Future Airport Surface Wireless Communication Network," *Proceedings of IEEE Aerospace Conference*, 2009.
3 RTCA Paper No. 182-09/PMC-749, "Terms of Reference, SC 223 Airport Surface Wireless Communications," August 7, 2009.
4 RTCA, Minimum Operational Performance Standards (MOPS) for Aeronautical Mobile Airport Communication System (AeroMACS), RTCA DO-346, September 2014.
5 EUROCONTROL, "Evolution of the AM(R)S VHF Capacity in the ECAC States," *ACP WGC 6th meeting*, Toulouse, France, October 2003.
6 ICAO COCR, "Communications Operating Concept and Requirements for the Future Radio System" (COCR) Version 2.0, ICAO ACP Repository, May 2007, 2007.
7 J. M. Budinger and E. Hall, Aeronautical Mobile Airport Communications System (AeroMACS), in S. Plass, (ed.), *Future Aeronautical Communications*, InTech, 2011.
8 EUROCOTROL/FAA, Action Plan 17, "Future Communications Study Final Conclusions and Recommendations Report," Version 1.1, November 2007.
9 EUROCONTROL, "IEEE 802.16E System Profile Analysis for FCI's Airport Surface Operation," September 2009.
10 EUROCONTROL, "AeroMACS System Requirement Document," February, 2010. (Revised July.)
11 ICAO, ACP WG-S/2, "Aeronautical Mobile Aerodrome Communications System", Montreal Canada, October 2012.
12 ICAO, ACP-WG-S/WP-07, "AeroMACS Operations Concept," Montreal, Canada, October 2012.
13 ICAO, ACP WG-S/WP01 Draft AeroMACS SARPS, "Proposed Aeronautical Mobile Airport Communications System (AeroMACS) SARPS," September 2014.

14 ICAO, ACP WG-S/5, "Proposed Aeronautical Mobile Airport. Communications System. (AeroMACS) SARPS," September 2014.

15 EUROCONTROL, "Airport Surface Data Link (AeroMACS)", 2015.

16 ICAO, "Adoption of Amendment 90 to Annex 10, Volume III", April 2016.

17 NextGen – SESAR, "State of Harmonization Document," December 2014.

18 RTCA, "Aeronautical Mobile Airport Communications System (AeroMACS) Profile," RTCA-DO-345, December 18, 2013.

19 EUROKAE, "Aeronautical Mobile Airport Communications System (AeroMACS) Profile," 142/ED-222, November 2013.

20 WiMAX Forum, "Air Interface Specifications WiMAX Forum® Mobile System Profile" WMF-T23-001-R010v09, September 2010.

21 EUROCONTROL/FAA, Memorandum of Cooperation, "Action Plan 17 Future Communications Study Final Conclusions and Recommendations Report," Version 1.1, November 2007.

22 U.S. Government Publishing Office, "Electronic Code of Federal Regulations, e-CFR," Title 1, Chapter I, Subchapter D, Part 90, November 23, 2016.

23 IEEE, IEEE 802.16–2009 Standard for Local and Metropolitan Area, Part 16: Air Interface for Broadband Wireless Access Systems, May 2009.

24 Mobile WiMAX Standard Référence Document, DRAFT-T23-004-R010v02-C_SRD, 2011.

25 RTCA, "Aeronautical Mobile Airport Communications System Profile," DRAFT-T24-002-R010v01-H_TPA, 2011.

26 IEEE, IEEE 802.16–2009 Standard for Local and Metropolitan Area, Part 16: Air Interface for Broadband Wireless Access Systems, May 2009.

27 RTCA, "RTCA Documents Definitions," 2013. Available at www.rtca. org/documents.

28 EUROCAE, "Minimum Operational Performance Standards (MOPS) for the Aeronautical Mobile Airport Communication System," 143/ED-223, October 2013.

29 EUROCAE, "Minimum Aviation System Performance Standards (MASPS) AeroMACS" 146/ED-227, September 2016.

30 ARINC Airlines Electronic Engineering Committee, "AEEC Project Initiation/Modification (APIM)," March 8, 2014.

31 ICAO, "Manual on the Aeronautical Mobile Airport Communications System (AeroMACS)" Doc 10044, 2016.

32 Federal Aviation Administration, "Engineering Brief No. 97, Guidance for AeroMACS Installation by the Airport Operator," June 3, 2016.

33 WiMAX Forum First Aviation Event, Washington DC, September 2013.

34 C. A. Wargo and R. D. Apaza, "Application Survey for the Future AeroMACS," *Proceedings of IEEE ICNS*, pp G7-1–G7-10, 2011.

35 ICAO, ACP-WG-S/5 WP-10, "DL: UL OFDM Symbol Ratio in AeroMACS TDD Frame," July 2014.

36 R. J. Kerczewski, R. D. Apaza, and R. P. Dimond,"AeroMACS System Characterization and Demonstrations, "*Proceedings of IEEE Aerospace Conference*, 2013.

37 ICAO, "Testbed for Prototype C-band Airport Surface Wireless Network," June 2006.

38 ICAO, "Status of AeroMACS Prototype Mobility Test and Evaluation," March 2012.

39 W. Hall J. Magner, N. Zelkin, S. Henriksen, B. Phillips, and R. Apaza, "C-Band Airport Surface Communications System Standards Development – AeroMACS Prototype Test Results Summary," *2012 ICNS Conference*, April 2012.

40 EUROCONTROL, "IEEE 802.16E System Profile Analysis for FCI'S Airport Surface Operation," September 2009.

7

AeroMACS Networks Characterization

7.1 Introduction

Chapter 6 explored the ideas and incentives behind the creation of AeroMACS as a standard-driven technology, its standardization process, its testing and evaluation based on measurements conducted over a prototype test bed, and the general evolution of the technology. In this chapter, we study AeroMACS as a short-range high-aggregate-data-throughput broadband wireless communications system and concentrate on the detailed characterization of AeroMACS Physical layer (PHY) and media access sublayer (MAC) features. Subsequently, the AeroMACS architecture and network reference model (NRM) are discussed. A brief discussion on functionalities and position of the AeroMACS network within the broader contexts of the airport network and the Aeronautical Telecommunications Network (ATN) follows. AeroMACS is planned to be an all-IP network that supports high-rate packet-switched ATC and AOC services for efficient and safe management of flights, while providing connectivity to the aircraft, operational support vehicles, and personnel within the airport area.

Chapter 6 and previous chapters have made the case for selection of IEEE 802.16-2009 technology for AeroMACS networks. In addition to satisfying a major objective of making the most of existing technologies, as opposed to establishing a proprietary standard, the technical characteristics of WiMAX systems closely match the communication requirements and constraints of the airport surface channel. In particular, NLOS propagation mode in the airport channel is supported by provision of OFDM, inherited by AeroMACS from IEEE 802.16 standards and WiMAX technology.

The IEEE 802.16-2009 standard, however, contains a large number of options and features and as such, as a whole, represents a technology complex rather than a single interoperable communication system.

AeroMACS: An IEEE 802.16 Standard-Based Technology for the Next Generation of Air Transportation Systems, First Edition. Behnam Kamali.
© 2019 the Institute of Electrical and Electronics Engineers, Inc. Published 2019 by John Wiley & Sons, Inc.

WiMAX Forum is charged with the task of constructing interoperable technologies by assembling a subset of optional and mandatory protocols of the IEEE 802.16 standards, through what is known as WiMAX System and Certification Profiles. In other words, the WiMAX Forum downscaling of the scope of an extensive suite of protocols of a massive standard enables the construction of interoperable technologies. AeroMACS is a WiMAX-driven network, specifically based on WiMAX Forum Mobile System Profile Release 1.0 version 0.9 of the IEEE 802.16-2009 standard [1]. The IEEE 802.16-2009 standard stipulates the air interface of combined fixed and mobile broadband wireless access, specifying PHY and MAC layers and leaving the higher layers unspecified to be selected from various available network technologies such as IP and Ethernet. This provision essentially separates the access service network from connectivity service network and, therefore, enables network design flexibility. The medium access control sublayer is designed to support multiple PHY specifications that are applicable to various operational environments. AeroMACS as a telecommunications network has an architecture that is defined by IEEE 802.16-2009 standard, however, an implicit objective is to integrate AeroMACS, as a subnetwork, into an IP-based aeronautical network (ATN), in other words AeroMACS can be considered as an all IP network. We now explore the PHY and MAC layers features and functionalities of AeroMACS networks.

7.2 AeroMACS Physical Layer Specifications

AeroMACS physical layer protocols and functionalities are based on PHY Layer specifications of the IEEE 802.16-2009 standard. The key PHY layer feature of the IEEE standard is OFDMA that may be scaled to accommodate several channel bandwidths from 1.25 to 20 MHz (see Chapter 4), supporting various levels of requested data throughput and capacity. AeroMACS channel bandwidth at this stage is chosen to be of 5 MHz, however, a possible extension to 10 MHz is currently under investigation [2]. The Standard supports both time-division duplexing (TDD) and frequency-division duplexing (FDD) schemes, however, AeroMACS network has adopted TDD, for the reasons that were outlined in Chapter 6. The AeroMACS specifications are, therefore, based on the Common Part TDD of WiMAX profile [3]. With TDD, the DL and UL segments of the OFDMA frame can be adaptively selected for the provision of a dynamic allocation of DL and UL resources, leading to efficient support of asymmetric DL/UL traffic. It should be noted that only a single channel is required for this duplexing technique, that testifies to the spectral efficiency of the TDD scheme. Table 7.1 highlights the key PHY layer features of AeroMACS networks. The last three rows of Table 7.1 represent some cutting-edge PHY layer features of modern wireless networks such as AeroMACS, they will be explored in some detail later in this chapter. As Table 7.1 shows a number of modulation formats and channel coding schemes are supported by AeroMACS standard, two sections of this chapter are dedicated to these subjects.

Table 7.2 provides a quick review of some important PHY layer parameters of AeroMACS systems.

Table 7.1 Key AeroMACS physical layer features.

Feature/parameter	Description/quantity	Comments
Access technology	Orthogonal frequency division multiple access (OFDMA)	Scalable to accommodate several possible channel bandwidths
Duplexing scheme	Time division duplexing (TDD)	Supports asymmetric DL/UL traffic
Subchannelization	Partial usage subchannelization (PUSC)	Optimum for multiuser scenarios and channel estimation, provides high frequency diversity
Modulation formats	QPSK (4-QAM), 16-QAM, and 64-QAM	16-QAM is an option for UL
Channel coding	Convolutional codes (CC), convolutional turbo codes (CTC), repetition codes, Reed–Solomon (RS) codes	AeroMACS limits the rates for CC and CTC to 1/2, 2/3, and 3/4, repetition codes rates 1/2, 1/3, 1/6
Adaptive modulation and coding (AMC)	Dynamic link adaptation, through the selection of modulation and coding combination that best match the channel conditions	AMC enables transmission of more bits per symbol leading to higher overall throughputs and greater spectral efficiencies
Multiple-input multiple-output (MIMO)	AeroMACS applies MIMO exclusively for spatial diversity	MIMO is applied in AeroMACS exclusively for enhancement in reception performance
Fractional frequency reuse (FFR)	Adaptive selection of frequency reuse factor (FRF) in AeroMACS cells	A technique used for mitigation of co-channel or co-subchannel interference in OFDMA networks

Table 7.2 Important AeroMACS physical layer parameters.

Parameter	Quantity/value	
Channel bandwidth	5 MHz (10 MHz under investigation)	
Fast Fourier transform size	512 (1024 under consideration)	
Frame duration	5 ms divided into DL and UL subframes	
Sampling factor	28/25	
Sampling frequency	5.6 MHz	
OFDM symbol duration	102.9 μs	
OFDM symbols per frame	48	
Subcarrier spacing	10.94 kHz	
Number of data subcarriers	DL: 360	UL: 272
Number of subchannels	DL: 15	UL: 17
Cyclic prefix duration factor	1/8	

In the same fashion that WiMAX technology is established by assembling a subset of mandatory and optional protocols of IEEE 802.16e standards, AeroMACS chooses a subset of functionalities and features of WiMAX needed and appropriate for airport surface communications, which further scales down the scope of IEEE 80.16-2009 standard. Table 7.3 highlights the differences between WiMAX and AeroMACS's main PHY layer features and parameters.

Table 7.3 Physical layer differences between WiMAX and AeroMACS.

Feature/parameter	WiMAX	AeroMACS
Bandwidth	5, 10, and 20 MHz	5 MHz
Duplexing method	TDD and FDD	TDD
Subchannelization	PUSC, FUSK, AMC	PUSC
MIMO	Advanced antenna techniques, beamforming, spatial diversity	Spatial diversity
Modulation	BPSK is optional	BPSK is not supported

7.2.1 OFDM and OFDMA for AeroMACS

In Chapter 4, we explored the virtues of OFDM as a modulation and multiplexing technique, and described its associated scalable multiple access technology, OFDMA. Perhaps, the single most important characteristic of OFDM/OFDMA multicarrier schemes vis-à-vis the multipath fading environment is their capability to combat the effects of frequency selective fading and the associated ISI, which appears once the transmission speed has increased beyond certain level. In this manner, the design and implementation of costly and complex adaptive equalization filters are avoided or at least simplified. This is of particular interest in AeroMACS networks, since the propagation mode variations in an airport surface channel could create harsh fading, wide delay spread, and tap correlation distortions.

As the general OFDM/OFDMA system characteristics reveal (see Chapter 4), the physical resources are divided among users following a specific subchannelization procedure. The IEEE 802.16-2009 standard supports multiple subchannelization formats, however, as Table 7.2 indicates, AeroMACS applies partial usage subchannelization (PUSC) scheme exclusively. This is a mandatory subcarrier allocation that is used in every AeroMACS frame. With PUSC data regions are assigned to subchannels following a pseudorandom permutation, in order to guarantee high-frequency diversity [2].

7.2.2 AeroMACS OFDMA TDD Frame Configuration

Since the duplexing configuration for AeroMACS is TDD, the uplink and downlink subframes are transmitted at different times, while they use the same radio channel. The TDD frame has a fixed duration that is divided between a single DL subframe and a single UL subframe. The frame is further divided into an integer number of physical slots (PS), which help to partition the bandwidth. The split between UL and DL is adaptive and is a system parameter controlled at higher layers within the AeroMACS network. Figure 7.1 shows the AeroMACS TDD frame structure.

Parameter n can be calculated from Equation 7.1 [4].

$$n = \frac{f_s \times T_F}{4} \tag{7.1}$$

Where, f_s is the sampling frequency and T_F is the frame time duration. The time duration of physical slots depends upon channel bandwidth and sampling factor, and therefore, when these parameters are selected for AeroMACS, slot time duration is set to a fixed value. Consequently, an AeroMACS OFDM symbol is composed of constant number of slots once frame duration time and sampling frequency are determined. The

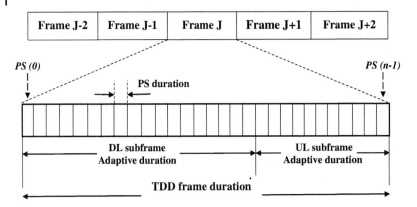

Figure 7.1 AeroMACS TDD frame structure with variable time division between DL and UL subframes.

number of OFDM symbols in an AeroMACS frame depends on the FFT size, CP duration time, sampling factor, and channel bandwidth, and therefore, has a fixed value once the value of these quantities are determined. According to IEEE 802.16-2009 [5], the sampling factor is set in direct relation to channel bandwidth, as Table 7.4 suggests.

The channel bandwidth for AeroMACS is 5 MHz, corresponding to FFT size of 512 points. The sampling factor according to Table 7.2 is 28/25, and the cyclic prefix duration time is 1/8 of the useful OFDM time length. Having all these values at disposal, the time duration of an AeroMACS OFDM symbol may be calculated as shown in Equation 7.2 [6].

$$T_{\text{S-OFDM}} = \frac{N_{\text{FFT}}}{\eta_{sam} \times BW}\left(1 + \frac{T_{cp}}{T_S}\right) = \frac{512}{28/25 \times 5 \times 10^6}\left(1 + \frac{1}{8}\right) = 102.9\,\mu s$$

(7.2)

In order to calculate the number of OFDM symbols in a 5 ms long frame, one needs to determine the useful transmission time within the frame time by subtracting the sum of transmit time gap (TTG) and receive time gap (RTG) (see Chapter 4) from 5 ms, and dividing the resultant time by

Table 7.4 Sampling factor selections for various channel bandwidth [5].

Channel bandwidth (BW)	Multiple of 1.75 MHz: 1.75 MHz, 3.5 MHz, and so on	Multiple of 2.5 MHz: 2.5 MHz, 5 MHz, 10 MHz, 20 MHz	Otherwise
Sampling factor: η_{sam}	8/7	28/25	8/7

$T_{\text{S-OFDM}} = 102.9\,\mu s$, as shown in Equation 7.3.

$$N_{\text{OFDM-S}} = \frac{T_F - RTG - TTG}{T_{\text{S-OFDM}}} = \frac{0.005 - 6 \times 10^{-5} - 103.7 \times 10^{-6}}{102.9 \times 10^{-6}} = 47$$

$$(7.3)$$

The value of RTG is given by the Standard and TTG and RTG are related by Equation 7.4.

$$\text{TTG} = T_F - \text{RTG} - N_{\text{OFDM-S}} \times T_{\text{OFDM-S}} \qquad (7.4)$$

The 47 OFDM symbols are divided between DL and UL subframes. At the beginning of each frame, the BS announces the number of OFDM symbols transmitted in the DL and UL subframes in the corresponding DL-MAP (media access protocol) and UL-MAP messages.

As we mentioned earlier, an important advantage of TDD format over FDD scheme is the ability of TDD to lend itself to asymmetric DL/UP link transmission. It is actually possible to dynamically change the DL to UL ratio in the TDD frame configuration between contiguous BSs with no interruption in the service. A BS conceptually can, at the beginning of a frame, switch to a different DL/UL configuration as the serviced MSs are notified on a per frame basis. Furthermore, contiguous AeroMACS base stations may adopt their ratio configuration independently, as the BSs belonging to contiguous cells use different radio channels and there is no risk of additional interference. Insofar as handover process is concerned, there is no complications as the mobile stations acquire the information regarding DL/UL ratio of the target cell during synchronization [6]. It should be mentioned, however, that current COTS products do not support adaptive frame configuration. In other words, TDD frame division into DL and UL subframes is assumed to be fixed to a particular DL/UL ratio for each BS. The only mechanism currently available to support alteration in DL/UL ratio is re-configuring the BS for a different DL/UL ratio; this, however, requires an interruption in the communications.

The IEEE 802.16-2009 standard [4] and AeroMACS MOPS [7,8] do not place any limitation on how TDD frames may be split between DL and UL subframes, as long as integer number of OFDM symbols are assigned to each subframes. The AeroMACS profile [9], however, allows a finite number of ways that 47 OFDM symbols may be split between DL subframe and UL subframe. These options are presented in Table 7.5.

Table 7.5 specifies that all these options are mandatory and are supported by BSs, SSs, and MSs, and in case IEEE 802.16j-based relays are in the future of AeroMACS, the RSs need to support this TDD feature. A reasonable setting is to have 29 OFDM symbols for the DL and 18 OFDM symbols for the UL. However, the individual setting is dependent

Table 7.5 DL/UL ratio allowed for AeroMACS.

Number of OFDM symbols in DL	Number of OFDM symbols in UL	Nodes	Status
35	12	BS, SS/MS	Mandatory
34	13	BS, SS/MS	Mandatory
33	14	BS, SS/MS	Mandatory
32	15	BS, SS/MS	Mandatory
31	16	BS, SS/MS	Mandatory
30	17	BS, SS/MS	Mandatory
29	18	BS, SS/MS	Mandatory
28	19	BS, SS/MS	Mandatory
27	20	BS, SS/MS	Mandatory
26	21	BS, SS/MS	Mandatory

on the service provider. Valid values can be taken from the WiMAX Forum™ Mobile System Profile Specification TDD Specific Part [10].

7.2.3 AeroMACS Modulation Formats

The process of digital modulation is more complex than that of analog modulation in many respects, however, the primary advantage of digital modulation is the simplicity of the recreation of the baseband signal at the receiver. A digital modulation signal set consists of a finite number of symbols (the alphabet) that are transmitted through the channel at a specified rate. At the receiver, based on some well-defined algorithm, the signal detector is to make "the best possible choice" as to which one of these symbols has been sent by the transmitter at any given time. Statistically speaking, the estimation process involved in the demodulation of analog-modulated signals is reduced to a merely decision procedure. In contrast to handful number of methods that are available in the analog modulation domain, the catalogue of digital modulation techniques is rich with several classes of modulation schemes that may or may not even have analog counterpart. One reason for this profusion is the fact that in digital modulation it is possible to vary more than one attribute of the sinusoidal carrier signal to form the modulated waveform. For instance, amplitude and phase of the carrier may be changed, this creates families of powerful digital modulation schemes such as quadrature amplitude modulation (QAM). The IEEE 802.16-2009 standard supports QAM digital modulation techniques with three different orders (noting that QPSK may be considered as 4-QAM).

Table 7.6 Modulation schemes supported by WiMAX and AeroMACS networks.

Mode	WiMAX DL	WiMAX UL	AeroMACS DL	AeroMACS UL
Optional	BPSK	BPSK	None	64-QAM
Mandatory	QPSK, 16-QAM, 64-QAM	QPSK, 16-QAM, 64-QAM	QPSK, 16-QAM, 64-QAM	QPSK, 16-QAM

The concept of link adaptation, to be discussed in the Section 7.2.5, generally means the dynamic adjustment of some PHY layer parameters such as modulation order and error correction coding rates according to noise and fading conditions of the channel. The IEEE 802.16-2009 standard, as well as WiMAX and AeroMACS networks, supports link adaptation in the form of adaptive modulation and coding. To this end several modulation schemes are available in the Standard. Table 7.6 provides the list of modulation schemes supported for DL and UL of WiMAX and AeroMACS Networks.

7.2.3.1 How to Select a Modulation Technique for a Specific Application

We digress to address an important issue critical to all digital communications systems, including WiMAX and AeroMACS. In general, a modulation/detection (demodulation), that is, a MODEM technique is selected for a given application based on evaluation of various performance measures of the technique, and how closely they match the characteristics of the transmission channel and the requirements set forth by the user. Critical performance measures for the selection of a MODEM system for a specific application are presented in the following paragraphs.

1) An important measure of performance for various MODEM techniques is the average probability of error in the detection process. This is normally provided either by a mathematical equation or as a graph that plots bit error rate (BER) versus expected value of bit-energy-to-noise power spectral density (PSD) ratio, E_b/N_0, measured in decibels. In regard to error probability benchmark, often time, MODEM techniques are compared on the basis of the value of E_b/N_0 they require to achieve a specified BER value, typically 10^{-5}. For instance, BER equation for QPSK modulation (which is also equal to BER equation for BPSK) with coherent detection (matched-filter detection), and the assumption of Gray coding signal arrangement is given by Equation 7.5.

$$P_{QPSK} = P_{BPSK} = Q\left(\sqrt{\frac{2E_b}{N_0}}\right) \tag{7.5}$$

This MODEM technique requires approximately 9.6 dB of E_b/N_0 at the front end of the receiver to deliver BER of 10^{-5}. To make a comparison, let us consider coherently detected binary frequency-shift keying (BFSK) MODEM system, whose BER is given by Equation 7.6.

$$P_{BFSK} = Q\left(\sqrt{\frac{E_b}{N_0}}\right) \tag{7.6}$$

Using Equation 7.6, it can be shown that for a BFSK system to deliver a BER of 10^{-5}, the required E_b/N_0 is 12.6 dB. It is, therefore, concluded that QPSK and BPSK performs better than BFSK by 3 dB, at this BER. This marks the superiority of coherently detected QPSK and BPSK over that of BFSK in error performance, or equivalently, in power efficiency.

2) A second important criterion for comparison of various modulation schemes is *spectral efficiency (bandwidth efficiency) η*, defined as the ratio of data rate to channel bandwidth, measured by bits per second per hertz. Bandwidth efficiency quantifies the amount of data that may be transmitted in a specified bandwidth, within a given time. It therefore reflects how efficiently the allocated spectrum is utilized. Assuming null-to-null bandwidth, that is, frequency spacing between the nulls bounding the main lobe of the modulated signal power spectral density, and the knowledge of baseband pulse shaping technique, one can come up with simple equations for η. For an M-ary modulation scheme with ISI-free raised-cosine baseband pulse shaping method, with roll-off factor α, the spectral efficiency of the modulated signal can be calculated from Equation 7.7 [11].

$$\eta = \frac{\log_2 M}{1 + \alpha} \quad \text{bits/s/Hz} \tag{7.7}$$

It should be noted that the ideal ISI-free Nyquist pulse is a special case of raised-cosine pulse in which the roll-off factor $\alpha = 0$.

3) Bandwidth efficiency is not the only important spectrally related aspect of a modulation scheme. The null-to-null bandwidth is a measure of required channel bandwidth for the modulated signal. However, null-to-null bandwidth alone may not provide adequate information regarding all aspects of spectral behavior of a modulation technique. Another form of bandwidth that reflects spectral compactness of a modulated signal is power containment bandwidth (PCB). Power containment bandwidth is defined as a band of frequencies that contains a specified percentage of the signal power. For instance, the 90% PCB is the width of the frequency band that encloses 90% of the power of the modulated signal. However, another relevant spectrum-related parameter of a modulated signal is the rate at which the

sidelobes of the signal PSD fall off, which is an important issue in regard to interference into adjacent radio channels. For instance, the sidelobes of QPSK spectrum falloff at a rate of 6 dB/Octave whereas minimum shift keying (MSK, a continuous-phase modulation method) modulation's sidelobes fall off at a rate of 12 dB/Octave [12]. In order to curtail the effect of adjacent channel interference, the modulated signal is oftentime filtered before transmission. Nonetheless sharp spectral falloff is a fine trait for a modulation scheme to be employed for wireless applications.

4) Hardware or software complexity/cost of implementation of the modulator/detector system is another criterion for selection of a MODEM technique. Hardware/software availability in the form of COTS VLSI circuits, for instance, that might be used to implement the MODEM system, is also an issue to be considered while selecting a MODEM technology for an application [12].

7.2.3.2 General Characteristics of Modulation Schemes Supported by AeroMACS

As Table 7.5 illustrates, modulation schemes available in AeroMACS standard consists of three square M-QAM modulations with $M = 4$, 16, and 64, where 4-QAM is equivalent to QPSK. Quadrature amplitude modulation schemes are widely applied in data communications and digital radio systems, and particularly with OFDM and OFDMA systems (see Chapter 4). QAM waveforms are constructed by changing the amplitudes of two orthogonal carrier components of the same frequency, in accordance with variation in the baseband signal sequences, and transmitting the sum of the two components simultaneously. The general expression for QAM signals is given by Equation 7.8.

$$s(t) = I(t)\cos\left(2\pi f_c t\right) + Q(t)\sin\left(2\pi f_c t\right) \tag{7.8}$$

Where $I(t)$ is the "in phase" component that determines the amplitude of carrier $\cos\left(2\pi f_c t\right)$ and $Q(t)$ is the "quadrature component" defining the amplitude of the carrier $\sin\left(2\pi f_c t\right)$.

An expression for bit error probability (BER) of M-QAM modulation, with coherent detection and Gray coding arrangement of the signal set, is provided in Equation 7.9 [11].

$$P_B = \frac{2\left(\sqrt{M} - 1\right)}{\sqrt{M}\log_2\sqrt{M}} Q\left(\sqrt{\frac{6\log_2\sqrt{M}}{M - 1} \times \frac{E_b}{N_0}}\right) \tag{7.9}$$

An advantage of M-QAM modulation schemes is their high spectral efficiency, particularly when AMC schemes are in place. In other words,

Table 7.7 Some defining attributes of AeroMACS modulation schemes.

Modulation format	QPSK	16-QAM	64-QAM
BER: Coherent detection with Gray coding arrangement	$P_B = Q\left(\sqrt{\dfrac{2E_b}{N_0}}\right)$	$P_B = \dfrac{3}{4}Q\left(\sqrt{\dfrac{4E_b}{5N_0}}\right)$	$P_B = \dfrac{7}{12}Q\left(\sqrt{\dfrac{2E_b}{7N_0}}\right)$
E_b/N_0 required for BER $= 10^{-5}$	9.6 dB	13.4 dB	17.8 dB
Spectral efficiency with raised-cosine filter roll-off factor α	$\dfrac{2}{1+\alpha}$ bits/s/Hz	$\dfrac{4}{1+\alpha}$ bits/s/Hz	$\dfrac{6}{1+\alpha}$ bits/s/Hz
Power spectral density (PSD) [13]	$4E_b\left[\dfrac{\sin(2\pi f T_b)}{2\pi f T_b}\right]^2$	$10E_0\left[\dfrac{\sin(4\pi T_b f)}{4\pi T_b f}\right]^2$	$42E_0\left[\dfrac{\sin(6\pi T_b f)}{6\pi T_b f}\right]^2$
Null-to-null bandwidth	$2R_s = R_b$	$2R_s = R_b/2$	$2R_s = R_b/3$

when M is selected adaptively, the M-QAM modulation scheme can carry more data bits per symbol as M is increased.

Table 7.7 summarizes some important properties of AeroMACS's three modulation schemes.

The parameter shown by E_0 in the Table 7.7 denotes the energy of the waveform with lowest amplitude, R_s is the symbol rate for QAM modulations, and R_b is the bit rates expressed in bits/second. In general, R_s and R_b in an M-ary modulation scheme are related by $R_b = R_s \times \log_2 M$. Table 7.7 demonstrates that the price that is paid for higher spectral efficiency of QAM modulation schemes is the requirement for greater levels of signal energy.

Figure 7.2 illustrates signal constellations for 4-QAM (QPSK) and 16-QAM modulation formats. Figure 7.3 demonstrates the same for 64-QAM modulation scheme. Along with each constellation diagram, bit sequence-to-symbol (waveform, corresponding to signal points in the constellations) assignments according to Gray coding are shown. For the QAM modulations supported by AeroMACS, constellation diagrams are arranged in a square grid with adjacent symbols (signal points) having equal vertical or horizontal spacing. Gray coding[1] is widely applied in digital communications for binary data sequence to symbol (waveform) mapping.

---

1 The Gray coding, also known as reflected binary code (RBC), is an encoding process in which successive binary sequences are different in only a single bit.

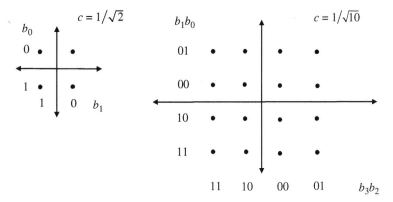

Figure 7.2 Constellation diagrams for 4-QAM (QPSK) and 16-QAM modulation formats with Gray coding bit-to-symbol mapping. (Adopted from Ref. [4] with permission of IEEE.)

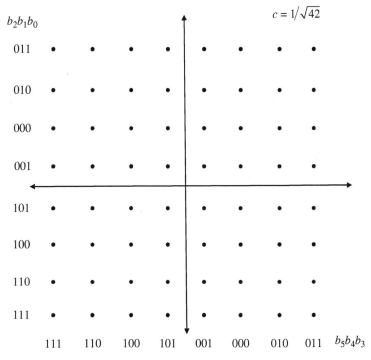

Figure 7.3 64-QAM signal constellation diagram with Gray coded bit-map arrangement. (Adopted from Ref. [4] with permission of IEEE.)

In all three constellation diagrams, b_0 denotes the least significant bit (LBS). In order to normalize the constellations, that is, equalize the average power in the signal sets, the constellations are multiplied by the factor c shown in each diagram.

Each signal point in a constellation diagram defines the amplitude and the phase of the sinusoidal waveform that is used for transmission of the corresponding binary sequence. As Figures 7.2 and 7.3 demonstrate, the data bit sequences assigned to adjacent symbols (signal points) differ in merely one binary digit. The significance of Gary coding arrangement manifests itself in the detection process and the related bit error rate. When a received waveform is detected in error, the likelihood that the symbol is erroneously mapped to one of the adjacent symbols is much higher than otherwise. This follows because of the Euclidean proximity of the adjacent symbols. Therefore, with Gray coding arrangement, the most likely erroneous symbol decisions result in merely a single bit in error [14].

7.2.4 AeroMACS Channel Coding Schemes

IEEE 802.16-2009 and its interoperable technology version WiMAX support two levels of error control coding. Primarily, WiMAX standards define an adaptive modulation coding scheme at the physical layer that employs several channel coding techniques. Second, there is a multilayer (link layer and transport layer) ARQ and HARQ error control included in the IEEE 802.16e and in mobile WiMAX standards. Our focus is on PHY layer coding method, and subsequently on how they can combine with various modulation formats for link adaptation in WiMAX and AeroMACS networks.

Channel coding is a mandatory procedure. The IEEE 802.16-2009 standard supports the coding process through four consecutive steps of randomization, FEC (forward error correction) encoding, bit interleaving, and repetition. This encoding process is followed by modulation and mapping into OFDMA subchannels. It should be mentioned that repetition may be applied only when QPSK modulation is used [4]. A brief description of FEC steps is provided here; for detail of randomization and bit interleaving, the reader is referred to the Standard's document [4].

7.2.4.1 Mandatory Channel Coding for AeroMACS

The mandatory FEC coding scheme, supported by the Standard, is a tail-biting nonsystematic nonrecursive (feed-forward) convolutional code (NSNRCC). The code is a (2, 1, 6) convolutional code with generator sequences given by $g_1 = (171)_{OCT}$ and $g_2 = (133)_{OCT}$. The encoder circuit

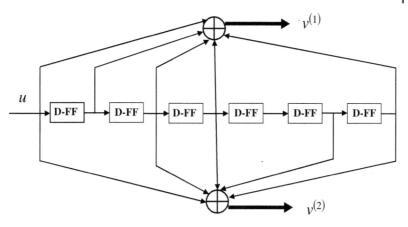

Figure 7.4 Encoder circuit for the basic (2, 1, 6) AeroMACS convolutional code.

for this code is shown in Figure 7.4. Generator sequences, once the Octal numbers are converted to their binary equivalent digit-by-digit, define the connection of the input sequence and encoder shift-register systems to each of the encoder outputs. For truncated convolutional encoding, after the data sequence has been entered the encoder, blocks of zero input are applied to the encoder to flush it out and return it back to the all-zero initial state, in preparation for next data sequence. This generates additional coded bits and reduces the actual code rate. A tail-biting convolutional encoding process ensures that the starting and ending states of the encoder is the same state that does not have to be necessarily the all-zero state. In this manner, the generation of additional coded bits are avoided [15].

The encoder is a sequential logic circuit with six D-flip-flops (DFF, delay element), each one provides one bit time-duration delay. Binary sequence u represents the input data and sequences X and Y are the coded sequences generated by the encoder. In order to create different rates, the code is punctured. Puncturing is a simple process of withholding certain encoded bits from being transmitted through the channel. This process increases the code rate but reduces the code error correction capabilities. Table 7.8 explains how the convolutional code words generated by encoder shown in Figure 7.4 are punctured for generation of various rates that are supported by WiMAX and AeroMACS standards. Binary sequences shown in Table 7.8 signify the transmission or withholding of coded bits, "1" means transmission and "0" implies withholding of the corresponding coded bit [4].

Table 7.8 Puncturing procedure for (2, 1, 6) convolutional code.

Code rate	1/2	2/3	3/4	5/6
d_{free}	10	6	5	4
Puncturing pattern for the first output	1 Intact	10 Every other coded bit is deleted	101 The middle bit out of every three coded bits is deleted	10101 Second and fourth bits, out of every five coded bits, are deleted
Punctured pattern for the second output	1 Intact	11 Intact	110 The third bit, out of every three coded bits is deleted	11010 The third and fifth bits, out every five coded bits are deleted
Notes	Supported by: WiMAX/ AeroMACS	Supported by: WiMAX/ AeroMACS	Supported by: WiMAX/ AeroMACS	Supported by: WiMAX

Table 7.8 reveals that as puncturing of the code words becomes deeper, the codes d_{free}[2] is reduced, which translates into decline in code's error correction capability, while, data throughput is increased.

7.2.4.2 Optional CC–RS Code Concatenated Scheme

In addition to the mandatory coding format, a number of optional coding schemes are supported by the Standard. One optional FEC scheme consists of cascading (concatenation) a Reed–Solomon (RS), as the outer code, and a convolutional code with Viterbi decoding, as the inner code. This is an interesting concatenation by virtue of the fact that the decoding errors out of the Viterbi decoder are "event errors," that is, burst errors, and RS codes are particularly powerful in correcting burst errors. The convolutional code is the (2, 1, 6) code whose encoder is shown in Figure 7.4. Major parameters of a typical applied RS code are highlighted in Table 7.9.

The overall encoding process is performed by first passing the data through the RS encoder. This code is a high rate (code rate in excess of 0.937) systematic code that means the encoder produces parity check words (bytes) followed by an exact replica of data words. The resultant RS codeword is passed through the convolutional encoder followed by the puncturing process, if required.

---

2 d_{free} is the minimum Hamming distance between any two code words of any length that is generated by a convolutional code, it is a key measure of the code's error correction capability [15].

Table 7.9 The characterization of the Reed–Solomon code for AeroMACS networks.

Item	Description	Comments
Finite field	$GF(256) = GF(2^8)$	Word length is 8 bits or 1 byte
Field generator polynomial	$p(X) = 1 + X^2 + X^3 + X^4 + X^8$	Not unique
Codeword and message block length	$N = 255, \quad K = 239$	Error correction capability: $T = \dfrac{N - K}{2}$
Code generator polynomial	$g(X) = \prod_{i=0}^{2T-1} (X + \alpha^i)$	If α is a primitive of $GF(256)$, the RS code is a narrow-sense and primitive
Code modification	May be shortened or punctured	Enables variable block sizes and error correction capabilities

7.2.4.3 Convolutional Turbo Coding (CTC) Technique

Another optional coding scheme is convolutional turbo coding. The encoder of CTC essentially consists of a constituent encoder, an interleaver, and a switch. The constituent encoder generates the coded bits using a double binary circular recursive (feedback) systematic convolutional code (CRSCC) [15] encoder. The general structure of the CTC encoder is depicted in Figure 7.5. The bits of the data to be encoded into CTC are alternately fed to inputs A and B, starting with most significant bit (MSB) of the first data byte being fed to A. The same data delivered to inputs A and B are fed to CTC interleaver that permutes the data bits. As Figure 7.5 demonstrates, the input sequence to the constituent encoder is provided by a switch that connects, alternately, to inputs A and B on one

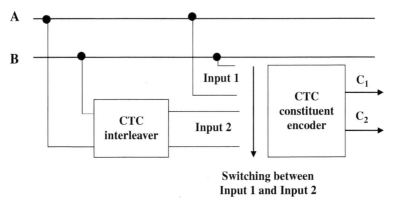

Figure 7.5 The general structure of the CTC encoder.

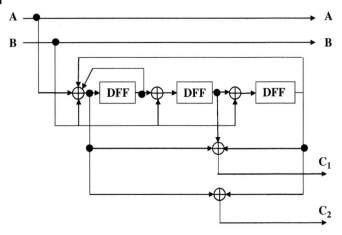

Figure 7.6 CTC constituent encoder circuit.

hand, and the output of the CTC interleaver, on the other hand. The parity check bits of coded sequence are shown in Figure 7.5 by C_1 and C_2. The circuit for the encoder of constituent CRSCC is shown in Figure 7.6. The constituent encoder is a linear feedback shift register system.

It should be noted that data links designated as A and B in Figure 7.6 are connected to the switch shown in Figure 7.5, that is to say they are alternately connected to actual data lines and the permuted version (interleaved) of the same data. The data sequence that is directly coupled to the CTC encoder through "Input 1," generates parity check sequence C_1 (called C_1 coding) and the interleaved sequence connected to the encoder via "Input 2" creates parity check sequence C_2 (called C_2 coding) [4].

Table 7.10 summarizes mandatory and optional channel coding techniques for AeroMACS networks.

Table 7.10 Channel coding schemes for AeroMACS.

Link	Mandatory	Optional
DL	NSNR convolutional code at rates 1/2, 2/3, 3/4, 5/6	CTC at rates 1/2, 2/3, 3/4, 5/6, Repetition codes at rates 1/2, 1/3, 1/6 RS-CC concatenation block turbo codes (BTC) LDPC codes
UL	NSNR convolutional code at rates 1/2, 2/3, 3/4	CC of rate 5/6 CTC of rates 1/2, 2/3, 3/4, 5/6, Repetition codes at rates 1/2, 1/3, 1/6 RS-CC concatenation TBC, LDPC

In digital communications systems, a coding scheme is generally selected based on required signal reliability, expressed in BER or frame error rate (FER), channel characteristics, throughput/goodput requirements, latency concerns, hardware/software complexity, and so on. If a wireless channel is infested with frequent deep fades, then coding schemes with burst error correcting capabilities are most suitable. This raises an important question with respect to the selection of channel coding techniques for AeroMACS networks. Do the mandatory NSNRCC scheme adequately address the performance requirements of AeroMACS, considering the airport channel peculiarities, interference concerns, and power limitations, or more powerful coding schemes, allowed in the WiMAX standard as optional choices, need to be employed?

7.2.5 Adaptive Modulation and Coding (AMC) for AeroMACS Link Adaptation

The unpredictability caused by random fluctuations of the airport mobile radio channel characteristics makes the standard approach of using a single bandwidth-efficient modulation scheme, which represents an optimum or near-optimum choice under various conditions on an airport surface environment, unrealistic. Same observation may be made in regards to choosing a channel coding technique. However, since a variety of modulation formats and coding schemes are supported by IEEE 802.16-2009 standard, it is feasible to develop techniques through which the most appropriate modulation and coding combination is selected to confront the propagation conditions of the communication channel. This is the essence of adaptive modulation and coding (AMC) concept in AeroMACS networks, which enables efficient link adaptation that best matches the channel parameters and user requirements. In fact, AMC is an important feature of many contemporary wireless telecommunications networks such as LTE, Advanced-LTE, and WiMAX. In plain language, when channel is clear and in good conditions, AMC algorithm assigns a high-order modulation format combined with a low redundancy (high rate) coding scheme (best case 64-QAM combined with NSNRCC of rate 3/4 for DL), while under the conditions of noisy channel and deeply faded signal, a lower order, therefore spectrally less efficient but robust modulation scheme, along with a low-rate coding format is assigned to maintain both connection quality and link stability, without any increase in signal power. In the worst-case scenario, QPSK modulation with CC or CTC of rate 1/2, perhaps along with repetition, may be applied. In this manner, high volumes of data are thrusted through when the channel is clear, and a conservative modulation coding combination is applied when the channel is affected by harsh fading and signal attenuation.

As our studies of cellular radio channel over the Extended MLS C-Band, presented in Chapters 2 and 3, demonstrate, the quality and strength of a received signal, either by a base station or a mobile unit, in an airport surface environment are influenced by a number of factors. These factors include BS–MS distance, propagation mode (LOS, near LOS, NLOS, or a combination thereof), path loss (much more intense in C-band than it is in legacy VHF band), large-scale fading (log-normal shadowing), small-scale fading (flat fading; frequency selective fading), whether or not multihop relays are applied, MIMO configuration, and so on. Earlier, we argued that different zones of an airport surface may bear different propagation conditions, ranging from runway areas with a relatively clear channel to ramp vicinities that are infested with harsh multipath effect. These zones, clearly, require different modulation-coding combinations to maintain link quality at an acceptable level.

It is observed that AMC is the methodology of link adaptation in AeroMACS systems. The important role of AMC and link adaptation in broadband networks is in achieving high spectral efficiency while maintaining received signal reliability (bit error rate, frame error rate, and so on) at an acceptable level as wireless channel constantly changes in time and frequency domains. In the previous sections, we explored various modulation and coding schemes that are supported by AeroMACS. The Standard allows for the modulation-coding combination scheme to change, if so needed, on a burst-by-burst (frame-by-frame) basis per link. The channel condition information based on which this transition is permitted is normally acquired from the channel quality feedback indicator that the MS can provide for the BS with feedback on the DL channel quality. For the UL path, the BS normally estimates the channel quality by observing the received signal quality.

AeroMACS standard allows for seven combinations of modulation and coding schemes in support of AMC adaptation. QPSK with code rate of 1/2 may be complemented with 4 different repetition provisions for higher level of error protection, which increases the number of possible modulation-coding combination to a total of 10 options. Table 7.11 summarizes these schemes along with data rates for DL and UL paths. Coding techniques listed in Table 7.11 are the mandatory tail-biting convolution codes and the optional convolutional turbo codes described in the previous sections. Data rate computation is based on 5 MHz channel bandwidth, PUSC subchannelization, frame duration of 5 ms, and other PHY layer parameters [16].

Specific methods of channel estimation based on which AMC adaptation decision may be made are described in Ref. [17]. Reference [18] proposes a couple of AMC methods for WiMAX networks. The first method, called "channel state technique," adapts the modulation-coding

Table 7.11 AeroMACS physical data rates with 5 MHz BW and PUSC.

Modulation	Coding schemes	Code rate	Repetition factor	DL data rate (Mbps)	UL data rate (Mbps)
QPSK	CC/CTC	1/2	6	0.53	0.38
QPSK	CC/CTC	1/2	4	0.79	0.57
QPSK	CC/CTC	1/2	2	1.58	1.14
QPSK	CC/CTC	1/2	1	3.17	2.28
QPSK	CC/CTC	3/4	1	4.75	3.43
16-QAM	CC/CTC	1/2	1	6.34	4.57
16-QAM	CC/CTC	3/4	1	9.50	6.85
64-QAM[a]	CC/CTC	1/2	1	9.50	6.85
64-QAM[a]	CC/CTC	2/3	1	12.67	9.14
64-QAM[a]	CC/CTC	3/4	1	14.26	10.28

a) Optional for UL.

combination using an estimation of the channel behavior based on attenuation coefficient. The second technique, named "error state technique," follows a different approach in which the received frame error rate is the basis for AMC adaptation decisions.

The original WiMAX standard defines AMC at the PHY layer to be used in the WiMAX profiles and applications; however, except for a suggested simple channel quality table lookup method, the detail of the algorithms to be used for the scheme are not specified. The implications are that different applications might need to develop their own adaptation algorithm to be applied in their particular circumstances for optimization with respect to critical system criteria such as Quality of service (QoS), performance, throughput, and latency.

It is clear that the selected modulation coding combination have direct bearing on system throughput and goodput. Depending on robustness of the coding scheme, different theoretical throughput values can be achieved. Although a number of coding-modulation combinations are available, as shown in Table 7.11, the most likely combinations are QPSK/16QAM-CC rate ½ for UL and 16QAM/64QAM-CC rate 1/2 for DL [10].

7.2.6 AeroMACS Frame Structure

The AeroMACS TDD frame structure is essentially the same as that of WiMAX TDD, shown in Figure 5.6 and described in Section 5.3.2.4. Accordingly, the AeroMACS frame has a time duration of 5 ms and starts up with a DL Preamble (prefix) that spreads over an entire OFDM symbol (time duration of 102.9 μs). This is followed by the frame control header (FCH). Like in the case of WiMAX frame, FCH contains information about the following DL MAP. Next DL MAP and UL MAP appear containing important management information, and essentially defining for the SS/MSs how the upcoming frame is used for either exchange of data or management information. These four elements, that is, preamble, FCH, DL MAP, and UL MAP show up in each DL subframe. On the UL subframe side, ranging opportunities is scheduled for the SS/MS for the purpose of synchronization with the BS and for bandwidth request that might be needed by SS/MS. The rest of the time–frequency resources are utilized for transmission of user data [10].

7.2.7 Computation of AeroMACS Receiver Sensitivity

According to IEEE 802.16-2009 standard, the sensitivity for any WiMAX (AeroMACS included) receiver is defined as the required power level at the front end of the receiver to secure a BER of 10^{-6}. The Standard provides Equation 7.10 for calculation of WiMAX and AeroMACS receiver sensitivity [5].

$$R_{SS} = -114 + (SNR)_{R_X} - 10\log_{10}(R) + 10\log_{10}\left(\frac{f_s \times N_{Used} \times 10^6}{N_{FFT}}\right)$$
$$+ (L)_{IMP} + NF$$

$$(7.10)$$

Where R_{ss} is the receiver sensitivity in dB_m, $-114\,dB_m$ is the thermal noise power, referenced to 1 MHz bandwidth and temperature of 300 K. $(SNR)_{R_X}$ is the receiver SNR required at the demodulator input to produce the desired BER of 10^{-6}. Clearly, the required value of $(SNR)_{R_X}$ depends on applied modulation scheme and coding rate. R is the repetition factor, f_s is sampling frequency in Hertz, N_{FFT} is the FFT number of points (FFT size, number of subcarriers). N_{Used} is the number of subcarrier in use for data and pilot transmission that is equal to N_{FFT} minus the number of guard band subcarrier and DC subcarrier, $(L)_{IMP}$ is the implementation loss that accounts for items such as channel estimation errors, tracking errors, quantization noise, and phase noise; the assumed value for this loss factor is 5 dB. Finally, NF is the receiver noise figure that is assumed to be about 8 dB.

As was noted above, the required value of $(SNR)_{R_X}$ depends on the modulation and coding combination. Table 7.12 lists the required values of $(SNR)_{R_X}$ for convolutional codes of different rates combined with various modulation schemes available in the Standard. Table 7.12 also

Table 7.12 AeroMACS receiver SNR and receiver sensitivity.

Modulation scheme	Code rate (CC)	Repetition factor	Receiver SNR $(SNR)_{R_x}$ in dB	Receiver sensitivity in dB$_m$
QPSK	1/2	2	5	−92.37
QPSK	3/4	1	8	−89.37
16-QAM	1/2	1	10.5	−86.37
16-QAM	3/4	1	14	−83.87
64-QAM	1/2	1	16	−80.37
64-QAM	2/3	1	18	−76.37
64-QAM	3/4	1	20	−74.37

highlights the receiver sensitivity for AeroMACS system of channel bandwidth equal to 5 MHz, calculated from Equation 7.10, for selected values of repetition factors. The corresponding parameter values to be used in Equation 7.10 for 5 MHz AeroMACS system are given by [19];

$$f_s = 5.6 \times 10^{-6}, N_{FFT} = 512, N_{Used} = 420, (L)_{IMP} = 5\,dB, NF = 8\,dB$$

7.2.8 Fractional Frequency Reuse for WiMAX and AeroMACS Networks

As we have described in Chapters 4 and 5, WiMAX and AeroMACS multiple access technique, OFDMA, groups large number of orthogonal subcarriers to form user's subchannels. When neighboring cells use the same subchannels for signal transmission "co-channel or co-subchannel interference" occurs. As we have deliberated in Section 5.3.2.8, to mitigate the effect of this type of interference WiMAX applies fractional frequency reuse (FFR) scheme. FFR is a technique that, essentially, renders frequency reuse factor (FRF) to be adaptive, rather than taking on a fixed value as in classical cellular networks. We explored a relatively complex case of FFR for WiMAX networks in Chapter 5, here we investigate the simplest possible form of fractional frequency reuse format that may be suitable for application in AeroMACS networks. In this FFR scheme, proposed by WiMAX Forum, the cells are primarily divided into "inner" and "outer" regions. Inner regions are closer to the base stations and less susceptible to co-channel or co-subchannel interference. For these regions, the FRF is one, in other words, all subchannels are accessible to all users. However, for the outer cell regions, the subchannels are divided in the same way as they are in traditional cellular networks (see Chapter 2), that is, disjoint sets of sub-channels are allocated to adjacent cell edges. In other words, in outer zones

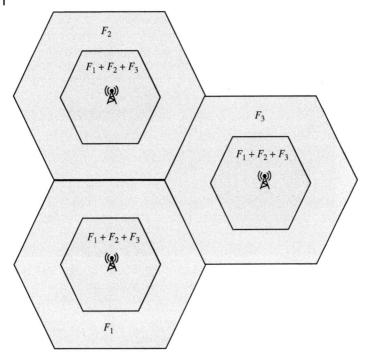

Figure 7.7 Illustration of a simple FFR scheme that may potentially be applied in AeroMACS.

the user has access only to a fraction of available subchannels. Figure 7.7 demonstrates how subchannels are divided between various zones according to this elementary FFR scheme. The subchannels are divided into three disjoint subsets, designated as F_1, F_2, and F_3. The user in inner zones has access to all subchannels in all three subsets, while only one subset is accessible to MSs in each cell or sector outer zone. With this simple configuration, the full frequency reuse is maintained for users in the inner zone with maximum spectral efficiency, and fractional frequency reuse is implemented for users in outer areas that mitigates the effect of co-subchannel interference and assures connection quality and throughput for these users [20].

This is a static form of adaptive FRF, many new algorithms for dynamic adaptive FRF schemes have been devised recently [21].

7.2.9 Multiple-Input Multiple-Output (MIMO) Configurations for AeroMACS

In Chapter 5 multiple-input multiple-output (MIMO) antenna systems, along with the concept of space–time coding (STC), for general WiMAX network were explored in some detail. In this section, we briefly discuss applications of MIMO configurations in AeroMACS systems.

Currently, AeroMACS networks implement MIMO with the sole intention of enabling improvement in reception quality and performance, and is not employed for the purpose of increasing the network throughput and spectral efficiency [22].

On the downlink, the basic two-antenna rate-1 (Matrix A) is the only planned STC–MIMO configuration for implementation in AeroMACS. This means that the base station may optionally implement one or two transmission chains. With two transmit antennas at the BS, the MIMO system provides a simple transmit diversity that is associated with Alamouti's STC scheme (see Chapter 5). Currently, no MIMO configuration is planned on the uplink, which indicates only one transmitter chain is assumed at the MS.

The simple MIMO configuration (proposed for AeroMACS DL) with two transmitting antennas is defined by the Matrix A given in Equation 7.11.

$$A = \begin{bmatrix} S_1 & -(S_2)^* \\ S_2 & (S_1)^* \end{bmatrix} \tag{7.11}$$

In this matrix, the row index indicates antenna number, and column index designates the OFDMA symbol number. Therefore, the top row shows what is emitted from the first transmit antenna, and the bottom row demonstrates the same for the second transmit antenna. Accordingly, the first channel uses the first antenna to transmit S_1 and the second antenna to transmit S_2. Similarly, second channel uses the first antenna to send $-S_2^*$ and the second antenna to send S_1^*. The signal detection at the receiver follows the algorithm described in Section 5.3.2.7. This amounts to a simple space–time transmit diversity (STTD) [5].

7.3 Spectrum Considerations

Studies to estimate the bandwidth required to carry AeroMACS applications were initiated in the United States and the Europe within the 2003–2004 time frame. In the United States, the effort was led by the FAA in collaboration with NASA Glenn and MITER Corporation. An early NASA/FAA study estimated the FAA's existing and anticipated data requirements for instrument landing systems, radar systems, runway visual range, visual aids, and A/G communications [23]. The highest requirements for wireless communications from airlines and port authorities included communications with ground maintenance crews and airport security. NASA Glenn later conducted a study to estimate additional bandwidth requirements to accommodate wake vortex sensing and the overhead associated with security provisioning features of the IEEE 802.16 standard [24]. MITER Corporation in a series of studies, conducted for the FAA, from 2004 to 2008, established estimates of the aggregate data rate requirements for a high-data-rate airport surface wireless network [16]. These studies addressed potential bandwidth requirements for Phase 1 (through 2020) and Phase 2 (beyond 2020).

The bandwidth requirements for a proposed mobile and fixed applications using an IEEE-802.16-based system were estimated for both low-density and high-density airports. The highest total aggregate data capacity requirements for fixed and mobile applications was based on large airports (e.g., DFW, ATL) with a terminal radar approach control (TRACON) ATC facility, at the time, and not with an ATC tower (ATCT) [16].

As we elaborated in Chapter 6, WRC-07 approved the addition of an aeronautical mobile route service (AM(R)S) allocation within the 5091–5150 MHz band to the ITU-R international table of frequency allocations, through which prior limitations in the so-called microwave landing system (MLS) extension band for "support of navigation/surveillance functions only" was removed. The new designation provided protected spectrum for safety and regularity of flight applications, which lead to the creation of AeroMACS. The WRC-07 specifically limits the new allocation to communications with aircraft when wheels are in contact with the airport surface, exclusively.

Table 6.3 provides information on the entire operating band of frequencies (5000–5150 MHz), currently allocated spectrum (59 MHz bandwidth within 5091–5150 MHz spectrum), as well as center frequencies for AeroMACS channels. The reference center frequency is 5145 MHz, and the frequency band has been channelized for 5 MHz bandwidth. All preferred channel center frequencies are defined downward from 5145 MHz in 5 MHz incremental frequencies. Accordingly, the currently allocated AM(R)S band allows for 11 AeroMACS channels whose center and end frequencies are listed in Table 7.13.

No guard band has been stipulated between these channels; however, 1.5 MHz of band on each end of the 5091–5150 MHz spectrum has been left unused to separate AeroMACS band from the adjacent bands. In addition, the band of 5000–5030 MHz is a potential frequency spectrum that may be allocated for AeroMACS in the future, which could provide additional 6 AeroMACS channels. Accordingly, the design of the AeroMACS systems takes this potential additional spectrum into consideration in developing equipment that can operate not only in the original designated band, but also in the 5000–5030 MHz band [25].

It should be noted that, although Table 7.13 lists the end frequencies for each channel and defines the boundary of the allocated spectrum, the

Table 7.13 Currently allocated AeroMACS channels, frequencies are in MHz.

Channel designation	1	2	3	4	5	6	7	8	9	10	11
Channel center frequencies	5095	5100	5105	5110	5115	5120	5125	5130	5135	5140	5145
Channel lower and upper end	5092.5	5097.5	5102.5	5107.5	5112.5	5112.5	5122.5	5127.5	5132.5	5137.5	5142.5
Frequencies	5097.5	5102.5	5107.5	5112.5	5117.5	5122.5	5127.5	5132.5	5137.5	5142.5	5147.5

behavior of the signal spectrum around the edges of the spectrum depends upon the channel spectral mask that is applied to AeroMACS signals. Figure 7.8 illustrates the original notational AeroMACS channel plan. The figure is adopted from Ref. [25] with permission. The figure designates the 11 available AeroMACS channels in this band by their center frequency locations. The C-band beyond 5150 MHz is allocated for nonaviation applications.

An important issue, related to AeroMACS allocated spectrum, is the fact that the spectrum is shared with other applications, including non-geostationary satellite feeder links. Spectrum sharing and the existence of in-band coallocated applications present an additional challenge in the design of AeroMACS networks. The next section is dedicated to the discussion of interference to coallocated applications.

Like in any wireless communications application, the main spectrum-related factors that affect the design and development of AeroMACS systems are spectral restrictions defined by the allocated spectrum, power limitations vis-à-vis interference to in-band and adjacent band applications, and the efficient use of available spectrum within these constraints [25].

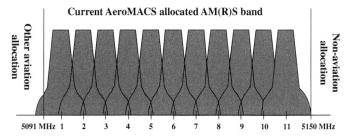

Figure 7.8 AeroMACS original notational channel plan with 5 MHz channel spacing. (From Ref. [25] with permission from IEEE.)

7.4 Spectrum Sharing and Interference Compatibility Constraints

The 5091–5150 MHz ARM(S) band allocated for AeroMACS is shared with nongeostationary (low Earth orbit, LEO) mobile satellite services (MSS), on a coprimary basis, for earth-to-space feeder uplinks. The Globalstar constellation is the primary existing system operating such links in this band. This coallocation requires that the total aggregate power from AeroMACS installations must not exceed interference thresholds for the feeder link receivers on Globalstar spacecraft, placing a significant limitation on the total system capacity of AeroMACS [26].

In 2005, MITRE Corporation was tasked by the FAA to conduct an investigation on whether precautionary measures are needed to be taken for protection of MSS feeder uplinks from radio frequency interference (RFI) due to AeroMACS signals emitted from airports. The interference may be coming from both UL, that is to say from MS/SS, and DL, that is, from AeroMACS base stations installed on the airport surface. The objective was to establish practical limits on AeroMACS transmissions from airports so that the threshold of interference into Globalstar feeder links is not exceeded. The basic assumptions in this investigation included the existence of 497 towered airports in the contiguous U.S. continent. This was considered to be the primary source of interference into Globalstar feeder links, with multiple BSs presumed for larger airports such that the radius of the coverage area for each BS is kept below 2 km. The BS antennas are configured into three sectors. Five OFDMA channels of 10 MHz wide was assumed for AeroMACS (this investigation was conducted before allocated bandwidth for AeroMACS channels was determined to be 5 MHz). The complete list of AeroMACS and MSS LEO satellite parameters and constraints based on which this preliminary investigation was conducted is provided in MITER report [27]. The important finding of this study is, under the conditions defined in the final report of the study, that the sharing of MLS Extended band between AeroMACS and MSS feeder uplinks is feasible, provided that the MSS–LEO satellite interference threshold power of −157.3 dBW (corresponding to a 2% increase in the satellite receiver's noise temperature) is not exceeded [27]. In summary, the MITER report concludes that under the 2% interference criterion and under the assumptions that are made in the report, AeroMACS-based WiMAX technology can share the aeronautics C-band spectrum (5091–5150 MHz) with LEO–MSS feeder uplink without any need for interference mitigation measures. However, the report calls for validation of its results either through simulation or field measurements. The report also recommends that as IEEE 802.16 profiles for aeronautics are developed, the AeroMACS/MSS compatibility analysis should be updated to reflect the new AeroMACS characteristics changes.

Later, NASA Glenn created a global AeroMACS–MSS interference model using Visualyse Professional Version 7 software from Transfinite Systems Limited,[3] assuming 6207 airports on the planet. It was assumed that base station transmission occurs in all 11 5 MHz AeroMACS channels shown in Figure 7.8. The propagation model was assumed to be free space brand according to ITU-R Rec. P.525. All in all, 19 scenarios with variations in airport size (small, medium, and large) antenna distribution, antenna beamwidth, and antenna tilt were simulated. The maximum simulated cumulative interference power at the low Earth orbit hot spot for these variations was used to establish transmitter power limits [28].

The most realistic scenario of all was corresponding to the case in which the 6207 airports were divided into three size categories of large, medium, and small. In the United States, 35 airports were identified as large and 123 as medium. In Europe, 50 were designated as large and 50 as medium. The rest of the airports in the United States, Europe, and the rest of the world were identified as small. It was assumed that the large airports would use all 11 AeroMACS channels shown in Figure 7.8, medium airports would use only 6 channels, and small airports would use just 1 channel. Channel power levels for base stations were deemed to be limited to 1711, 855, and 285 mW for large, medium, and small airports, respectively. This amounts to a transmission power ratio of 6/3/1 for large/medium/small airports. Simulations were also used to determine MS/SS transmission limits. The MS/SS antenna model was based on the antenna system employed for mobile measurements conducted at the NASA-CLE (Cleveland Hopkins International Airport) AeroMACS Test Bed (see Chapter 6). The power transmission ratio for this case was assumed to be 8/4/2 for MS/SS operating in large/medium/small airports, respectively. This translates into transmission power level limits of 664, 332, and 83 mW for large/medium/small airports, respectively. For the BS simulations, the elevation angle of the transmitter antennas was assumed to be zero. It should be possible to increase the transmission limits with a downward transmitter angle tilt (base station installed in the control tower). However, the results indicate that it will be more important to limit the power transmission from subscribers. One approach to increase the allowable transmission power would be to utilize antennas with reduced gain at high elevation angles [26].

---

3 http://transfinite.com/content/Professional.html.

7.5 AeroMACS Media Access Control (MAC) Sublayer

Physical layer represents the "lowest layer" (Layer 1) of the standard seven-layer Open Systems Interconnection (OSI) model defined by International Standards Organization (ISO), or lowest layer in Internet Protocol (IP) defined by Internet Engineering Task Force (IETF). The data link layer (DLL), which contains the sublayer media access control (MAC) is the layer 2 in the OSI model, which is essentially responsible for transfer of data packets between adjacent network nodes. In other words, DLL is the protocol layer that facilitates interfacing of the PHY layer and layer 3, the network layer. Consequently, DLL is in charge of interfacing the hardware of physical layer to the software of higher layers.

A key functionality of the MAC sublayer is to provide an addressing mechanism and channel access method with an associated QoS attached to it, for the purpose of facilitating communications between nodes existed in a single network or between nodes of different networks. The "MAC address"[4] is a unique identifier assigned to network interfaces by MAC sublayer protocols. MAC address distinguishes different network interfaces and is widely used in many network technologies. For the majority of IEEE 802-based networks, including IEEE 802.16 networks, MAC addresses are used as network addresses.

The AeroMACS MAC layer is compliant with that of IEEE 802.16-2009, as is explained in Chapter 5. Figure 7.9 provides MAC "protocol reference model" for IEEE 802.16-2009 Standard-based technologies, including WiMAX and AeroMACS.

As Figure 7.9 illustrates, the AeroMACS MAC protocol stack consists of three parts of service-specific convergence sublayer (CS), common part sublayer (CPS), and security sublayer. The functionalities of various blocks shown in Figure 7.9 are more or less the same as corresponding ones illustrated in Figure 5.2. The MAC CPS is responsible for all MAC-specific functionalities, including resource allocation, scheduling and connection establishment and maintenance, modulation and coding combination selection, and QoS management (see Section 5.4). The CS is generally in charge of mapping external data from the upper layer, for example, IP or Ethernet, into appropriate service data units (SDUs). The security sublayer provides means for secure communications, and hence it supports functionalities such as authorization, authentication, key exchange, encryption, ARQ, and HARQ.

Multiple users access the channel in accordance with TDMA–OFDMA scheme. The smallest time–frequency resource unit that can be assigned

---

4 MAC address is known by several other designations in the literature: "physical address," "hardware address," "burned-in address."

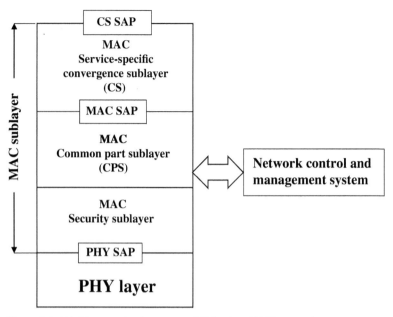

Figure 7.9 MAC "protocol reference model" for AeroMACS networks.

to each user is a slot. A slot is composed of a group of adjacent subcarriers divided into 48 data subcarriers. In AeroMACS systems, DL and UL slots have different compositions. DL slots are formed by arranging two subcarrier clusters (each cluster consists of 14 subcarriers) per OFDM symbol assembled within two symbols with a total of 56 subcarriers, of which 48 subcarriers are data subcarriers and the remaining 8 subcarriers are pilot subcarriers. Uplink slots, on the other hand, are composed by assembling 6 tiles, each tile consists of 4 subcarriers from 3 OFDM symbols, resulting in a total of 12 subcarriers of which 8 are data and 4 are pilot subcarriers [2].

The IEEE 802.16-2009 standard stipulates a point-to-point connection-oriented link. This implies that each SDU received from an interfaced higher layer is mapped to a unique unidirectional "service flow"[5] with a specific set of associated QoS parameters. Depending upon the nature of the higher layers connected to AeroMACS MAC layer, there are several interfacing possibilities. The MAC CPS operates in a point-to-multipoint environment. The DL resources are used by AeroMACS base stations exclusively, whereas UL resources are shared by subscriber and mobile

---

5 Service flow is a term that is used, in the jargon of IEEE 802.16 standards, for simply a BS–MS connection with an associated QoS.

stations. Downlink transmissions are normally of broadcast mode, that is, all MS/SSs receive downlink transmissions. The MAC header of each MAC PDU carries a "connection identifier (CID)" with it. Based on the content of CID, the MS or SS determines whether or not the MAC PDU is addressed to it. It should be emphasized that a key character of IEEE 802.16 standard is the availability of transport connections that allows for the incorporation of a desired QoS at the MAC layer [2].

7.5.1 Quality of Service for AeroMACS Networks

Quality of service in wireless networks, or in any telecommunications network for that matter, means, at least conceptually, the probability that the network fulfills the promise of a given traffic agreement. QoS manifests itself in the ability of network elements to offer some level of assurance that the network's promised traffic and service requirements would be met. Therefore, QoS is a measure of how reliable and consistent a network operates.

As was discussed in Chapter 5 in connection with general WiMAX networks, the key concepts related to QoS are service flow and bandwidth grant provisions. A service flow is a one-way MAC SDU with a defined set of QoS parameters and an associated service flow ID (SFID). A provisioned service flow possesses SFID but it may not carry traffic, rather it might be waiting to be activated for usage. An admitted service flow goes through the process of activation. An active service flow is an actuated service flow with all required resources allocated to it. Normally, upon an external request for a service flow, the BS/MS will check for required resources to satisfy the requested QoS. The use of service flows is the main mechanism for provision of a requested QoS [29].

The IEEE 802.16 standards and their driven technologies, such as WiMAX and AeroMACS, provide broadband access for various services requiring different QoS. The MAC sublayer schedules traffic flows and allocates bandwidth such that the QoS requirement for each service is met. The QoS can include one or more measures that are known as "*QoS parameters*." Some common WiMAX/AeroMACS QoS parameters are outlined and described further.

- *Maximum sustained traffic rate* (MSTR): This represents the maximum allowable data transmission rate for the service, expressed in bits per second. MSTR defines the maximum bound but not a guaranteed rate that is always available.
- *Minimum reserved traffic rate* (MRTR): This parameter specifies the minimum information rate reserved for a service flow, and it is expressed in bits per second. This is the minimum average rate.

- *Maximum latency tolerance* (MLT): This parameter specifies the maximum latency between the reception of a frame by the BS or SS/MS on its network interface and the forwarding of the packet to its RF interface. MLT is normally expressed in milliseconds.
- *Jitter tolerance:* Packets arrive at the destination with different latencies, and this creates packet delay variation (PDV) or as it is commonly known, delay jitter. Jitter tolerance defines the maximum delay variation that is tolerated for the connection and it is expressed in milliseconds.
- *Request/transmission policy:* This QoS parameter allows for specifying certain attributes of the service flow. For instance, it allows the selection of an option for PDU formation.
- *Traffic priority:* Traffic priority specifies the priority assigned to a service flow. To define priority a value between 0 and 7 is assigned to each service flow. Given two service flows that have identical QoS parameters and different priorities, the higher priority service flow will be given lower delay and higher buffering performance [30].
- *Maximum burst rate:* Maximum number of OFDM bursts transmitted per second.
- *Scheduling algorithms:* They include techniques such as earliest deadline first (EDF), weighted fair queue (WFQ), round robin (RR), deficit round robin (DRR), and first-in, first-out (FIFO).
- *ARQ type:* This entails the selection of an ARQ technique.
- *SDU type and size:* This is related to the size and type of service data units to be used.

Another item directly related to QoS is *"bandwidth request* (BR) *mechanism."* This is an important element, not only on account of its direct bearing on QoS, but also because of its connection to the most precious resource in communications systems, namely, the spectrum. In order to enhance bandwidth efficiency while provisioning a desired level of QoS, wireless mobile networks must adopt well-organized bandwidth request,[6] grant,[7] and polling mechanisms. In WiMAX and AeroMACS networks, the base station is in charge of bandwidth allocation. The DL bandwidth assignment is managed exclusively by the DL scheduler at the BS. The UL bandwidth allocation is also managed by the BS through the *resource request and grant* process, which is initiated upon the reception

---

6 Request refers to the mechanism in which the MS/SS informs the BS that it needs uplink bandwidth allocation. Request is normally made on a stand-alone bandwidth request, but it may be optionally made as a piggyback on an uplink data burst [31].
7 Each bandwidth grant is addressed to the MS/SSs' basic CID. Usually bandwidth grant takes a nondeterministic pattern, so some SSs may happen to get less frequent grants than expected [31].

of the request from a SS/MS unit. The IEEE 802.16-2009 supports a pair of bandwidth request modes (mechanisms), a contention mode and a contention-free (polling[8]) mode. In the contention mode, the MS sends bandwidth requests during a contention period, and the BS resolves contention by using an exponential backoff strategy. In the contention-free mode, the BS polls each MS/SS, and in reply, an MS/SS could send its bandwidth request. Contention mode in connection with scheduling classes (to be discussed in the next section) of extended real-time polling service, non-real-time polling service, and best effort have the options of making BRs via both mechanisms, depending on the scheduling decision made by the BS.

It is imperative to distinguish between *QoS parameters,* as listed already, and *QoS metrics.* QoS metrics are in fact performance-related entities such as *throughput, goodput, latency, jitter,* and *packet loss rate.*

As was mentioned earlier, WiMAX/AeroMACS systems define a set of QoS parameters over the air interface connections. Combined with QoS mechanism over the backbone wireline or wireless networks (e.g., IP, Ethernet), it would be feasible to have a predictable end-to-end quality of service.

7.5.2 Scheduling, Resource Allocation, and Data Delivery

Scheduling schemes are data handling mechanisms enabling a fair distribution of resources among network users. Scheduling algorithms are implemented at the CPS of the MAC layer with the objective of efficient assignment of network resources to users. Resource allocation in communication systems is often formulated as an optimization problem. Two common approaches are the minimization of total transmit power by placing a constraint on traffic data rate, and the maximization of data rate by imposing a restriction on total transmit power. For fixed data rate applications, the first approach is more appropriate, whereas the second option is more suitable for bursty packet environments. The techniques corresponding to the latter option are collectively referred to as "*rate adaptive resource allocation* (RARA) *algorithms,*" which are more suitable for use in AeroMACS systems. SESAR has conducted a feasibility study on four different RARA algorithms for AeroMACS, as outlined here [19].

1) Maximum sum rate produces maximum throughput.
2) Maximum fairness provides equal data rates for all users.

---

8 Polling refers to the process in which the BS allocates bandwidth for the MS/SSs, for the purpose of allowing them to make bandwidth requests. In addition to polling individual MSs, the BS may issue a broadcast poll by allocating a request interval to the broadcast CID, when there is insufficient bandwidth to poll the stations individually.

3) Proportional rate constraints stipulates proportional rates for various users.

4) Proportional fairness scheduling attempts to maintain a balance between competing users by providing at least a minimal service to all users, while trying to maximize the overall throughput.

These algorithms present possible techniques for implementation of resource allocation scheduling. The first three algorithms mentioned in the list are of the type that try to instantaneously achieve their respective objectives, whereas the last one tends to realize its objectives gradually and over time. This provides the latency as an additional parameter in the trade-off list. Based on this study, SESAR recommends that proportional fairness scheduling is probably the best scheduling algorithm for the AeroMACS application, particularly in light of the fact that Communications Operating Concept and Requirements (COCR) had already concluded that latency requirements were not stringent for the most critical AeroMACS safety functions.

In the particular case of OFDMA systems, since several users are allowed to communicate simultaneously during the same symbol period, there is a need for selection or development of an algorithm, such as the ones explored in the previous paragraph. This algorithm serves to determine the user scheduling on how to allocate the subcarriers among the users, and how to determine the correct power level setting for each user per burst [19]. The IEEE 802.16-based WiMAX system is designed to accommodate fixed and mobile nodes while maintaining differentiating QoS for various offered services. In order to support a broad range of applications, AeroMACS (WiMAX) defines five categories of scheduling services that may be supported by the BS MAC scheduler for information transmission over any given connection. These services, sometimes referred to as *QoS classes*, support different applications ranging from real-time applications such as voice transmission, to non-real-time applications such as file transfer. Since these services are required to provide different QoS, they need different bandwidths that are defined by MAC scheduler. There are five QoS classes that are supported by AeroMACS.

- *Unsolicited grant services* (UGS): Originally designed to support real-time constant bit rate traffic, as such this service provides opportunity for fixed rate transmission at regular time interval with no need for polling. Among QoS parameters defined for UGS are jitter tolerance, minimum reserved traffic rate, maximum latency, and the unsolicited grant interval. This implies that UGS service flow is offered UL resources periodically, without requesting them. Examples for UGS commercial applications are T1/E1 transport and VoIP.

- *Real-time polling services* (rtPS): Designed to support variable rate traffic. In this service, the BS polls the MS/SS/RS periodically to provide request opportunities. Among QoS parameters defined for rtPS are maximum latency, minimum reserved traffic rate, polling interval, and traffic priority. In this scheme, the SS/MS/RS is polled at the fixed interval, and the SS/MS may use these polls to request bandwidth. An example of rtPS-based communications application is transport of MPEG audio and video steaming.
- *Non-real-time polling services* (nrtPS): This is appropriate for delay-insensitive applications such as FTP (file transfer protocol). This service allows the SS/MS/RS to use contention request and unicast request for bandwidth bid. Unicast request opportunities are offered regularly in order to ensure that the SS/MS/RS has a chance to request bandwidth in congested network environments. QoS parameters associated with this scheme are minimum reserved traffic rate, maximum sustained traffic rate, and traffic priority.
- *Best effort services* (BE): This service along with nrtPS are meant for lower priority services, as such BE does not specify any minimum service requirement. Similar to nrtPS, BE offers contention and unicast request opportunities but not on a regular basis. In this scheme, there are no grantees neither for QoS nor even for data delivery, therefore, no QoS parameter is defined for it. Bandwidth request ranging opportunities are provided to the user through the transmission of bandwidth request ranging codes, and when this code is received by a BS it would poll the corresponding SS/MS [10]. Data packets are carried as space becomes available. Delays may be incurred and jitter does not pose a problem. This service is normally used for Internet services, such as email and browsing in WiMAX networks. Another related commercial communications application example that takes advantage of this scheduling mode is Hypertext Transport Protocol (HTTP).
- *Extended real-time polling services* (ertPS): This service provides real-time service flows that generate variable-sized data packets, therefore, it requires a dynamic bandwidth allocation on a periodic basis. In effect, this scheduling algorithm combines the efficiencies of UGA and rtPS. An example of a ertPS-based commercial communications application is VoIP with silence suppression. Another typical example that uses ertPS scheduling scheme is Skype.

The five QoS classes (or scheduling techniques) outlined here, in fact define data delivery methods available through MAC layer of AeroMACS standard. Each service flow, as described previously, has specific set of QoS parameters that are initialized at connection setup. This facilitates the selection of different data delivery strategies such as UGS, BE, and rtPS.

For the downlink path, similar QoS classes may be applied, their names are slightly different, although they have comparable QoS parameters. There is no need for polls or ranging opportunities for the DL. The QoS classes, or data delivery services, for downlink are, real-time variable rate (RT-VR) service, non-real-time variable rate (NRT-VR) service, and BE service.

Generally speaking, requirements and needs of a particular application determine the specific QoS class that is suitable for the application. The IEEE 802.16-2009 standard does not specify how different data delivery services and scheduling strategies are implemented, they are left to the vendors [10].

7.5.3 Automatic Repeat Request (ARQ) Protocols

Automatic repeat request techniques, as was described in Chapter 5, have been widely applied in telecommunications and computer networks for error control. ARQ techniques are equipped with effective error detecting schemes. Normally, a cyclic redundancy check (CRC) code, which requires significantly lower number of redundancy bits than that of a typical FEC scheme, is employed. At the receiver, each received code word is examined for error detection using the CRC code. If no error is detected, the receiver accepts the code word and positively acknowledges (ACK) the successful reception of the code word to the transmitter through a feedback (return) channel. However, when errors are detected, the transmitter is informed with a negative acknowledgment (NACK) message sent by the receiver (through the return channel) and thus "instructed" to retransmit the same code word. The retransmission continues until the code word is received successfully [15].

The ARQ techniques are applied in AeroMACS/WiMAX networks with two objectives in mind, synchronization of data flows between transmitting and receiving nodes and error control. Data flow synchronization is enabled through the acknowledgement mechanism that is imbedded in ARQ system. The principal error control functionality of the ARQ technique in WiMAX and AeroMACS systems is in recovery of lost packets. ARQ and HARQ[9] (Hybrid ARQ) protocols are implemented in the MAC security sublayer of AeroMACS networks. Since the airport surface environment includes a large population of aircraft and other

---

9 Hybrid automatic repeat request system is an error control scheme that combines forward error correction (FEC) and ARQ techniques. In HARQ techniques, the FEC is exploited to detect and correct up to certain number of transmission errors. When the number of errors is beyond the error correcting capability of the FEC the ARQ scheme is invoked [15].

mobile and stationary terminals, with a combination of unicast, multicast, and broadcast traffic, it has been proposed in the aviation literature that packet-level coding techniques may be required for AeroMACS networks. In particular, SANDRA Project (seamless aeronautical networking through integration of data links, radios and antennas) has conducted investigation related to the necessity and feasibility of application of packet coding in AeroMACS [32]. Packet coding is beneficial whenever large amounts of information have to be broadcast to large number of mobile and stationary terminals, essentially complementing ARQ and HARQ protocols. However, the packet-level coding protocols reside in higher layers of OSI (or IP protocols)-layered protocol architecture such as application layer, and is beyond the scope of the coverage of this text.

When a service flow needs to be protected by ARQ, CRC is added to each MAC PDU that is carrying data for the service flow. WiMAX and AeroMACS apply CRC-32 code's standard generator polynomial is given in Equation 7.12.

$$g(x) = x^{32} + x^{26} + x^{23} + x^{22} + x^{16} + x^{12} + x^{11} + x^{10} + x^8 + x^7 + x^5$$
$$+ x^4 + x^2 + x + 1 \qquad (7.12)$$

CRC is a linear cyclic block code and the polynomial in Equation 7.12 is used to calculate CRC bits (i.e., parity check bits) to be appended to the payload of MAC PDU [4].

Some notes regarding construction and transmission of MAC PDUs are in order at this point. A MAC PDU consists of a fixed six-byte long header containing the connection CID, a variable payload, and a fixed 4-byte long CRC sequence. A MAC SDU may consist of one or more PDUs. PDUs may be cascaded for a single transmission in either DL or UL directions, this is known as *concatenation*. *Fragmentation* is a process in which a MAC SDUs are divided into smaller MAC PDUs that allows for efficient use of available bandwidth relative to QoS requirements of a connection's service flow. *Reassembly* is the reverse of fragmentation, that is to say placing of fragments together to regenerate the original SDU. Fragmentation and reassembly capabilities are mandatory for WiMAX and AeroMACS networks. The authority of traffic fragmentation is defined when the connection is created by MAC service access point (SAP) [4]. With respect to automatic repeat request procedure, two types of connections are supported by the IEEE 802.16-2009 standard.

1) Non-ARQ connections: For this type of connection, the fragments are transmitted in sequence only once. The sequence number assigned to each fragment enables the receiver to reassemble the payload and to detect the absence of any fragments, that is, lost fragments. Upon the

detection of any loss, the receiver simply discards all the received PDUs until it receives a first fragment again or a nonfragmented PDU.

2) ARQ-enabled connections: For this type of connections, fragments are formed for each transmission by concatenating sets of ARQ blocks with adjacent sequence numbers. The block sequence number (BSN) value carried in the fragmentation subheader (FSH) is selected to be the BSN for the first ARQ block[10] appearing in the segment [4]. The segment may be transmitted more than once in correspondence with the procedure defined by the applied ARQ type.

The IEEE 802.16-2009 standard offers four different types of ARQ schemes. First is the go-back-n (cumulative) ARQ format. Second is the selective repeat ARQ (also known as selective reject ARQ) technique. The others are two ARQ methods that are combinations of go-back-n and selective repeat schemes.[11]

The size of a basic ARQ information unit is at least 4 bytes and at most 12 bytes. The basic components of an ARQ information unit are the CID and the BSN. The CID identifies the transport connection and the BSN may vary depending upon the ARQ type. ARQ blocks are central to IEEE 802.16 ARQ protocols in the sense that each incoming SDU from a higher layer is logically divided into a number of ARQ blocks. Each ARQ block so formed gets a block sequence number. The size of each ARQ block, in a given scenario, is fixed and is specified by ARQ_BLOCK_SIZE. An exception to this fixed size rule is made for the last ARQ block associated with a SDU, or a SDU whose length is smaller than ARQ block size.

WiMAX and AeroMACS technologies support different types of acknowledgement message formats.

- Acknowledgment type 0: Corresponding to selective ACK entry, contains up to four fixed length acknowledgment maps. The length of such a map is 16 bits, where each bit indicates whether or not the corresponding ARQ block has been received error free. When using the selective ACK entry option, the BSN corresponds to the first bit of the following acknowledgement map. This type of an acknowledgement is only applicable if no less than 16 ARQ blocks have been received with no prior transmitted acknowledgment.
- Acknowledgment type 1: Corresponding to cumulative ACK entry, uses the BSN to cumulatively acknowledge all ARQ blocks received. Type 1 feedback message essentially points to the last consecutive ARQ block that was received successfully. Type 1 acknowledgement message

---

10 A MAC SDU is logically partitioned into ARQ blocks whose length is specified by the "TLV" (type/length/value) parameter [4].

11 For an extensive coverage of the theory of ARQ schemes, please refer to Ref. [15].

has a fixed length of 4 bytes. Types 0 and 1 are basic acknowledgement schemes. Type 1 requires less overhead, however, it is less efficient. Type 0 generally offers a better performance in terms of delay and throughput, while type 1 has a reduced feedback dimension [2].

- Acknowledgment type 2: Corresponding to cumulative with selective ACK entry is in fact a combination of acknowledgment type 0 and type 1. BSN in type 2 is interpreted as cumulative acknowledgement, and the first bit of the following map is set to 1 and the remaining bits of the map are the same as type 0.
- Acknowledgement type 3: Corresponding to cumulative with block sequence ACK entry is a combination of type 1 and a series of sequence ACK maps. In this case, the BSN acknowledges all correctly received ARQ blocks cumulatively. The sequence ACK map contains either two sequences with each 6 bits long, or three sequences each with a length of 4 bits. Thereby, each sequence specifies a number of consecutive BSN entries, with the first sequence starting at the cumulative BSN plus one, which is always a negative acknowledgment [4,10].

Since type 2 and type 3 acknowledgements are combinations of type 0 and type 1, they are very similar, where the difference between them appears to be in the selective part. Type 2 directly joins type 0 and type 1, where the feedback message points to the last received error-free ARQ block in sequence and composes a bitmap indicating which of the ARQ blocks that followed the sequence are error free, which allows for selective retransmission. Type 3 is different from type 2 only in the composition of the bitmap that acknowledges the entire sequences of ARQ blocks through a suitable encoding [2].

The WiMAX Forum™ Mobile System Specification requires ARQ acknowledgment type 1, type 2, and type 3 to be implemented, while ARQ acknowledgment type 0 is optional. AeroMACS profile, however, requires that type 2 and type 3 should be mandatory and type 0 and type 1 be optional.

The performance quality of various types of acknowledgements are assessed by considering many factors, including feedback intervals, error patterns, computational complexity, and the size of the ARQ blocks.

7.5.4 Handover (HO) Procedures in AeroMACS Networks

Handover, or handoff (HO), is the ability of an MS, SS, or even a mobile RS to terminate communications with one BS and recommence it with a different BS without interruption in the ongoing session. Handover is a key mobility attribute differentiating the cellular networks from traditional precellular mobile networks (see Chapter 2). The proper implementation of HO is a critical issue for all cellular networks, including

WiMAX and AeroMACS. For fully developed and deployed AeroMACS networks with aircraft speed on the airport surface reaching up to 370 km/h, the HO must be optimized in context the of careful and accurate cellular network planning.

The HO protocols in AeroMACS are directly dependent upon two processes of *scanning* and *ranging*. Scanning is the process by which the MS/SS acquires a DL channel from one of the reachable BSs. Scanning takes place upon initial network entry or while a handover process is in progress. Ranging is the process by which the MS/SS, after having acquired a DL channel and information on the channel parameters, adapts its time offsets and power adjustments to be aligned with the BS.

Scanning process for the purpose of performing HO, in fact starts before the handover process is even triggered. A MS/SS, while being served by a BS, periodically receives information from the serving BS on the status of neighboring BSs as potential future target stations for handover. When the MS triggers scanning it makes a request that data transfer in both DL and UL paths be stopped so that the scanning process can be conducted. During the scanning process the MS listens selectively only to the BSs from which it has received information [33].

The AeroMACS standards are required to support HO between different AeroMACS BSs during aircraft movement, and under the condition that the connection is degraded with the current serving BS. The same applies to all other mobile or stationary stations on airport surface that are connected to AeroMACS network. The IEEE 802.16-2009 standard supports three handover mechanisms.

1) MS initiated handover
2) MS initiated handover with scanning
3) BS initiated handover

The AeroMACS MOPS and Profile (see Chapter 6) require that the two MS-initiated handover mechanisms, listed here, be supported by all AeroMACS systems [33].

In regard to which one of the two MS-initiated HO procedure should be used, the profile leaves that as a design option.

7.5.4.1 MS-Initiated Handover Process

MS-initiated handover is the simplest AeroMACS handover process. In this HO technique, the MS receives "Neighbour Advertisement" information from its serving BS, and chooses the candidate target BS based on this information and not on actual measured-based channel conditions. Figure 7.10 is a diagram that describes the MS-initiated HO process and exchanges that take place between the MS and the two BSs involved in the process, that is, the serving BS and the target BS.

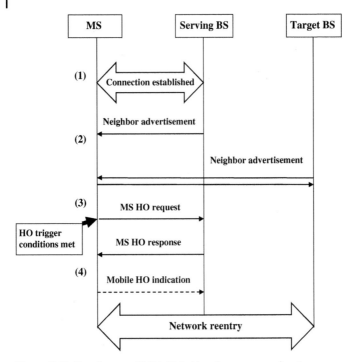

Figure 7.10 The diagram of MS-initiated handover process showing message flows [33].

The numbers on the left side of Figure 7.10 point to the order of sequence of exchanges taking place in the MS-initiated handover process, as described in the following paragraphs.

1) Successful completion of initial network entry and service flow establishment.
2) Acquiring network topology: neighbor advertisement
   a) The serving BS and the target BS send "Neighbour Advertisement" message. An AeroMACS BS broadcasts information about the network topology using the message, which is an AeroMACS layer 2 management message. This message provides channel information about neighboring BSs.
   b) The serving BS sends Neighbour Trigger TLV[12] in Neighbour Advertisement message. The Neighbour Trigger TLV defines several important factors: the type of metric (received signal strength

---

12 TLV stands for type, length, value. In the context of AeroMACS MAC sublayer, it is a tag that is added to each transmitted parameter that describes the type of parameter, the length of the encoded parameter, and its value.

indicator (RSSI) or carrier to interference and noise ratio (CINR)), function (metric of neighbor BS is greater/less than absolute value, or than the sum of serving BS metric and relative value), action (respond on trigger with MS HO request), and value of the trigger.

3) Handover decision
   a) When the trigger conditions are met, the MS sends one or more MOB_MSHO-REQ to the serving BS.
   b) The serving BS sends MS HO respond message.
   c) The MS optionally sends MS HO indicator to the serving BS with final indication that it is about to perform HO.

4) Handover process initiation.

AeroMACS supports the application of hard HO procedure that is initiated by the MS with the involvement of scanning process as well. In a typical HO procedure, the BS sends the scanning request to the MS when the received signal level is lower than a certain threshold. This request is sent in an unsolicited manner and contains all the parameters useful for scanning (scanning interval, report mode, report period, first scanning frame, etc.). The MS performs the scanning and then sends the result report back to the BS. The decision to start the HO procedure is made by the BS. Reference [33] provides a detail description of the second AeroMACS handover procedure, that is, MS-initiated handover with scanning.

The performance goodness of various handover procedures may be measured or compared on the basis of few HO-related parameters, among them are handover delay, interruption time, mean packet loss, and probability of handover failure [2].

7.6 AeroMACS Network Architecture and Reference Model

The IEEE 802.16-2009 [4] and WiMAX Forum network architecture [34] define the architecture for AeroMACS networks. The WiMAX architecture guarantees intervender and internetwork interoperability. As we have pointed out earlier, the AeroMACS standard specifies only PHY layer and MAC sublayer protocols and leaves the higher layer protocols unspecified for greater flexibility. Nonetheless, AeroMACS is envisioned to be an all-IP network that decouples the access network from IP connectivity, thereby enabling modularity.

7.6.1 AeroMACS Network Architecture

Figure 7.11, adopted from Ref. [2], illustrates a general architectural diagram of an AeroMACS network.

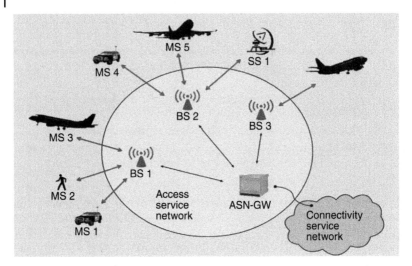

Figure 7.11 An AeroMACS network architectural diagram. (Adopted from Ref. [2] with permission from IEEE.)

As Figure 7.11 demonstrates, the logical entities in an AeroMACS network are as outlined next.

- Mobile stations (MS): moving nodes such as aircraft, service and emergency vehicles, and fuel and catering trucks.
- Subscriber stations (SS): stationary (fixed) nodes such as control tower, radar systems, airline equipment, sensors, and weather stations.
- Base stations (BS): the central logical entities of the AeroMACS network representing access points to the network and implementing some key air interface and access functionalities, such as radio resource management (RRM), UL and DL scheduling, and HO control. The physical BS can be sectorized, where sectors operate like logical BSs, each one operates on an angular sector with distinct channel assignment.
- Access service network (ASN): this is composed of a complete set of functionalities required for providing MS/SSs with radio access to the AeroMACS network. In other words, ASN is essentially an access network infrastructure to which MSs and SSs are connected. Radio-related functions of ASN are the responsibility of BSs that are logical and physical parts of ASN.
- Connectivity service network (CSN): this entity contains a set of network protocols and functionalities that deliver IP connectivity services (within the airport intranet or internet) to the AeroMACS user. The CSN may consist of several network components such as routers, AAA, and internetworking gateways. In short, the CSN provides connectivity for AeroMACS network to other networks.

- ASN gateway (ASN-GW): another important component within ASN that is essentially an AeroMACS router, representing the link between the ASN and the CSN.

It should be noted that a functional ASN needs to contain at least one base station and one ASN-GW for connecting the AeroMACS network to outside networks. Some of the entities shown in Figure 7.11 are characterized in more detail in the context of AeroMACS NRM in the next section.

7.6.2 AeroMACS Network Reference Model (NRM)

Network reference model (NRM) for the AeroMACS network, or any other wireless network for that matter, is a logical representation of the network architecture, identifying functional entities and reference points over which interoperability is achieved between functional entities. In particular, the NRM describes the functional blocks and reference points of the access service network, connectivity service network, and their interconnection.

The general approach for the design of AeroMACS network architecture, like that of WiMAX network, is grounded on several principles. First and foremost, the architecture is based upon the functional decomposition concept. The required features are divided into functional entities with no specific implementation assumptions for their physical realization. The architecture specifies well-defined reference points (RPs) between these functional entities. This is to ensure multivendor interoperability. Second, a modular and flexible approach is adopted to guarantee that a broad range of physical/logical implementation options are acceptable for deployment. However, another important feature of the AeroMACS architecture, as was mentioned earlier, is the decoupling of access network and its support technologies from the IP connectivity network. This enables the unbundling of access infrastructure from the IP connectivity protocols.

Each of the entities MS, SS, BS, ASN, and CSN, shown in Figure 7.11, represent a grouping of functional entities. Reference points are conceptual links connecting different functions of various functional entities, which are not necessarily physical interfaces. They become physical connections when functional entities that they interface reside in different physical devices. These functions make use of various protocols associated with an RP. Figure 7.12 illustrates a visual depiction of AeroMACS NRM, identifying overall interoperability reference points between various functional entities.

The AeroMACS NRM contains several interoperability RPs, as shown in Figure 7.12. Two protocols associated with an RP shall be able to

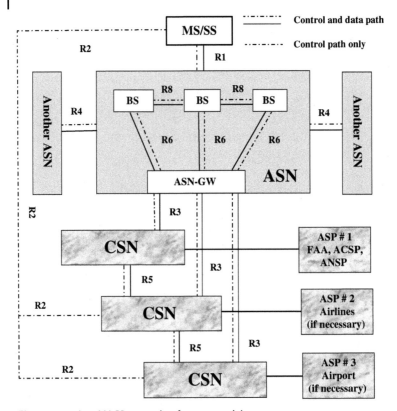

Figure 7.12 AeroMACS network reference model.

originate and terminate in different functional entities, that is, a given protocol associated with an RP may not always terminate in the same functional entity.

It is noted that reference points R1 through R5 are inter-ASN RPs and the rest are, by and large, intra-ASN reference points. The reference point R1 is the interface between the MS/SS and the ASN, or in fact a BS. The connection is provided directly, or in the case of relay-augmented AeroMACS networks, it is provided through an IEEE 802.16j-based relay station. Table 7.14 provides description for different RPs that are defined in AeroMACS NRM in accordance with information provided in AeroMACS Manual [33].

Three connectivity service networks (CSN) are identified in Figure 7.12. Each one hosting a particular ASP (Application Service Provider), representing the major stakeholders of the AeroMACS network. The first set of ASPs are the FAA, ACSP (Aeronautical Communication Service Provider), and ANSP (Air Navigation Service Provider). ANSP is the entity that manages the flight traffic in a region or in a country. ANSPs have their networks deployed to support the air traffic applications. ANSP networks belonging to various regions are interconnected to each other and thus providing a global network for data link services [33].

Note that Reference Point 7 is not listed on Table 7.14. This is for the reason that RP7 has been deleted by the IEEE, while the original numbering is kept intact.

The AeroMACS network reference model allows multiple implementation options for a given functional entity, while achieving interoperability among these different realizations. Interoperability is based on the definition of communication protocols and data plane[13] treatment between functional entities to achieve an overall end-to-end function, such as security or mobility management. Thus, the functional entities on either side of a reference point (RP) represent a collection of control and bearer plane end points. In this setting, interoperability may be verified based only on protocols exposed across an RP, which depends on the end-to-end function, and/or capability realized [33].

The MS/SS and BS, shown in Figure 7.12, are specific AeroMACS entities whose functions are described by the IEEE 802.16-2009 Standard. The MS/SS manages the user and control planes of the PHY layer and MAC sublayer of the user node, while BS does the same for the access network. The AeroMACS BS is a logical entity that embodies a full instance of the MAC and PHY layers in compliance with the AeroMACS specifications.

The ASN defines a logical boundary and represents a convenient way to describe an aggregation of functional entities and corresponding message flows associated with the access services. The ASN also represents a boundary for functional interoperability with AeroMACS clients, Aero-MACS connectivity service functions, and aggregation of functions embodied by different vendors. Mapping of functional entities to logical entities within ASNs as depicted in the NRM is informational.

An ASN shares reference point R1 with the MS/SS, R3 with a CSN, and R4 with another ASN. The ASN consists of at least one BS and one ASN-GW. A BS is logically connected to one or more ASN-GWs. The R4 reference point is the only RP for control and bearer planes for interoperability between similar or heterogeneous ASNs [33].

According to AeroMACS architecture, a generic ASN-GW is responsible to carry out a number of tasks and functionalities, a partial list of which is presented here [33].

---

13 Data plane refers to the part of a network that carries user traffic. Data plane is also known as user plane, forwarding plane, or bearer plane.

Table 7.14 AeroMACS major reference points descriptions [33].

RP	End points	Description
R1	MS/SS/RS-ASN	Consists of the protocols and procedures between an MS/SS and a BS as part of the ASN air interface specifications. IEEE 802.16e/802.16j air interface protocols implementation, PHY and MAC to support MS/SS/RS to communicate with BS [35]
R2	MS/SS/RS-CSN	A nondirect-protocol logical interface for authentication, authorization[a], IP host configuration management, and mobility management. This reference point is logical in that it does not reflect a direct protocol interface between MS/SS and CSN. R2 may support IP host configuration management running between the MS/SS and the CSN
R3	ASN-CSN	Control protocols to support AAA, mobility management capabilities, policy enforcement, tunneling, and other data plane methods to transfer IP data between ASN and CSN. Some of the protocols foreseen on this RP are RADIUS (Remote Authentication Dial-In User Service) and DHCP (Dynamic Host Configuration Protocol)
R4	ASNGW-ASNGW	A set of control and data path protocols originating/ terminating in an ASN-GW that coordinates MS mobility between ASNs and ASN-GWs. It is the only interoperable RP between ASN-GWs of one or two different ASNs.[b] In fact, R4 is the only interoperable RP between similar or heterogeneous ASNs
R5	CSN-CSN	Consists of a collection of control plane and bearer plane protocols for internetworking between the CSN operated by the home NSP and that operated by a visited NSP. This reference point will only exist between CSNs that have an institutional or business relationship that requires such internetworking
R6	BS-ASNGW	A set of control and data path protocols for communication between BS and ASNGW within a single ASN. The data path consists of intra-ASN and inter-ASN tunnels between the BS and ASNGW. The control protocols include function for data path establishment, modification, and release control in accordance with MS mobility
R8	BS-BS	A collection of intra-ASN control message flows between BSs to ensure fast and seamless handovers. The data plane includes protocols that allow inter-BS communication protocols and data transfer. Additionally, protocols that control data transfer BS/ABSs involve in handover of a certain MA/AMS

a) "The authentication part of R2 runs between the MS/SS and the CSN operated by the home NSP; however, the ASN and CSN operated by the visited NSP may partially process the aforementioned procedures and mechanisms" [33].
b) Note that R4 is not shown in AeroMACS NRM of Figure 7.12, it is normally included when connecting ASNGWs, for cases in which there are multiple of ASNGWs.

- Connectivity of MS with AeroMACS layer 2
- Establishment of IP connectivity between the MS and the CSN
- Network discovery and selection of the AeroMACS subscriber's preferred NSP
- Allocation of subscriber IP address by referring to the DHCP server for network establishment and DHCP messages forwarding
- AAA proxy/client, AeroMACS ASN-GW triggers the exchange of susceptible subscriber information and transfers AAA messages of AeroMACS subscriber visited NSP for authentication, authorization, and accounting to the Home NSP [33].

As it has been pointed out earlier, CSN is the network that provides end-to-end connectivity to AeroMACS subscribers with network entities and enables the provision of services by AeroMACS ASPs. CSN main functionalities are AAA server and DHCP server. The applicable reference points are R3 (CSN with ASN) and R5 (between two CSN) [33].

7.7 Aeronautical Telecommunications Network Revisited

In this section, we explore the role of the AeroMACS network within the larger context of the ATN, sometimes referred to as the "Aviation Internet" (see Chapter 1). An airport network, discussed in the next section, may be viewed as a subnet of the greater ATN, and AeroMACS may be considered as an access network for the airport network to connect to ATN.

ICAO has developed a new standard, and has recently created a technical manual, for ATN based on Internet protocol suite, ATN/IPS, to replace the legacy OSI-based ATN. The ATN/IPS has adopted the same four-layer model as defined in the Internet-layered protocol architecture, namely, the link layer, the Internet protocol (IP) layer, the transport layer, and the application layer. The ATN/IPS standard does not adopt any specific link layer protocol, as this is a local or bilateral issue which does not affect overall interoperability [36].

The ATN consists of multiple independent networks, such as ANSP network, airlines network, airport service provider network, with separate administrative domains interconnected to each other to achieve an overall safety communication infrastructure. These networks predominantly operate based on ATN/IPS (IPv6) protocols, and are closed networks, that is, sufficiently isolated from the public Internet. Therefore, the overall aeronautical network can be viewed as islands of closed networks interconnected over public infrastructure to form a closed internet for aeronautical purposes that is invisible to the public Internet. This is shown in Figure 7.13.

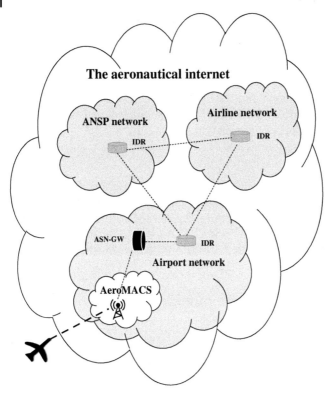

The aeronautical internet

ANSP network

Airline network

IDR

IDR

ASN-GW

IDR

Airport network

AeroMACS

Figure 7.13 Aeronautical telecommunications network (Aviation Internet).

The autonomous networks within ATN, utilizing the IPS standards and protocols, are interconnected through inter-domain routers (IDR), such as BGP (Border Gate Protocol). In order to maintain privacy vis-a-vis public Internet, these BGP routers are not exposed to other routers in public domains. Secured links, either based on VPNs (virtual private network[14]) or dedicated telecommunication lines are deployed to interconnect the BGP routers in the aeronautical network as shown in Figure 7.13. Since AeroMACS acts as an access network for the airport domain, ASN Gateway of AeroMACS network is connected to the core BGP router in the airport network [33].

BGP-4 with extensions is adopted for interdomain routing. The transmission control protocol (TCP) and user datagram protocol (UDP) are

---

14 A virtual private network is a network that is constructed using public media, normally the Internet, to connect to a private network.

adopted for connection-oriented and connectionless services at the transport layer [33].

It should be noted that the AeroMACS networks, capable of packet data service provision, offer mechanisms to transport both ATN/IP- and ANT/OSI-based messaging.

7.8 AeroMACS and the Airport Network

An airport network is an infrastructure interconnecting a number of service networks for the overall air traffic and flight safety management of the airport area. An airport network is not necessarily dedicated to a single airport, it may actually cover multiple regional airports. Figure 7.14 provides a logical representation of the overall network at an airport [33].

Figure 7.14 represents a logical portray of an airport network with all auxiliary networks and services connected to it. ARINC is a service provider in North America, and STIA[15] provides the same type of services in Europe. The role of major components interconnected in an overall airport network to the core airport network of Figure 7.14 is briefly described at this point [33].

- The networking platform that supports various airport services is the *airport network*, which interconnects several other airport service networks, as shown in Figure 7.14. The airport network may provide infrastructure for Airport Service Providers (ASPs) to register and host their airport services in global or site-local domains. Global services are available worldwide, while site-local services are provided within the airport network only.
- *AeroMACS* network, in the context of the larger airport networking infrastructure, plays the key role of providing mobile connectivity to aircraft to access the core airport network. AeroMACS dynamic connections can handle subscriptions for mobile users, and ensures authenticity and privacy needed for aircraft's safety communications. AeroMACS network may also offer fixed link services to interconnect with various networks within the airport domain.
- *ANSP* manages the flight traffic in a region or in a country, as such ANSP networks support air traffic applications. Various regional ANSP networks are interconnected to provide a global network for data link services.
- *ASPNs* can be considered as private enterprise networks. An airport normally has multiple of ASPNs. ASPNs can host the servers or

---

15 STIA is the acronym for the French name of a company called "Société Internationale de Télécommunications Aéronautiques."

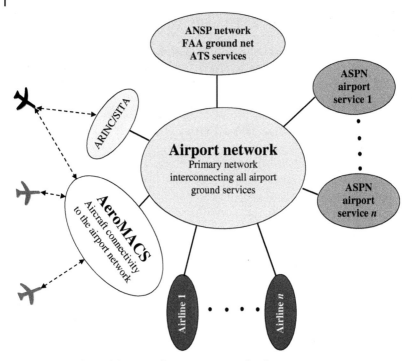

Figure 7.14 A logical diagram of an airport network infrastructure.

applications that are required to be accessed by external entities in the perimeter network, while the internal network elements can be kept inside a private LAN.

In conclusion, the key role of an AeroMACS network, within the greater airport network, is essentially providing seamless connectivity for the aircraft, while on the surface of the airport, to enable access to all services available in the airport network. These services include ANSP network services/applications, as well as airline or service partner operational centers to access data link applications. AeroMACS may also be used to extend private networks that are owned by different ASPs within airport premises or to interconnect networks within airport regions.

7.9 Summary

In this chapter, we have explored PHY layer specifications and MAC sublayer protocols of AeroMACS networks. AeroMACS main PHY layer feature is its robust and multipath resistant multiple access technology,

OFDMA, inherited from IEEE 802.16-2009 standard. This allows 5-MHz OFDMA channels within the allocated ITU-regulated aeronautical C-band of 5091–5150 MHz. The duplexing method is TDD, enabling asymmetric UL and DL signal transmission. Adaptive modulation–coding (AMC) is another key physical layer feature of AeroMACS network. AMC allows for a proper combination of a modulation scheme (selected from QPSK, 16-QAM, 64-QAM) and a coding scheme (selected from convolutional codes, turboconvolutional codes, repetition codes, RS codes), depending on the channel condition. MIMO and smart antenna systems are another PHY layer feature of AeroMACS networks. The chapter also discusses the AeroMACS MAC layer. In particular, scheduling, QoS, ARQ system, and HO procedure are described in some detail. AeroMACS network architecture and NRM are discussed. The last part of the chapter highlights the position and the role of the AeroMACS network within the larger contexts of the airport network and the global ATN.

AeroMACS is the AP17 proposed data communications link for the airport surface delivering an ATN/IP high data rate radio link to enable air traffic services (ATS), which includes the safety critical communications involving aircraft and air traffic management, Aeronautical Operational Control, that are directly linked to the safety and regularity of flight, and airport authority communications that affect the safety of flight vis-a-vis airport surface vehicles and ground services.

AeroMACS is based on a commercial telecommunication standard, the IEEE 802.16-2009 and its interoperable version; the WiMAX technology. This enables the benefits of economy of scale, the possibility of upgrading as the mother standard is evolved, and the use of COTS equipment. Additionally, the adoption of an established standard reduces the time and the cost of initial development, implementation, and the equipment required on the airport surface and onboard of aircraft.

A high point of IEEE 802.16 standards and its derivative technologies, WiMAX and AeroMACS, is the fact that the standard only specifies and standardizes the network air interface and radio link for various entities such as SSs, MSs, BSs, and RSs. The air interface rules include PHY and MAC layers protocols. This leaves the higher layers of the protocol stack open to other bodies like the Internet Engineering Task Force to fill in and complete the standard for end-to-end communications.

In summary, AeroMACS is a broadband end-to-end IP network that provides worldwide interoperability and integration of critical communications for ANSPs, airspace users, and airports. It supports the exchange of wide range of applications, such as voice, video, data, in an efficient, safe, and secure manner. Finally, AeroMACS is an aviation data link and is the first new pillar of a wider future aviation communications infrastructure envisioned by NextGen and SESAR projects. AeroMACS presents an important test case for the aviation industry in leveraging on commercial communication developments and technologies; pooling synergies between ANSPs, airports and airlines; handling the security issues of future aviation communications infrastructure; and implementing IP data link on-board aircraft [33].

References

1 WiMAX Forum®, "Air Interface Specifications, WiMAX Forum® Mobile System Profile WMF-T23-001-R010v09, Approved, 2010.

2 G. Bartoli, R. Fantacci, and D. Marabissi, "AeroMACS: A New Perspective for Mobile Airport Communications and Services," *IEEE Wireless Communications*, 20(6), 44–50, 2013.

3 WiMAX Forum®, "Mobile System Profile Specification: Release 1.5 TDD Specific Part," WMF-T23-002-R015v01, August 2009.

4 IEEE, IEEE 802.16–2009 Standard for Local and Metropolitan Area, Part 16: Air Interface for Broadband Wireless Access Systems, May 2009.

5 IEEE, IEEE 802.16–2009 Standard for Local and Metropolitan Area, Part 16: Air Interface for Broadband Wireless Access Systems, May 2009.

6 ICAO ACP-WG-S/5 WP-10, "DL: UL OFDM Symbol Ratio in AeroMACS TDD Frame," July 2014.

7 RTCA, Minimum Operational Performance Standards (MOPS) for Aeronautical Mobile Airport Communication System (AeroMACS), RTCA DO-346, September 2014.

8 EUROCAE, "Minimum Operational Performance Standards (MOPS) for the Aeronautical Mobile Airport Communication System," 143/ED-223, October 2013.

9 RTCA, "Aeronautical Mobile Airport Communications System (AeroMACS) Profile, RTCA-DO-345, December 18, 2013.

10 M. Ehammer, E. Pschernig, and T. Gräupl, "AeroMACS; An Airport Communications System," *Proceedings of 30th Digital Avionics Systems Conference*, pp. 4C1-1–4C1-16, 2011.

11 S. W. Ellingson, *Radio System Engineering*, Cambridge University Press, 2016.

12 B. Kamali and R. L. Brinkley, "Reed-Solomon Coding and GMSK Modulation for Digital Cellular Packet Data Systems," *Proceedings of IEEE PACRIM Conference on Communications, Computers, and Signal Processing*, vol. 2, pp. 745–748, 1997.

13 F. Xiong, *Digital Modulation Techniques*, Artech House, 2000.

14 J. G. Proakis and M. Salehi, *Digital Communications*, 5th edn, McGraw Hill, 2008.

15 S. Lin and D. J. Costello, *Error Control Coding: Fundamentals and Applications*, 2nd edn, Prentice Hall, 2002.

16 I. Gheorghisor, "Spectral Requirement for Wireless Broadband Networks for the Airport Surface," *Proceedings of IEEE ICNS Conference*, pp. 1–10, 2008.

17 A. F. Molisch, *Wireless Communications*, 2nd edn, Chapter 19, John Wiley & Sons, Inc., New Jersey, 2011.

18 D. Marabissi, D. Tarchi, R. Fantacci, and F. Balleri, "Efficient Adaptive Modulation and Coding Techniques for WiMAX Systems," *Proceedings of IEEE ICC*, pp. 3383–87, 2008.

19 SESAR, "AeroMACS System Requirements Document," 2010.

20 WiMAX Forum, "Mobile WiMAX – Part I: A Technical Overview and Performance Evaluation", 2006.

21 Y.-J. Choi, C. S. Kim, and S. Bahk, C "Flexible Design of Frequency Reuse Factor in OFDMA Cellular Networks," *Proceedings of IEEE ICC*, 2006.

22 RTCA, "Aeronautical Mobile Airport Communications System (AeroMACS) Profile, RTCA-DO-345, December 18, 2013.

23 R. D. Apaza, *Wireless Communications for Airport Surface: An Evaluation of Requirements*, IEEE, 2004.

24 ICAO, "*Bandwidth Requirements and Channelization Approaches for Future Airport Surface Communications in the 5091–5150 MHz Band*", ACP-WGF 15/WP23, presented by Robert Kerczewski, June 2006.

25 E. Hall, J. Budinger, R. Dimond, J. Wilson, and R. Apaza, "Aeronautical Mobile Airport Communications System Development Status," *Proceedings of IEEE ICNS Conference*, pp. A4-1–A4-15, 2010.

26 R. J. Kerczewski, B. Kamali, R. D. Apaza, J. D. Wilson, and R. P. Dimond, "Considerations for Improving the Capacity and Performance of AeroMACS," *Proceedings of IEEE Aerospace Conference*, pp. 1–8, 2014.

27 I. L. Gheorghisor, Y.-S. Hoh, and A. E. Leu, "Analysis of ANLE Compatibility with MSS Feeder Links," MITER CAASD Report MTR090458, 2009.

28 J. D. Wilson, "Simulating Global AeroMACS Airport Ground Station Antenna Power Transmission limits to Avoid Interference with Mobile Satellite Service Feeder Uplinks, NASA/TP-2013-216480, April 2013.

29 B. Li, Y. Qin, C. P. Low, and C. L. Gwee, "A Survey on Mobile WiMAX", IEEE Communications Magazine, December 2007.

30 A. F. Bayan and T.-C. Wan, "A Review on WiMAX Multihop Relay Technology for Practical Broadband Wireless Access Network Design," *Journal of Convergence Information Technology*, 6(9), 363–372, 2011.

31 D. Shwetha, H. J. Thontadharya, and J. T. Devaraju, "A Bandwidth Request Mechanism for QoS Enhancement in Mobile WiMAX

Networks," *International Journal of Advanced Research in Electrical, Electronics, and Instrumentation Engineering*, 3(1), 6881–6888, 2014.

32 SANDRA, "Report on Long Term AeroMACS Evolution" D6.2.3, 2012.

33 ICAO, "Manual on the Aeronautical Mobile Airport Communications System (AeroMACS)" Doc 10044, 2016.

34 WiMAX Forum, "WiMAX Forum Network Architecture – Stage 2: "Architecture Tenets, Reference Model and Reference Points," February 2009.

35 WiMAX Forum®, "Network Architecture: Architecture Tenets, Reference Model and Reference Points Base Specification," Release 2.2, 2014.

36 ICAO, "Manual on the Aeronautical Telecommunication Network (ATN) using Internet Protocol Suite (IPS) Standards and Protocol", 2nd edn, DOC 9896, January 1, 2015.

8

Aeromacs Networks Fortified with Multihop Relays

8.1 Introduction

The basic concept of relay-augmented cellular networks is the application of small form-factor relays to support either the expansion of the base station coverage area, or to improve the network capacity and throughput, or both, without any change in the network radio resource distribution arrangements. Typically, the relay-based configuration is rolled out in the early stages of network deployment to meet the objectives of coverage over a larger area at lower cost than the more expensive alternative of integrating additional base stations into the network [1].

Over the past few decades, several classes of multihop wireless networks have been developed and standardized. Among them are various wireless *ad hoc* networks such as Bluetooth networks, wireless sensor networks, relay-based wireless multihop networks, and wireless mesh networks. Each of these communications systems has its own characteristics that are defined by the main network attributes such as mobility, scalability, power constraints, and form factor, leading to different protocol design scenarios.

Within the same time frame, it gradually became apparent that for the future generations of wireless networks to be able to deliver ubiquitous broadband content, the network is required to provide acceptable level of coverage in all parts of the covered area at significantly higher bandwidth per subscriber. In order to achieve this objective at operating frequencies of AeroMACS networks, or any other network operating over C-band and beyond, the network must significantly reduce the distance between the MS/SS and base stations (BS) antennas. The notion of redefining the wireless network architecture to support the use of multihop relay stations (RS) has been shown to be a promising idea to cope with this challenge. On the one hand, multihop relays are small, cost effective, and easy to install; enabling rapid deployment in the airport environment for

AeroMACS: An IEEE 802.16 Standard-Based Technology for the Next Generation of Air Transportation Systems, First Edition. Behnam Kamali.
© 2019 the Institute of Electrical and Electronics Engineers, Inc. Published 2019 by John Wiley & Sons, Inc.

provision of coverage in all corners of an airport. On the other hand, relays do not require any dedicated backhaul equipment, as they receive their resources from network base stations through the same means used for the access services. Relay stations standardization was commenced by the IEEE in mid 2000s. The results of the standardization activities were published as an amendment to the IEEE 802.16-2009, namely, the IEEE 802.16j Amendment.

In this chapter, the focus is on relay-augmented AeroMACS networks with relay station specification conformant to the IEEE 802.16j amendment. Multihop relays are not currently included in the component arsenal of AeroMACS systems, however, a strong case can be made in favor of IEEE 802.16j-base WiMAX technology for AeroMACS, particularly in regard to potential expansion and future evolution of the airport that require modification in the airport surface communications system. As in the case LTE-based 5G cellular networks and the future communications standards for vehicular systems and ITS (intelligent transportation systems), our deliberation in this chapter demonstrates that there is very little doubt that multihop relays will have to be included in the future expansions of AeroMACS networks.

We first revisit the IEEE 802.16j amendment. This is followed by an extensive coverage on characteristics, applications, classifications, and usage scenarios of IEEE 802.16j-based multihop relays. A case is then made for the need to make use of multihop relays in AeroMACS networks. This is presented by examining many reasons that AeroMACS networks would benefit from incorporation of multihop relays. The fortification of any wireless network with multihop relays is not without a cost. The price to be paid in this case manifests itself, by and large, in the form of added system and protocol complexities and by introducing some adverse effect on network operational behavior, particularly with respect to latency. These challenges are also addressed.

8.2 IEEE 802.16j Amendment Revisited

Owing to practical shortfalls arising from early implementation of IEEE 802.16e-based WiMAX networks, the need for some modification and amendment to the Standard was recognized early on. In particular, initial field trials of mobile WiMAX products have shown that IEEE802.16e systems provide poor QoS around WiMAX cell boundaries for indoor users, as well as in outdoor areas severely shadowed by man-made structures and natural obstacles. For instance, even with application of advanced signal processing techniques such as OFDM, MIMO, and AMC the projected data rates require SINR (signal-to-interference-plus-noise ratio) levels at the front end of the receivers that are difficult to obtain at the WiMAX cell boundaries or in the shadowed areas.

To address this challenge, as was mentioned earlier, the IEEE composed an amendment to the IEEE 802.16-2009 standard, designated as IEEE 802.16j, in which multihop RS may be used as an extension to a base station and transport traffic between the BS and the subscriber station (SS) and/or mobile station (MS). The new standard specifies a set of technical issues in order to enhance the legacy standard (IEEE 802.16-2009), with the main objective of supporting the relay concept. Referring to the approval of IEEE 802.16j amendment project, the IEEE states the following.

"The approval of this project aims to enable exploitation of such advantages by adding appropriate relay functionality to IEEE Std 802.16 through the proposed amendment. The multihop relay is a promising solution to expand coverage and to enhance throughput and system capacity for IEEE 802.16 systems. It is expected that the complexity of relay stations will be considerably less than the complexity of legacy IEEE 802.16 base stations. The gains in coverage and throughput can be leveraged to reduce total deployment cost for a given system performance requirement and thereby improve the economic viability of IEEE 802.16 systems. Relay functionality enables rapid deployment and reduces the cost of system operation. These advantages will expand the market opportunity for broadband wireless access. This project aims to enable exploitation of such advantages by adding appropriate relay functionality to IEEE Std 802.16 through the proposed amendment" [2].

Thus, in relay-fortified cellular networks base stations are complemented with relay stations instead of additional BSs. An RS communicates with the BS through a wireless channel and may operate without additional carrier frequency [3]. This eliminates the need for a wired or a dedicated wireless (microwave) connection to the backhaul network and significantly reduces the installation and operation cost, in comparison with the alternative of employing additional BSs. By replacing the direct link between a BS and an SS/MS in a poor coverage area with two links with better channel quality, as we will see in this chapter, the overall network capacity will increase dramatically [4].

The IEEE 802.16j amendment introduces the multihop relay as an optional deployment. Consequently, in relay-augmented network, the BS needs to be upgraded in order to integrate relay stations into its UL and DL transmission protocols. This type of base station is referred to as multihop relay base station (MRBS). In other words, a legacy BS is replaced by a MRBS and as many RSs as needed. The new expanded cell is called multihop relay cell (MR-cell). MRBS manages all communications resources within a MR-cell through a centralized or decentralized procedure. Resource management of SS/MS may be carried out directly through the BS or via radio links to a relay station. Exchange of traffic data and control signals between the SS/MS and MRBS may be direct, or may be relayed by the RS thereby extending the coverage and enhancing performance of the system in areas where RSs are deployed. Each RS is under the supervision of an MRBS. When more than two hops take place in the transmission path, traffic and control signals between the last RS in the chain (called access RS) and MRBS may also be relayed through intermediate RSs. The RS may be fixed in location, it may be nomadic, or in the case of an access RS, it may be mobile [2]. The allocation of bandwidth and other resources for RSs and SS/MSs may be controlled in one of the two modes, centralized or noncentralized (distributed) scheduling. In centralized scheduling mode, the bandwidth allocation for an RS's subordinate SS/MSs is determined at the MRBS, whereas in distributed scheduling mode, the bandwidth allocation of an RS's subordinate SS/MSs are determined by the RS in cooperation with the MRBS. In other words, in distributed scheduling RSs share resource allocation responsibilities with their superordinate MRBSs.

The IEEE 802.16j working group was charged with the task of generating a standard for WiMAX mobile multihop relay (MMR) network. The standard specifies a set of technical issues in order to enhance the legacy standards (in this case IEEE 802.16-2009) with the main objective of supporting the multihop relay concept. Key technical issues that were discussed in the IEEE 802.16j working group included new frame structure to support relay stations, centralized versus distributed control, centralized versus distributed scheduling, radio resource management, power control, call admission and traffic sharing policies, QoS based on network-wide load balancing and congestion control, and security management. The sections that follow are dedicated to address several of these issues.

In the context of broadband AeroMACS networks, small form-factor relays are used to support the expansion of the AeroMACS base station coverage, and/or to augmenting the network capacity, without any change in the radio resource distribution arrangements of the network. In short, the fundamental goals of multihop relay-augmented AeroMACS networks are the provision of very high-rate data services with a coverage area that expands beyond the footprint of the legacy AeroMACS cell, while improving the aggregate network throughput.

8.3 Relays: Definitions, Classification, and Modes of Operation

Multihop relaying for extension of radio outreach is an old concept that has become revitalized through the recent emergence of broadband radio access technologies. Before embarking on characterization and classification of relays and relay-augmented wireless networks, few words are in order regarding the benefits of employing relays in a wireless network. First and foremost, among these benefits is the fact that the IEEE 802.16j-defined relay deployment strategy presents a cost-effective, low-complexity, and easy-to-install-infrastructure alternative for wireless network radio outreach extension in a variety of situations. Second, the relays can provide capacity improvement and throughput enhancement in areas that are not sufficiently covered by the serving BSs. In the majority of usage models, however, relays are deployed to satisfy a combination of the two objectives just mentioned. Another important result of deployment of a relay-fortified wireless infrastructure, in comparison with the all-BS architecture, is the reduction of aggregate output power of the cellular network. For the AeroMACS application, this translates into less IAI (inter application interference), that is, interference into coallocated applications such as MSS feeder link. In what follows, a list of major benefits, usage models, and applications of relay-augmented cellular networks is provided.

1) Extension of radio coverage into areas severely shadowed by buildings or natural obstacles.
2) Enhancement of coverage and throughput on or beyond WiMAX/ AeroMACS/LTE 5G (Or any other cellular network for that matter) cell footprint boundary.
3) Improvement in radio coverage inside a building or a high-rise complex.
4) Providing coverage in a dense urban area.
5) Providing coverage in a rural area.
6) Providing temporary coverage and temporary capacity upgrade.
7) Increasing network capacity to support intense usage areas: "hot spots."
8) Resolving special coverage challenges: "coverage hole filling."
9) Extension of radio access to moving vehicles such as buses, trains, aircraft.
10) Providing radio coverage in tunnels.

In summary, the main aspects for present and future usage of relays are coverage extension, capacity and throughput enhancement, support for mobility at all levels, cost efficiency, and improvement in frequency planning [1].

A variety of multihop wireless networks have been developed in the past couple of decades or so, among them are *ad hoc* networks, short range access networks such as Bluetooth, sensor networks, and mesh networks. More recently, standard-based WMAN and WWAN multihop networks based on relays, such as the ones based on IEEE 802.16j (WiMAX), IEEE 802.16m WiMAX, and LTE/A-LTE, have appeared and rapidly gaining attention for various applications. While IEEE802.16m amendment supports multihop relay architecture and mesh networking topology (as well as a host of other 4G and 5G desirable traits), IEEE 802.16j is solely dedicated to cell coverage extension and throughput/capacity enhancement of WiMAX networks through the application of multihop relays. A major difference between the 802.16j and 802.16m standards is that the latter is not constrained by legacy issues, and therefore, it is at liberty to design an entirely new radio access system. It is also worth mentioning that IEEE 802.16j has the most comprehensive relay categories compared to other standards. This means that the relay categories available in IEEE 802.16m and LTE-A consist of a subset of those provided by IEEE 802.16j. In all broadband cellular RS-augmented standards, the main concept is to complement BS with less complex, less costly, and easier-to-install relay stations instead of adding new BSs to the cellular network.

8.3.1 A Double-Hop Relay Configuration: Terminologies and Definitions

The IEEE 802.16j amendment presents a variation of the IEEE 802.16-2009 standards in which the relay-based multihop network capability is incorporated. This amendment specifies PHY and MAC layer enhancements to IEEE 802.16e, enabling the operation of relay stations over licensed bands without requiring any modifications in subscriber and mobile stations, and with full backward compatibility with IEEE 802.16-2004 and IEEE 802.16e. As was alluded to earlier, in relay-augmented networks small form-factor relays are used to support the expansion of the BS coverage, and/or the improvement of the network capacity/throughput, without any change in the radio resource distribution arrangements of the network.

Accordingly the IEEE 802.16j-2009 amendment offers an option on deployment of multihop relays as a cost-effective method of extending the radio outreach and/or network performance improvement, or both, in a cellular AeroMACS network [5]. Traffic and signaling between MRBS and SS/MS may be routed through "access RSs" or via a direct link between MRBS and the SS/MS. Figure 8.1 shows a simple two-hop relay configuration.

The physical channels connecting MRBS and RSs are called *relay links*, and the channel between an access relay and a SS/MS is termed *access link*, as illustrated in Figure 8.1. The MRBS may directly deliver the traffic

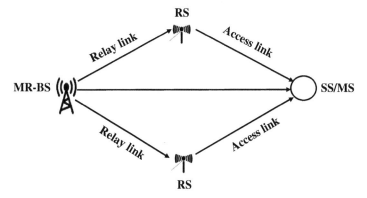

Figure 8.1 A two-hop relay system consisting of a MRBS and two RSs.

signal to the SS/MS, in which case the channel is referred to as *direct link*. In more complex multihop relay networks for which more than two hops may occur, the signaling between the MRBS and an access RS may be relayed through intermediate RSs. In those cases, the link between MRBS and the access RS, which may include several consecutive relay links and intermediate RSs, is called *relay path*. The protocols for the access link, including those related to mobility, remain the same as the ones in IEEE 802.16-2009, however, new functionalities are specified on relay links to support the multihop features [6].

Since a key feature of multihop relay architecture is the replacement of single-hop links with poor-quality channels, with multihop links with better quality and greater efficiency, a higher overall system capacity can be achieved. Moreover, multihop communications can support spatial reuse, which can in turn boost the overall system capacity as well. And since RSs are backward compatible with 802.16-2009 SS/MSs, while they are significantly less complex than 802.16-2009 BSs, an operator could deploy a network at a lower cost for providing wide coverage while delivering a required QoS to users. *Hence low cost and ease of deployment are the keys to the business case for 802.16j technologies.*

The IEEE 802.16j standard specifies two features of air interface between RS and MRBS as minimum requirements. First, the IEEE 802.16j is fully backward compatible with IEEE 802.16e. This implies that all SS/MS-BS air interface protocols are supported by IEEE 802.16j network with no need for any upgrade in the SS/MS. Second, the RS devices need to support all the licensed bands allocated for IEEE 802-16j-based systems. The network topology that is supported by the RS is limited to point-to-point, that is, mesh topology is not supported at this juncture.

The allocation of bandwidth and other resources for RSs and SS/MSs may be controlled using one of the two modes: *centralized* and *noncentralized*, that is, *distributed* scheduling. In centralized scheduling mode, the bandwidth allocation for an RS's subordinate SS/MSs is determined at the MRBS; whereas in distributed scheduling mode, the bandwidth allocation of an RS's subordinate SS/MSs are determined by the RS in cooperation with the MRBS, in other words, RSs share resource allocation responsibilities with their superordinate MRBSs.

8.3.2 Relay Modes: Transparent versus Non-Transparent

Depending upon whether or not a relay station has the authority to manage network resources, or in more specific terms if the relay can generate cell control messages, two relay modes are defined in IEEE 802.16j and IEEE 802.16m, as well as in LTE-A standards. They are designated as transparent relay stations (TRS) and Non-transparent relay stations (NTRS). The NTRS assigns and manages radio resources within its own "cell." On the other hand, TRS shares cell IDs and control messages with its MRBS. In other words, the major difference between these two modes is whether or not the relay is allowed to transmit frame header information that contains crucial scheduling data needed to determine when a node can transmit or receive information. Therefore, the TRS exclusively transmits traffic data to the subordinate MS/SS, and the framing information is directly communicated from the MRBS to the MS/SSs. This implies that TRSs can only operate in centralized scheduling format and cannot be used for radio outreach extension. The NTRSs, on the other hand, transmits both traffic data and control signals to its subordinate MS/SSs, thus they may operate either in centralized or distributed mode and may be used for extension of radio coverage.

From a slightly different perspective, the distinction between NTRS and TRS can be made by observing that NTRSs operate as a miniature BS for the MS/SSs that are connected to them, that is, the non-transparent RS transmits management messages and forwards traffic data while a transparent RS only exchanges traffic data with MS/SSs based on the frequency allocation information obtained from the MRBS. That is to say that the MS/SSs that are physically connected to a TRS are not aware of its existence, and therefore the name of transparent relay. Another major difference between these two modes of relays is that frequency reuse is allowed in non-transparent relays but not in transparent relays. This is another reason that nontransparent relays are suitable for radio range extension. Figures 8.2 and 8.3 compare the functionalities of transparent and non-transparent relays in a broadband WiMAX network [7]. Since a fundamental reason for considering IEEE 802.16j amendment for Aero-MACS applications is the expansion of radio coverage with minimal effect on overall power increase, non-transparent is the mode of choice for

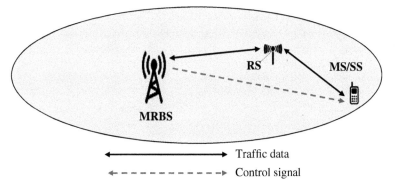

Figure 8.2 Transparent relay station (TRS).

AeroMACS, although transparent relay mode might be needed for throughput improvement.

In addition to complication in scheduling process, some of the disadvantages of using non-transparent mode include the required overhead for control signaling, the transmission delay of control messages, and frequent intra-BS handover [8]. Table 8.1 makes a comparison between TRS and NTRS in terms of their advantages and disadvantages for AeroMACS applications.

The choice of relay mode, in essence, is dictated by the application for which the relay is employed. For AeroMACS networks, the major applications of relays will be radio coverage extension to severely shadowed areas and to nodes that are located far outside of airport territory, and capacity enhancement at certain location of an airport surface. For the first application NTRSs are required, however, for the second, TRSs represent the mode of choice, although NTRSs may also be used for this purpose.

Transparent mode can only operate in centralized scheduling format and therefore they cannot support mobile relays. Moreover, they may not be

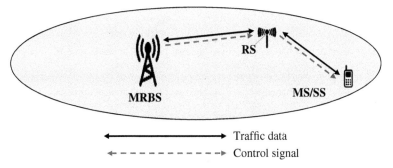

Figure 8.3 Nontransparent relay station (NTRS).

Table 8.1 Advantages and disadvantages of using TRS and NTRS in AeroMACS.

Relay mode	Advantages	Disadvantages
Transparent relay (TRS)	1) Low complexity: needs no framing no scheduling 2) Increases throughput which may be translated to improved QoS and/or capacity enhancement within the cell footprint 3) Can generate diversity gain 4) Ideal for cooperative communications within the framework of WiMAX 5) Straightforward handover 6) No additional overhead is needed 7) Since TRs do not transmit faming information there will be no inter relay interference 8) Ability to lower latency by using direct relay zone	1) Unable to extend radio outreach 2) Cannot be implemented in multihop topology, they can only be used for the last hop in a multihop communications system 3) TRS and MRBS must use the same subchannels that could potentially increase the interference level in the network 4) Channel quality measurement with TRS is not possible 5) Mobile relays cannot be supported as they are required to operate on distributed scheduling
Nontransparent relay (NTRS)	1) The main virtue of this mode is capability of extending radio coverage into shadowed areas and to nodes far outside of the BS cell footprint 2) Provides flexible solution and enables several types of scheduling and RRM approaches 3) May be used for modest increases in throughput and capacity 4) Supports multihop topology 5) Capable of operating in both centralized and distributed scheduling schemes 6) Enables straightforward cell splitting scheme 7) The NTRS and MRBS can use different subchannels that can control interference	1) Far more complex than TRS, to the extent that they can affect the cost of the network 2) Complex RRM and frequency reuse schemes are needed when NTRS are incorporated since MRBS and all of its subordinate relays are transmitting simultaneously 3) They require guard interval that increases the overhead 4) For NTRS, the transmission of framing information can create high inter-relay interference, thus not much capacity improvement can be achieved with this relay mode. 5) NTRS has fixed allocation for access and relay links, this is a major drawback for IP networks.

used for extension of radio coverage. However, TRSs are ideal for cooperative communications within the framework of WiMAX and AeroMACS networks, as Table 8.1 indicates. The main application of the TRS is to facilitate throughput/capacity increase within a BS cell footprint. Transparent relays are of low complexity, therefore, they are cost efficient.

Non-transparent mode may operate in centralized or distributed scheduling schemes. Perhaps the main virtue of the NTRS is its capabilities in extension of radio coverage area of the cell footprint into severely shadowed areas. This makes NTRSs suitable candidates for AeroMACS applications where the objective is to extend radio coverage without adding a new BS and without requiring a reconfiguration of the entire AeroMACS network that is already in place. NTRSs essentially act like a "mini BS" and therefore they are of high complexity and expensive to the extent that they can affect the total cost of the system. NTRSs may be used in any fashion in a multihop relay-augmented wireless network, that is, they do not need to be the last hop in a multihop transmission system. It should also be noted that since NTRSs transmit framing information, they can create high interrelay interference and thus not much capacity improvement can be achieved with this relay mode.

As Table 8.1 indicates, transparent and non-transparent relays have different advantages and disadvantages. Conceptually, there can be cases in AeroMACS networks that call for simultaneous application of both TRS and NTRS in an airport. The IEEE 802.16j standard provides very little detail of how this can be handled/realized [1].

8.3.3 Time Division Transmit and Receive Relays (TTR) and Simultaneous Transmit and Receive Relays (STR)

Non-transparent relay stations are further classified into time division transmit receive (TTR) and simultaneous transmit and receive (STR) relay.

Since the NTRS, from the perspective of the MS/SS, functions like a serving base station, it may provide most of the network capabilities offered by a standard IEEE 802.16e base station. For instance, an NTRS can generate control signals consisting of preamble, FCH, UP MAP, DL MAP and so on. Additionally, it can handle mobility tasks such as handover, as well as network entry and MS ranging. Hence, NTRSs can serve MS/SSs that are beyond the reach of the MRBS, that is, the NTRS may be used to extend the radio outreach of the MRBS to severely shadowed areas as well as extending the coverage to nodes that are located far beyond the boundary of the cell footprint. The NTRS can also provide a modest capacity enhancement as well.

Depending on assigned radio channels to the relay link and the access link, two types of non-transparent relays may be available: TTR and STR [9].

The TTR relay has captured most of the standard developing groups' attention, perhaps, because of its less complex structure than that of the STR relays. Both TTR and STR are supported by IEEE 802.16j standard. The TTR relay, sometimes termed as *single radio RS* also known as *inband relay*, communicates with its subordinate and superordinate nodes using the same radio channel, that is, the same OFDM subchannel is used in both relay link and access link, on a time division basis. TTR supports both single-frame and multiframe structures. The STR, also known as *dual radio RS* and *outband relay*, communicates with its subordinate and superordinate nodes using different OFDM subchannels. In other words, outband relays use different OFDM subchannels, or the same subchannel but without sharing other radio resources on relay and access link. The STR relays are capable of simultaneous signal transmission and signal reception, that is, transmitting signal to MRBS while listening to a SS/MS on the UL, and transmitting to SS/MS at the same time that they receive signal form MRBS on the DL. It has been claimed that for double-hop architecture (which is the most probable architecture for AeroMACS, should IEEE 802.16j-based WiMAX technology be adopted for AeroMACS) MRBS capacity with TTR relay mode is slightly higher than that of STR relay mode, however, that conflicts with another claim in the literature [10]. According to 802.16j standard, the TTR relay station shares the time frame for handling relay link and access link. During the DL subframe, the RS performs both functions of receiving payload from the superordinate station and transmitting payload to the subordinate station. In the opposite direction, during the UL the RS performs both functions of receiving payload from the subordinate station and transmitting payload to the superordinate station. Since the TTR relay cannot transmit and receive simultaneously it performs both functions on a time-sharing basis [10].

8.3.4 Further Division of Relay Modes of Operation

Based on their operational specifications, the TRS and NTRS are further categorized in IEEE 802.16j. Figure 8.4 shows various relay categories supported by the Standard. It should be mentioned that other technologies (IEEE 802.16m, LTE-A Release-10) support only a subset of categories shown in Figure 8.4 [11]. Relay operation in IEEE 802.16j standard is based on decode-and-forward procedure (to be explained later) with both TDD and FDD techniques supported.

An important feature of relay-fortified wireless networks lies in the fact that multihop communication supports spatial reuse, resulting in overall network capacity improvement [12].

Detailed descriptions of inband (NTRS–TTR) and outband (NTRS–STR) relays are provided in Section 8.3.9 [13].

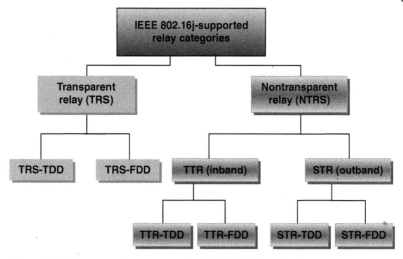

Figure 8.4 Relay categories supported by IEEE 802.16j standard.

8.3.5 Relay Classification Based on MAC Layer Functionalities: Centralized and Distributed Modes

Relays may be classified according to their PHY layer and MAC layer functionalities. We have already discussed that in terms of PHY layer processing; relay stations are classified as transparent relays (or Type II relays in LTE terminology) and non-transparent relays (or Type I relays in LTE jargon). Insofar as MAC functionalities are concerned, RSs can be characterized on the basis of their *scheduling arrangements* and *security capabilities*. In these respects, the RS may operate in *centralized* or *distributed* modes. Distributed mode with respect to scheduling means that the RS is capable of scheduling network resources in coordination with MRBS, otherwise the RS operates in centralized mode. RS in distributed scheduling mode creates DL-MAP and UL-MAP for allocation of bandwidth to its subordinate SS.

Same can be said about security, that is, the RS can be in distributed or centralized mode with respect to security arrangements. A TRS always operates in centralized mode with respect to both scheduling and security. In this case, bandwidth allocation and other scheduling procedures are carried out by the MRBS. An NTRS in distributed scheduling and security mode may provide radio outreach extension, higher bandwidth efficiency, as well as throughput enhancement in a WiMAX network. In centralized scheduling mode, all information related to the access link (bandwidth request, channel measurement, etc.) is forwarded to the MRBS for generation of proper DL-MPA and UL-MAP by the MRBS on behalf of the RS. This incurs latency in the network. On the other hand, a RS with distributed scheduling mode can process the information and generate proper DL/UL-Maps by itself [14].

8.3.6 Physical Classification of IEEE 802.16j Relays: Relay Types

Depending on their operational mode, mobility, scheduling, as well as other factors relay stations may be classified into several categories. With respect to mobility, relays are classified as fixed, nomadic, or mobile. Relays may be designed to perform different operations on the received signal. On this basis relays supported by IEEE 802.16j standard may be classified into three categories.

1) *Amplify and forward* (AF) *relays* (*nonregenerative relays):* This is perhaps the simplest type of relay, which acts like an analog repeater, that is, it receives the signal either from the BS or MS/SS, amplifies the signal, and retransmits them to the next node. There is a short latency associated with this type of relay. The downside, however, is that no error is rectified with this relay, and both signal and noise are amplified simultaneously, hence, the performance of this type of relay is not appealing, particularly at low SNR. There are two types of AF relays: fixed and adaptive (variable gain). In adaptive AF, relay's gain depends upon channel state, thus channel instantaneous gain information is required for the operation of this device.

2) *Decode and forward* (DF*), also known as detect and forward, relays:* This type of relaying can be viewed as a protocol for wireless cooperative communications. In this case, the relay receives the data, decodes it, and checks for any possible transmission error. It then performs error correction through CRC and ARQ system. It is not clear, at this point from the Standard, what would take place when the signal is an OFDM signal with a modulation/coding combination assigned to it by the AMC algorithm. Does the relay decode and demodulate the whole "burst profile" to detect and correct any transmission error? That would create excessive latency particularly in relation to decoding process. Or is it that the detection takes place with respect to some CRC codes and the relay does not have to decode the signal according to the coding scheme that has been used in the burst profile. In that case what would be the mechanism that adds the CRC-ARQ scheme? Conceptually, this type of relay receives the signal, decodes it, corrects the errors, then encodes the signal and retransmits it to the next node. This is a complex relay that can correct errors but incurs latency. DF relays are also of fixed or adaptive types.

3) *Demodulate and forward relays (D&F)*, also known as estimating and forwarding (ES) relay: This relay detects (demodulates) but does not decode the signal. As such the complexity of this relay system, the power consumption, and the latency incurred is reduced compared to the "decode and forward" relay. This type of relay essentially detects and quantizes the signal before forwarding it to the next node. In this case, some detection error may be introduced to the signal [15].

Relays may be applied on a time division basis where resources are available in the time domain. For instance, transmission or reception for relays occurs at different time slots, or various relays connected to a BS, access the medium at different assigned times. Relays may be managed over "frequency domain," that is, the relays operate on different radio channels and resources are made available on a frequency division basis. STR NTRSs operate on this mode and are capable of simultaneous transmission and reception on both up link and down link. It is also possible for relays to operate in a hybrid fashion. More recently, relays are being widely proposed for cooperative communications in which diversity is created through the transmission of the same signal to a single receiver via several relays [16–17]. Latency, signal overhead, and spectral efficiency are all affected by the selection of whether the resources are available on time-division or frequency-division basis.

8.3.6.1 Relay Type and Latency

It is clear that the selection of type of relay (AF, DF, D&F) has a direct bearing on latency. For instance, in DF relays the received signal has to be detected (demodulated) and decoded, and then reencoded, remodulated, and retransmitted. Each of these signal processing parts require certain processing time that contributes to latency. In particular, the decoding of a convolutional turbo code may take considerable amount of time, relatively speaking. On the other hand, an AF relay receives the signal, amplifies it, and then retransmits it. This incurs very little delay compared to that of DF relay case.

The overall latency (as well as signal overhead) related to an IEEE 802.16j systems is determined by the following factors.

1) The type of relay that is selected (AF, DF, D&F)
2) The scheduling procedure (centralized and distributed)
3) The number of hops in the network
4) The method of dividing the resources: time division or frequency division
5) The mode of relay (TRS, NTRS)
6) When NTRS is selected the further selection of TTR versus STR affects the latency as well

8.3.7 Modes of Deployment of IEEE 802.16j Relays in Wireless Networks

There are two deployment approaches, or methodologies, with respect to WiMAX/AeroMACS multihop relay networking: *Greenfield deployment* and relay *deployment for network evolution*. In the case of Greenfield deployment, the network designer utilizes relays as a basic network element, alongside other building block elements used in the network from its early planning and design stages. In such cases a systematic approach for the usage of relays in the network is adopted and optimum deployment conditions for the relays, such as the number of relays per base station, the distance between relays, the distance between relays and base stations, frequency reuse plan, and other parameters are analyzed, predefined, and finally placed in the network. In other words, in Greenfield approach the relays are initially deployed as a network building block for improvement of cell coverage with reduced number of otherwise required BSs for the coverage area. For Greenfield operators focusing on ubiquitous network connectivity, relay usage as a part of the initial rollout may result in network cost reduction. The method of deploying RSs may be considered as a cost-effective alternative to deploying low-powered BSs as RSs do not require backhauling.

In the second method, "relay deployment for network evolution," the main objective is to support the growth of an already deployed WiMAX/AeroMACS network due to continuous increase in number of users and usage of the network. In this case, the network that may include only BSs, or even a combination of BSs and RSs, is initially deployed to achieve a general coverage. As the traffic and the number of users increases or new uncovered or poorly covered areas come to light (in the case of AeroMACS, as new terminals and new buildings are added to an airport) and hence network capacity, coverage and throughput issues begin to emerge, new infrastructure consisting of RSs are added to the network. That is to say, RSs are used to upgrade the capacity, throughput, and radio outreach of the network. The network evolution approach minimizes the upfront investment associated with network development particularly when a new technology replaces an old one, and maximizes utilization of existing network assets [3].

For AeroMACS deployment over various airports, should IEEE 802.16j-based WiMAX technology be adopted, a combination of these two approaches seems to be the natural choice. The airport AeroMACS network may be initially deployed with IEEE 802.16j BSs and RSs and as the airport is expanded or the traffic is increased more RSs are added to the network.

8.3.8 Frame Structure for Double-Hop IEEE 802.16j TDD TRS

The physical layer WiMAX TDD frame structure containing transparent relays, more or less, includes the same framing and management/control information as does the original IEEE 802.16e TDD frame, given in Figure 5.6, which is reproduced in Figure 8.5 for quick reference. Table 8.2 describes the control/management information part of the frame.

The frame structure for IEEE 802.16j-based dual-hop transparent relays (TRS) is shown in Figure 8.6. The following remarks are in order for the frame structure of a dual-hop TRS IEEE 802.16j networks, at this juncture.

1) The MS/SS receives all of control data (preamble, FCH, DL MAP, UL MAP) directly from MRBS.
2) The IEEE 802.16j TRS shares the time frame to handle relay and access links. During DL subframe the RS receives the payload from the MRBS, or any other superordinate station, and transmits the payload

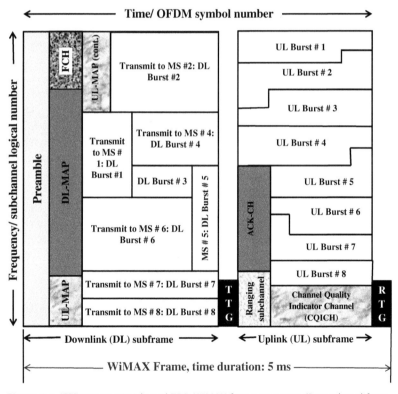

Figure 8.5 IEEE 802.16-2009-based TDD WiMAX frame structure. (Reproduced from Ref. [18] with permission of IEEE.)

Table 8.2 Control and management parts of WiMAX TDD frame.

Preamble	This is the first OFDM-DL symbol of the frame and is used by the MS for BS identification, time synchronizing of the MS to the BS, and channel estimation
FCH	A special management burst that follows the preamble, it provides the frame configuration information such as MAP message length and coding scheme, as well as usable subchannels
DL-MAP	Defines for each MS the location of their corresponding burst, that is, their OFDM subchannel allocation as well as other control information in the DL subframe addressed to each MS/SS
UL-MAP	This is a burst that defines for each MS/SS its subchannel allocation in the upcoming UL subframe, as well as other control information
UL Ranging	The UL ranging subchannel is allocated for MSs to perform closed-loop time, frequency, and power adjustment as well as bandwidth requests
UL CQICH	The UL CQICH subchannel is allocated for the MS to feedback channel state information
UL ACK	The UL ACK is allocated for the MS to feedback DL HARQ acknowledgement

to the MS, or any other subordinate station (a dual function). Similarly, during the UL subframe the TRS receives payload from the MS/SS and transmits that to the MRBS.

3) Figure 8.6 describes the frame structure of IEEE 802.16j-based dual-hop MRBS/RS/MS. Like in the original IEEE 802.16e standard, the frame is divided into two time domain subframes separated by a guard time interval. The figure illustrates both MRBS and RS frame structures.

4) Insofar as MRBS frame makeup is concerned, the control part of DL subframe is identical to that of IEEE 802.16e. The payload part of DL subframe, however, is divided into two zones. In the first zone the MRBS transmits to subordinate MSs and RSs. In the second zone MRBS has several options as outlined here:
   a) Stay silent
   b) Transmit to MSs that are being served directly by MRBS. This alternative should be employed with caution since the RS is transmitting to MSs at the same time using essentially the same spectrum
   c) Participate in cooperative communication

5) The DL subframe of RS frame is divided into two time zones compatible with those of MRBS, but separated by a guard time. In the first zone, called the DL access zone, the RS receives from MRBS,

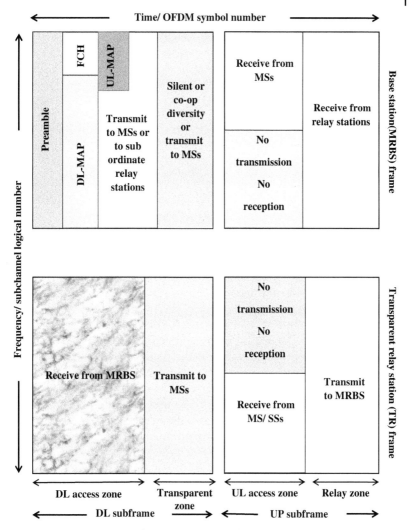

Figure 8.6 Frame structure for IEEE 802.16j-based dual-hop transparent relays.

whereas in the second time zone, known as transparent zone, the RS transmits the payload that it has received from the MRBS to the MSs that it serves.

6) The UL subframe is also divided into two zones in the time domain to match the division in the MRBS frame structure. The first zone is divided into two sectors in the frequency domain, compatible with that MRBS frame composition. In the first sector, the RS is silent, allowing the MRBS to use the same spectrum to receive from SS/MSs it directly

serves. In the second sector, the RS receives payload from the MSs that are communicating with MRBS through the TRS.

7) In the second time zone of the TRS UL subframe, designated as relay zone, the RS transmits to MRBS the payload that it had received from the MSs in the previous UL subframe.

8) The RS frame is naturally divided into DL and UL subframes as well.

8.3.8.1 The Detail of IEEE 802-16j Operation with Transparent Relays

1) The downlink course of actions
   a) MRBS transmits the frame control part (preamble, FCH, DL MAP, UL MAP) to all its subordinate MSs, whether they are directly connected to the MRBS or connected to TRSs. The MSs that are served by RSs find in their MAPs information related to the burst that are located in the transparent zone of the RS with modulation coding parameters that correspond to the RS-MS channel conditions. It should be noted that the MAP information communicated to the SS/MSs is related to the burst that was previously transmitted to the RS.
   b) MRBS transmits the frame control part of bursts to the subordinate RSs.
   c) MSs receive the control information that was addressed to them.
   d) RSs receive the control information that was addressed to them.
   e) MRBS transmits the payload part to its subordinate MSs, the direct served MSs, and RSs served in the DL access zone.
   f) RS receives the payload burst to be relayed in the access zone.
   g) RS transmits the payload burst to the relayed MSs in the transparent zone. Note that the burst information was received in the previous frame.
   h) Relayed MSs receive the payload in the transparent zone (without being aware that the payload came from a relay) after receiving the burst location in the MAP that was transmitted by MRBS [19].

2) The upload course of actions
   a) MRBS receives the payload parts of all of its directly served MSs in the UL access zone.
   b) MBRS receives the payload from all its subordinate relays in the UL relay zone. This zone contains burst that was received by the RS from relayed MSs in UL access zone at a previous frame.
   c) RS receives the payload to be relayed to MRBS from all of its subordinate MSs in the subzone of the UL access zone that is mirror image match to that of the MRBS subzone [19].
   d) RS transmits the payload burst to the relayed MSs in the transparent zone. The burst information that was received in the previous frame in the UL access zone.

8.3.9 The Frame Structure for TTR–NTRS

Figure 8.7 illustrates the frame structure of a dual-hop TTR–NTRS. As Figure 8.7 illustrates, the MRBS DL frame is divided into three parts, namely, transmission of control message, transmission to directly served MS/SSs, and transmission to TTRs. The MRBS UL subframe consists of two parts of receiving from directly served MS/SSs, and receiving from TTRs. The TTR DL subframe is similarly divided into three sections:

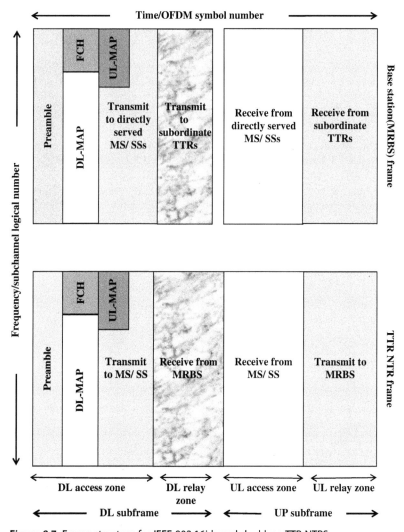

Figure 8.7 Frame structure for IEEE 802.16j-based dual-hop TTR NTRS.

transmission of relay's own control signal, transmission to subordinate MS/Sss, and receiving from MRBS. It is noted that the TTR has to change mode from a transmitter to a receiver. The TTR UL subframe consists of two sections, receiving from subordinate MS/Sss and transmitting to the MRBS. Again, a change of mode of operation is observed in the TTR.

Exploring Figure 8.7 reveals the following sources of degradations when using TTR–NTRS in the AeroMACS networks or any other wireless system.

1) In the TTR frame, both UL and DL subframes are divided between transmitting and receiving subsections. The transition between transmission and reception modes requires guard time interval that affects the capacity negatively, it may also affect the latency.

2) In the DL access zone, the relay transmits to subordinate MS/SS, and in the DL relay zone it receives from MRBS. It is clear that the signal transmitted to the MS/Sss is the one that was received from MRBS in the previous frame. This clearly demonstrates the occurrence of latency equal to the time duration of one frame, compared to the case that the BS directly transmits to the MS/SS.

3) The TTR does not listen to control part of MRBS message, the RS zone generates its own dedicated frame control part that includes midamble, R-FCH, UL R-MAP, and DL R-MAP. This is an additional source of overhead [20].

4) It should also be noted that, generally speaking, the TTR relay does not transmit the same burst that it has received from the MRBS to the target MS/SS. This is due to the fact that the conditions in MRBS-TTR and TTR-MS channels are not necessarily the same and the AMC algorithm creates a different burst for transmission over TTR-MS channel. Usually, the MRBS-TTR channel is a LOS link and therefore high-level modulation and high-rate coding schemes are used for it. That may or may not be the case for TTR-MS link.

8.3.10 The Frame Structure for STR–NTRS

STR relay possess a number of interesting features. STR is capable of transmitting and receiving simultaneously, that is, on the DL it can listen to the MRBS while transmitting to the MS/SS. Similarly, on the UL it can listen to the MS/SS while transmitting to the MRBS. Unlike TTR, STR does not require separate zones for access and relay links. STR features two antennas, one for communicating with its superordinate nodes that is pretty much like a backhaul link, and one for communicating with subordinate nodes such as another relay or the target MS/SS. The access and backhaul antennas have different requirements. The access antenna is normally omnidirectional or sectorized, although it conceptually could become directional toward a particular shadowed area. The backhaul antenna is always directional toward the MRBS. Transmitting antennas placed in close proximity can pose interference, thus arrangements must be made to prevent that. In order to overcome interference, one of the two options needs to be adopted: spatial separation between the antennas or spectral separation for backhaul and access links [19].

The frame structure for the dual hop STR–NTRS is illustrated in Figure 8.8. The following observations are made regarding this frame setup.

1.

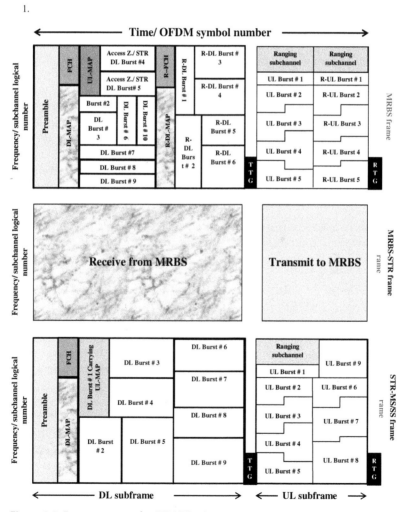

Figure 8.8 Frame structure for STR-NTS relay.

1) Unlike TTR relays, the STR relays do not have to divide their time resources into access zone and relay zone.
2) When the network incorporates both TTR and STR, the divided frame structure should be maintained.
3) In case the STR relays are used exclusively, the frame structure of 802.16j can be similar to that of IEEE 802e.
4) The frame structure part related to MRBS is normally left as TTR/STR compatible, therefore, the UL and DL subframes are divided into access and relay zones.
5) STR relays are allowed to communicate at both zones, while the TTR relays communicate with MRBS only through the relay zone.

It is noted that in STR frame, the entire time duration of subframes of antenna number one is employed to communicate with MRBS, while the whole time frame of antenna number two is utilized for communication with subordinate stations [21].

An important feature of STR relays is the fact that they can be implemented in different network layers. For instance, STR relays can be implemented in layer 2 as MAC relays, or in layer 3 as network relays. STR relays implemented in the MAC layer is compatible with IEEE 802.16j standards. When a MAC relay is implemented, the ASN-GW is not aware of any RS in the network, since all relay functionalities are realized in PHY and MAC layers.

8.3.10.1 STR Implementation in Different Layers

As it was pointed out earlier, the STR relay exhibits an important advantage over TTR. STR can be implemented in different network layers. The STR relay can be implemented using layer 2 as a MAC relay or using layer 3 as a network relay. This flexibility is important since the STR can adapt itself without great difficulties to different standards such as 802.16e, 802.16j, or 802.16m standard.

In MAC-based relay approach, the backhaul link between RS and MRBS is contained within MAC layer (L2). The MRBS behaves as if the MS is directly registered with it, therefore, in many respects, the ASN-GW does not realize that the MS is connected to an RS. A direct consequence of this is that no networking modification is required to support relay traffic, however, some alterations might be needed in RRM and other aspects. Although the NTRS operates as if it is a "mini BS" in relation to the MS it serves, from the networking point of view it does not behave like a BS. The GRE (Generic Routing Encapsulation) tunnel that is used to carry the service flows (to be defined in Section 8.4.4) is terminated at the MRBS and not the RS. The IEEE 802.16j MAC relay can support multihop relay with no network modifications.

The advantages of MAC-based STR relays are that no network modification is needed, lower latency for relay transport can be achieved, and easier to implement QoS constraints since RS-MRBS link uses dedicated MAC which can contain more information than is available over R6. On the other hand, problems arise since RS-MRBS link is not compliant with 802.16e MAC, and RS functionality does not build over existing WiMAX capability, therefore mobility and RRM need to be modified to support the relay. This is a critical drawback for the MAC-based STR that requires modification in MRBS. In other words, the MRBS system has to be modified in order for it to support IEEE 802.16e-based STR. It is much more complex to modify the IEEE 802.16e BS to support MAC-STR relay than it is to upgrade the BS to support network-layer-defined IEEE 802.16j STR. Likewise, upgrading a deployed IEEE 802.16e network to support relay is much easier if it does not involve radio modifications [9].

8.4 Regarding MAC Layers of IEEE 802.16j and NRTS

MAC sublayer, sometimes referred to as the brain of the wireless network, as we have indicated in the previous chapters, is the lower sublayer of the link layer, which bears control over addressing and channel access that facilitates resource sharing in a multiple access network. In this section, the focus is on MAC layers of IEEE 802.16j standard and non-transparent relay stations. From the MAC sublayer perspective, relays can operate in two scheduling modes of centralized and distributed. The same can be said about security arrangement, that is, relays can function in centralized or distributed security modes.

Three key aspects of MAC sublayer are considered: forwarding scheme, routing and path management mechanisms, and initial ranging and network entry techniques.

8.4.1 Data Forwarding Schemes

With the objective of maximizing network efficiency by aggregating traffic wherever possible, two forwarding schemes are defined in IEEE 802.16j standard: *tunnel-based* and *connection ID* (CID)-based schemes. In general, traffic aggregation has the following two benefits.

- Since less overhead information needs to be sent, overall system efficiency is enhanced.
- Management of signal forwarding is simplified, as groups of flows can be handled together.

An important IEEE 802.16j MAC sublayer functionality is to define a MAC message family; called R-MAC (Relay MAC), between MRBS and RS.

Since MSs are IEEE 802.16e-based systems, R-MACs are ignored by MSs or they do not concern them. The R-MAC essentially allows intermediate relays behave like relay entities. One of the R-MAC's major functions is to create a data forwarding scheme known as *tunneling connection*, which is specified by a special CID abbreviated as *T-CID*. A T-CID is a connection that traverses all the hops between a RS and the MRBS. A tunnel, characterized by a T-CID, has two specific end points and a QoS requirement, enabling connection to the MRBS for several different MSs that are being served by the same access RS. In tunneling, in the context of IEEE 802.16 standards, the PDUs with their associated CIDs are submitted to the MRBS MAC layer. The IEEE 802.16j PDUs and CIDs are encapsulated into R-MAC PDUs over the tunnel connection. A special header is attached at the beginning of the tunnel MAC PDU. This header is removed at the access RS. After the relay MAC header, several subheaders that provide the receiver with information about the payload (including but not limited to information related to QoS) are added. It should be noted that since intermediate relays cannot distinguish between the connections within the tunnel, QoS for the tunnel is determined by the highest level of QoS of all connections. The payload, which generally consists of a regular IEEE 802.16e MAC PDU, follows the subheaders. A CRC may be appended at the end. It is noted that tunneling is an option [1,22].

The second forwarding option is CID-based forwarding in which the packets are forwarded based on the CID of their destinations. In centralized scheduling, the MRBS sends a message to the RSs describing the relay link channel characteristics, including an extra field specifying the delay associated with each packet in either the DL or the UL. Thus, the RS knows in which frame each packet should be transmitted. This is necessary in order to meet the QoS requirements of each connection. In the distributed case, the RS has knowledge of the QoS requirements of each connection and can therefore make its own scheduling decisions [22].

8.4.1.1 Routing Selection and Path Management

Routing is an entirely different issue than forwarding in that the routing involves the selection of a suitable RS or set of RSs through which the signal is transmitted between the end nodes. In IEEE 802.16j, a route is composed of multihop paths between the serving MRBS and the destination MS/SS, therefore, one has to deal with routing path management issues. Although routing in such systems is tree-based, decisions must be made as to which RS should be associated with a particular SS/MS. Path management refers to issues relating to path establishment, maintenance, and release. Various path management techniques and algorithms have been proposed in the literature.

Insofar as routing techniques and path management are concerned, the IEEE 802.16j standard envisions routing decisions that are based on metrics such as radio resource availability, radio link quality, and traffic load at the RSs. However, the Standard does not indicate how the decision should be made, that is, the details of the path selection decision are left to the network designers and vendors. In any case path selection decisions are made at the BS based on information provided by the RSs. The path management mechanisms are then used to create the desired path. The Standard defines two approaches to path management as described further. The main difference between these two approaches is in the way the signaling information that manages the path is distributed in the system.

1) The Embedded Path Management uses a hierarchical CID allocation scheme in the system. The RMBS systematically assigns CIDs to its subordinate stations, such that the CIDs allocated to all subordinate RSs of any given station are a subset of the allocated CIDs for that station [5]. In this way, there is no specific routing table in each RS, and signaling to update path information is needed less frequently. This is a very simple approach to path management.

2) The Explicit Path Management uses an end-to-end signaling mechanism to distribute the routing table along the path. The MRBS sends the necessary information to the RSs involved in a path when a path is created, removed, or updated. Each path is identified by a path ID to which the CIDs are bound. This leads to a smaller routing table at the RSs and a reduction in the overhead required to update these tables. Optionally, the MRBS may include the QoS requirement associated with each CID to allow the RSs to make an independent decision regarding how to schedule the packet in distributed scheduling mode [5,22].

8.4.1.2 Initial Ranging and Network Entry

For the IEEE 802.16j networks, two different network entities might initiate network entry: the relays and MS/SSs. There are two different network entry procedures for these two entities. As IEEE 802.16j systems must maintain compatibility with legacy terminals, the network entry procedure as seen by the terminal must remain unchanged. However, there are some differences regarding how the MRBS and RSs deal with this procedure arising from the fact that the network needs to determine which node should be the access node for the MS. The initial ranging process in 802.16j systems varies, depending on the scheduling mode and mode of the relay (NTRS versus TRS), therefore, different ranging processes are needed as explained further.

1) MS initial ranging in TRS: The RSs monitor the ranging channel in the UL access zone and forward the ranging codes they receive to the MRBS.

The MRBS waits for a specified time for other messages with the same ranging code from other RSs and then determines the most appropriate path for the station, that is, direct or via an RS. If the direct path is chosen, the MRBS sends a response directly to the MS. Otherwise, the response is sent to the RS, which is then forwarded it to the MS.

2) MS initial ranging in NTRS: Due to legacy constraints, the MSs choose the MRBS or NTRS with the strongest preamble detected. This means that there is essentially no path decision to be made in this case, as the MS is communicating with a single RS. As the BS ultimately makes the network entry decision, the RS must communicate with the MRBS to ensure that the MS is permitted network entry. In the centralized case, this involves communicating all ranging information back to the MRBS, but in the distributed case the RS handles the ranging functions and simply makes a network entry query of the MRBS. The network entry process for RSs incorporates additional steps and defines a specific ranging process. More specifically, network entry is augmented with a neighborhood discovery and measurement process followed by a path selection algorithm to determine the most suitable access station for the RS.

3) RS initial ranging: During the ranging process the MRBS or NTRS, can determine if a node performing ranging is an RS, based on the used ranging code (a specific set of ranging codes are reserved for RSs). In this way TRS can easily ignore ranging performed by other RSs. Also, this enables priority to be given for RSs performing ranging. The rest of the initial ranging process for RSs is similar to that for MSs in non-transparent mode relay.

4) RS network entry: After the initial ranging, authentication, and registration processes, the MRBS may request the RS to determine the signal strength of each of its neighboring RSs and forward it to the MRBS. The MRBS can then determine the most suitable access station with which to associate the RS; based on traffic load, signal strength, and so on. The final stage of network entry is then the configuration of the RS parameters, including its operation mode, for example, TRS or NTRS, and scheduling mode.

8.4.2 Scheduling

Scheduling is one of the key tasks of MAC layer. Scheduling involves methods by which data flows are given access to system resources (e.g., processor time, communications bandwidth). This is usually done to distribute workload efficiently or to achieve a target QoS. The need for a scheduling algorithm arises from the requirement to perform multiplexing or multitasking. The scheduling algorithms' main quantities of concern are throughput and latency. In the case of WiMAX and Aero-MACS networks, scheduling means which MS should be served in a given WiMAX frame and what bandwidth (data rate) should be allocated for the MS. In multihop architecture much of MAC design deals with who decides which transceivers can transmit and how those decisions are communicated to the network. IEEE 802.16j MAC sublayer allows for both distributed scheduling and centralized scheduling.

In the original IEEE 802.16e, the BS makes the scheduling decisions, that is based on information gathered via several different techniques such as unsolicited bandwidth request, polling, and connection-based procedures. In this manner, the BS determines the optimal resource allocation. In multihop relay with NTRSs and distributed scheduling, however, some MAC intelligence is given to the relays. Hence, the IEEE 802.16j standards allow NTRSs to make decisions about resource allocation to their subordinate stations in coordination with the MRBS. The IEEE 802.16e allows for a connection-oriented network and the resource allocation to be carried out on per-connection basis. Each connection is given a MAC-based CID. Control messages are sent on special management CIDs. These CIDs are defined either as *basic connections* or as *primary connections* that are assigned during network entry process. These management connections can be assigned by NTRSs in what is known as local CID allocation mode. This is carried out in accordance with provisions of IEEE 802.16j amendment. The MRBS must grant a set of CIDs for this purpose. Transport connections, carrying higher layer data, are allocated by the MRBS.

The advantages of using distributed scheduling include prevention of heavy management traffic with the network, and allowing fast bandwidth request grants for high priority requests. Besides the distributed scheduling is the only feasible technique in a heavy multihop architecture. The down sides of applying distributed scheduling are the fact that it leaves no option for load balancing between relays, and the lack of centralized RRM can result in suboptimum resource utilization.

All TRSs must, and some NTRSs may, operate in centralized scheduling mode in which the MRBS allocates all resources, including scheduling according to IEEE 802.16e standards. There are no local CID capabilities and each RS provides the MRBS with special MAC message on channel quality information (CQI) of all its subordinate MSs and RSs. The MRBS provides each RS the MAP information that describes in detail the bursts to be transmitted or received in the following frame by each RS in the subnetwork toward its subordinate stations.

The advantages of centralized scheduling include enabling optimum RRM and network load balancing among the subordinate relays. However, the centralized scheduling procedure suffers from requiring heavy management traffic within the network, slow grant for high-priority bandwidth requests, and not being feasible in complex multihop networks.

8.4.3 Security Schemes

Like in the case of scheduling, security in IEEE 802.16j networks may be provided in the form of centralized or distributed. In *centralized security*, only the MRBSs and MSs may encrypt or decrypt MAC PDUs. In *distributed security* schemes, the access RSs are purveying to encryption and decryption keys so that the entire link from MRBS to access RS to MS/SS is secure.

8.4.4 Quality of Service (QoS) in Relay-Augmented Networks

Recall that QoS in wireless networks, or in any telecommunications network for that matter, means, at least conceptually, the probability that the network fulfills the promise of a given traffic agreement. Thus, QoS manifests itself in the ability of network elements (such as base stations, relays, routers, applications, hosts, and so forth) to have some level of assurance that its traffic and service requirements would be satisfied, therefore, QoS is a measure of how reliable and consistent a network operates.

The key elements related to QoS are *service flow* (SF) and *bandwidth grant* provisions. As describe earlier, a service flow is a one-way MAC service data unit (SDU) with a defined set of QoS parameters and an associated *service flow ID* (SFID). A *provisioned service flow* possesses SFID but it may not carry traffic, rather it might be waiting to be activated for usage. An *admitted service flow* goes through the process of activation. An *active service flow* is an actuated service flow with all required resources allocated to it. Normally upon an external request for a service flow, the BS/MS will check for required resources to satisfy the requested QoS. The use of service flows is the main mechanism for providing a requested QoS [23]. The five categories of scheduling schemes that are supported by the BS MAC scheduler for information transmission over any given connection are the same as the ones described in Section 7.4.1. QoS parameters are also the same as the one defined in Chapter 7, for example, MSTR, MRTR, MLT, jitter tolerance, and so on.

The AeroMACS/WiMAX-defined set of QoS parameters combined with QoS mechanism over the backbone wireline or wireless networks (IP or Ethernet networks), provide predictable end-to-end network QoS. When the network supports relays, the QoS analysis must also include two additional independent wireless links, that is, MS-RS and RS-MRBS links. The multihop relay QoS is essentially an issue of assigning individual QoS parameters to service flows, for each of the two wireless links. In the case of MAC-based relays, this can be done by observing the CID in each PDU, and associating each CID/SF with QoS parameters over the RS/MRBS link. For network-based relays the CID is not available, however, the SF can be deduced from the Generic Routing Encapsulation tunnel used for defining the data transport path. In this sense, MAC-based relay and network-based relay provide equivalent end-to-end QoS.

8.4.4.1 The Impact of Scheduling and Relay Mode on AeroMACS Network Parameters

Another key issue related to application of IEEE 802.16j relays in AeroMACS networks is the effect of the selected scheduling procedure and relay mode on major AeroMACS network parameters. We have discussed various relay modes earlier in this chapter. The IEEE 802.16j-based AeroMACS could apply either a centralized scheduling scheme or distributed scheduling procedure. This selection has direct bearing on signaling overhead, latency, and MAC protocol complexity. As a consequence, other network operations and functional parameters, such as throughput, overall delay, spectral efficiency, and QoS, are affected. In centralized scheduling, when a MS/SS approaches an RS, all information related to network entry and bandwidth request must be forwarded by the access relay to the MRBS for it to generate DL-MAP and UL-MAP. Additionally, packet scheduling, as well as network and connection managements procedures, has to be run by the MRBS that increases the signaling overhead and latency as well. On the other hand, in distributed scheduling schemes the access RS is allowed to handle the MS/SS network entry and bandwidth request, this clearly reduces latency. It is imperative to realize that latency directly influences the QoS, particularly for delay sensitive services of UGS, rtPS, and ertPS.

It may be that in centralized scheduling the downlink latency is reduced but uplink delay is increased, owing to the fact that the access RS has to pass the entry and bandwidth request to the MRBS on the uplink. However, on the DL all frame header and control information is generated by the MRBS and is forwarded to the RS along with traffic data, and thus the RS does not have to take time to generate new frame header and control signals, therefore, no latency is accrued in that sense.

It has been mentioned that the main factor in selection of relay mode is the particular application that is envisioned for the relay. For intracell (or cell boundary) throughput enhancement, TRS is preferred, whereas radio outreach extension calls for NTRS. With NTRS, it is possible to achieve radio coverage extension and a modest increase in throughput. This leads to the conclusion that if it is desired or necessary that a single type/mode relay be employed throughout the AeroMACS network, at least with respect to relay mode and scheduling procedure, NTRS with central scheduling is the optimum choice.

8.5 Challenges and Practical Issues in IEEE 802.16j-Based AeroMACS

We now examine some of the challenges that a system designer faces when considering the inclusion of multihop relays and IEEE 802.16j-based technologies in AeroMACS networks.

One of the challenges, and in fact a downside of using relay-fortified WiMAX, in AeroMACS networks, is the design of QoS mechanism and the scheduler that supports it. This is due to the fact that the MS-MRBS and MRBS-MS connections may involve more than a single link. In other words, relay links might be included in both uplink and downlink. These links, in general, have different channel conditions that make the scheduling process more complicated than the conventional WiMAX network. Clearly, as the number of hops increases, the scheduling process becomes more complicated.

8.5.1 Latency

The second problem that degrades the IEEE 802.16j-based AeroMACS is the additional latency that is created by multihop relays. The selection of the type of relay, that is, amplify and forward (AF) relay, detect and forward (DF) relay, and demodulate and forward (D&F) relay has a direct bearing on latency. In order to determine the overall latency related to an IEEE 802.16j network, the following factors need to be considered.

1) The type of relay that is selected
2) The scheduling procedure (centralized and distributed)
3) The number of hops or the maximum allowed number of hops in the network
4) The method of dividing the resources, time division or frequency division (for AeroMACS networks TDD is the duplexing scheme)
5) The mode of relay, TRS versus NTRS, also the selection of TTR or STR affects the latency

In fact, increased latency represents one of the prices that has to be paid for employing multihop relays in AeroMACS networks.

8.5.2 The Number of Hops

An important parameter in designing IEEE 802.16j-based WiMAX system is the number of hops, or the maximum number of hops that is allowed in the network. The Standard does not specify the optimum number of hops, perhaps for the obvious reason that different applications might require different "optimum number of hops." The number of hops, or the maximum number of hops, is a critical parameter as it has direct bearing on latency, throughput, and QoS in the network. Clearly, as the number of hops increases, so does the latency and the time it takes for the signal to get through the channel, resulting in a decrease in the effective system throughput. In regard to the number of hops in AeroMACS networks, a few important questions are posed at this point. Search for answers to these questions is a subject of future study in this area.

1) Should there be a maximum number of hops defined and accepted for the entire AeroMACS network, or at least for channels that include relay links?
2) Is there an "optimum" number of hops or "optimum maximum" number of hops for IEEE 802.16j-based AeroMACS?
3) Is it feasible to define a maximum and a minimum for the number of hops?
4) Isn't the optimum number, or optimum maximum number of hops in an AeroMACS network related to the airport size?

At a glance, it seems that it would make sense to assume a maximum number of two hops through AeroMACS network. This might be sufficient for main objectives of extending the radio coverage into nearby shadowed areas, or providing AeroMACS connection for nodes that are outside of airport area, and/or throughput improvement for particular locales on the airport surface. Limiting the number of hops to two would keep the complexity of relay-fortified AeroMACS network at a minimum, the required MAC protocol modifications would be very manageable (end-to-end QoS, latency, etc.), and it would minimize the escalation of the overall network latency. Hence, unless some compelling reasons arise, it should be assumed that any AeroMACS link may include at most a single relay.

Reference [24], based on research results presented in Refs [25] and [26], concludes that the number of hops in any multihop wireless network should not exceed three. The rational is that by increasing the number of hops beyond three, the achievable user's capacity in the network dramatically decreases. Moreover, an increase in the number of hops per connection increases the latency that would affect many network parameters negatively.

8.5.3 The Output Power and Antenna Selection

The output power of a transmitter is a key factor in determining the range of the transmission. A typical WiMAX base station transmits at power levels of approximately +43 dBm (10 W). A fixed CPE (customer premises equipment) station typically transmits at +23 dBm (200 mW). A mobile station integrated in a laptop computer or a PDA might be limited to a transmission level of 10–17 dBm (10–50 mW). The output power of the device determines not only the range but also the minimum size and the cost of the product. For AeroMACS base stations, output power levels on the order of 100 mW or so is assumed. This power restriction, owing to the need to control the level of IAI, that is, interference to coallocated application (Global Star feeder links), limits the range of MRBS and associated RSs.

In some literature, as a rule of thumb, it is assumed that the output power of a relay is 3 dB below that of a typical base station. This is to facilitate simulation runs for various measurements including but not limited to measurement of interference level to coallocated applications [3]. However, when discussing the required output power of an RS it is necessary to look at two separate links: the RS to MRBS link and the RS to MS link. Usually, the distance between the RS and the MRBS is greater than the distance between the RS and the MS; however, the RS to MRBS link is usually a LOS or near-LOS channel, and the RS to MS link typically is a NLOS channel. NLOS situations require a shadowing margin of about 15 dB, which is equivalent to four times distance difference of LOS cases. The power level of the RS has a major impact on its size, weight, cost, and the ability to operate in close proximity to people.

The selection of an antenna system is a critical part of multihop relay system design. For most applications, better performance may be achieved if antenna patterns are optimized separately for the relay link and the access link, that is, RS to MRBS and RS to SS/MS links. Usually, it may be beneficial to use a directional antenna pointing at the MRBS for the RS to MRBS link and use of omnidirectional antenna for the RS to MS link. In cases where the SS/MSs are located in a certain identified direction, it may be advantageous to use directional antenna for the RS/MSs links, as well. Alternatively, it is possible to use a single beam-switched antenna where the pattern of the antenna is determined by using one or more directional antennas chosen from a "bank" of directional antennas. A second choice with single antenna is the application of a beamforming technique, such as phase array, where the antenna elements are fixed and the antenna pattern is shaped by electrical means. In general, the use of directional antennas is preferred owing to three significant advantages: emission of the signal only in the specified direction, provision of additional antenna gain, and the elimination of interference from other directions.

8.6 Applications and Usage Scenarios for Relay-Augmented Broadband Cellular Networks

Relays of all types can improve the signal-to-noise-plus-inference ratio (SNIR) over the *access links*. In that sense, for a given level of signal reliability, it is possible to either reduce the overall signal power, which renders favorable effect on interference into coallocated application, or enhance the system throughput. For cases in which throughput improvement is the sole objective, the TRS is preferred for the reason of implementation of cost efficiency. It is noted that TRS cannot support radio coverage extension nor it can fully support MS mobility applications. However, it provides a cost-effective solution for fixed and nomadic applications.

In order to provide cost-effective simultaneous throughput enhancement and radio coverage extension in low-MS mobility cases, NTRS with centralized scheduling may be applied. The maximum number of RSs in the MRBS cell is limited by processing capabilities as well as increased complexity of MRBS scheduler. NTRS configuration with distributed scheduling may be applied for simultaneous throughput improvement and radio outreach extension in all environments. A distributed security and scheduling mode can achieve higher bandwidth efficiency with full support for MS mobility.

For AeroMACS applications, where the main objective for application of IEEE 802.16j-based WiMAX systems is radio coverage extension, while reducing interference to coallocated application, and NTRS in distributed mode might be appropriate; however, should the objective be just reduction in interference TRS might be applied as well.

Relay configurations range from a dedicated fixed outdoor, which can provide coverage to a femtocell, to nomadic and portable, to mobile relays such as those mounted on top of moving vehicles. An interesting case is a relay mounted on top of a vehicle (a bus, a train, or an aircraft) that can provide service to clients in the vehicle while wirelessly connected to a network and undergoes handover process from one BS (or RS) to another along the way. Relays, on a nomadic and portable configuration, may also be used on a temporary basis; providing additional capacity to certain locations where a heavy traffic is expected for a limited time. Relays can also be deployed to provide coverage in an area where emergency situation has risen, such is the case in airport incidents or rescue operation. In what follows, we provide an outline of applications and usage scenarios for RSs in a general WiMAX network, with a dedicated section for usage scenario in connection with AeroMACS.

8.6.1 Some Applications of Relay-Fortified Systems

Instances of applications of relay-augmented broadband wireless networks are briefly discussed in this section.

8.6.1.1 The European REWIND Project

The European "Relay-based wireless network and standard," REWIND (2007–2011) is a design and development project aimed at integration of relay stations into advanced OFDM-based wireless networks. The focus of REWIND project has been on WiMAX technologies. Therefore, should IEEE 802.16j be accepted for AeroMACS, REWIND Project studies on integration of IEEE 802.16j into cellular networks would have important referential value for AeroMACS system design and development.

The REWIND Project is dedicated to developing the next generation of broadband wireless networks based on relay-augmented WiMAX system (IEEE 802.16j Amendment). The posture of this project is to provide a ubiquitous wireless coverage to almost all applications, as it is expressed in its objective statement, *"The Relay based Wireless Network, will create through proliferation of low-cost, easy to install Relay Stations providing the necessary capacity and coverage to support new wireless applications with similar QoS currently available only through wired broadband service"* [27].

8.6.1.2 Vehicular Networks

Intelligent Transportation Systems (ITS) and broadband wireless access (BWA) technologies have rapidly converged into the so-called "vehicular networks" (VN). Vehicular networks provide a wide verity of services, including public safety communications, crash avoidance, and wireless access to the Internet for moving vehicles such as buses and trains. This requires roadside infrastructure consisting of BSs and perhaps RSs. Several technologies have been proposed for VN that includes IEEE 802.16j-based WiMAX networks. One of the challenging tasks in designing routing algorithm for VN is the accommodation for rapidly changing topology and high mobility of vehicles in the network. A critical issue is frequent path disruptions caused by high-speed vehicles leading to broken links that results in low throughput. This presents a challenge in providing a desired QoS to the user. MAC protocols for IEEE 802.16 can support the expected QoS for this application. Moreover, IEEE 802.16j has the advantage of providing extended coverage and improved network throughput by employing multihop relays. Furthermore, owing to high data rate, large network coverage, support for full mobility, relative inexpensive network deployment and maintenance, IEEE 802.16j stands as a very competitive alternative for this application [28–29].

A vehicular network based on IEEE 802,16j is configured as a point-to-multipoint system in which the MRBS serves a multiple of moving RSs. A moving RS, such as a train, serves a large number of SSs, such as passengers in the train. The BSs are, in turn, connected to the Internet through some Internet service gateway (ISG).

8.6.1.3 4G and 5G Cellular Networks

Relays have been long recognized as an important technological ingredient of IMT-Advanced technologies [30]. The attraction comes from the two main traits of multihop networks of cost effective radio outreach extension and throughput improvement. The IEEE has developed a new amendment to 802.16 standard, IEEE 802.16m, to accommodate the requirements of 4G cellular systems. In regard to 5G cellular networks, even though the 5G standards are in their early stages of development and not a great deal of information is available on their specifications at the time of preparation of this chapter, it is known that small cell (picocells and femtocells) configuration is one of the main technology components of 5G networks. It is plausible that the small cells will be supported by relay stations.

8.6.1.4 Cognitive Femtocell

Femtocells that can decode the BS control channel signal and make decision on their transmission opportunities; based on the BS scheduling protocols, are known as *cognitive femtocells*. The operation of such femtocells is similar to NTRSs as described in IEEE 802.16j amendment [31]. As it was mentioned earlier, the high data rates envisioned for mobile WiMAX networks may be difficult to realize for the projected data reliability and quality of service with conventional mobile WiMAX networks over the entire WiMAX cell footprint. Besides, in order to provide coverage to a large region, deployment of conventional WiMAX network infrastructure may be prohibitively expensive. Multihop relay configuration based on IEEE 802.16j provides a viable solution for both of these scenarios. Besides enhancement in capacity and extension of radio coverage; other major benefits that may be gained from the usage of relays are improvement in QoS, support for higher levels of mobility, decrease in cost of network installation and maintenance, and improvement in frequency planning.

8.6.2 Potential Usage Scenarios of IEEE 802.16j

In this section, we explore some usage scenarios and applications of IEEE 802.16j-conformant multihop relay systems that could potentially be exploited in airport surface environment.

8.6.2.1 Radio Outreach Extension

In this application, the RS is used to extend the coverage area of a WiMAX/AeroMACS cell beyond the original cell footprint. A RS can extend the coverage area of the cell in certain locations such as at the cell boundary or outside of the cell footprint in a variety of ways. Three such scenarios are shown in Figure 8.9.

Figure 8.9a illustrates a case in which extension of coverage area and/or stipulation of higher capacity or throughput at the cell boundary is

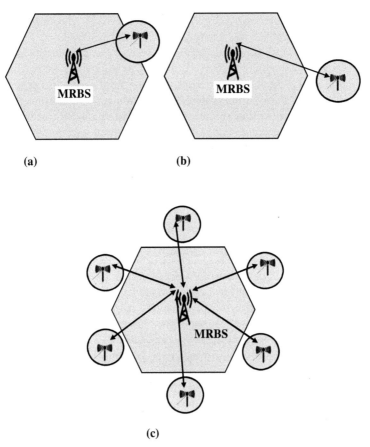

Figure 8.9 Three instances of radio outreach extension with multihop relays. (a) Improved coverage for cell boundary. (b) Extension coverage to an area outside of the cell footprint. (c) Radio coverage extension with relays and a single MRBS.

desired. Figure 8.9b demonstrates the case of coverage extension in an area outside of the cell footprint, a case that is sometimes referred to as *remote sector*. A remote sector relay could provide access to AeroMACS for airport assets that are located outside of the airport area. The latter configuration, shown in Figure 8.9c, is an example of providing coverage outside of the cell footprint, with deployment of as many RSs around the perimeter of a cell as needed in order to achieve larger coverage area with a single MRBS. This concept may be used as a design approach to provide radio coverage with minimum possible number MRBSs in a wireless cellular network. We recall here that "coverage" is basically determined by the power of the transmitter, the noise figure of the receiver, and receiver sensitivity [32].

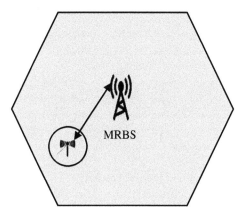

Figure 8.10 A "hole filling" scenario.

8.6.2.2 The Concept of "Filling a Coverage Hole"

In order to improve link quality for areas inside a WiMAX/AeroMACS cell footprint in which the signal is significantly attenuated, an RS can be deployed inside the particular service area of the cell. The excessive attenuation may be caused by shadowing or other effects. In this usage scenario, the RS is said to be filling a coverage hole. When using a relay as a "hole filler" it is recommended that handover between the coverage area of the MRBS and the coverage area of the RS should be avoided. Therefore, a TRS may be the suitable choice. However, in case of strong shadowing to the extent that frame control messages from the MRBS might be too weak to be detected by the MS/SS, a NTRS should be employed. Figure 8.10 portrays a "hole filling" scenario.

8.6.2.3 Relays for Capacity and Throughput Improvement

It is intuitively acceptable that usage of RSs inside a coverage area should increase the throughput and system capacity. Capacity increase in a given area can be achieved by placing a NTRS within the area. In this fashion, the NTRS essentially creates a smaller cell within the service area of the larger cell, and since each NTRS plays the role of a small BS, it contributes its capacity to the coverage area. Moreover, it is observed that MS/SSs with low-quality link to the MRBS may be enabled to have a better link quality connecting to the serving RS. This increase in link quality translates to higher throughput and therefore to a higher capacity.

8.6.2.4 The Case of Cooperative Relaying

Cooperative relaying enables improvement in spectrum efficiency and enhancement in reliability. It can substantially increase the throughput and provide *cooperative diversity.*[1] Multihop relays and MRBS can function together to provide cooperative diversity. The IEEE 802.16j amendment allows for generation of cooperative diversity through three mechanisms. In *cooperative source diversity*, the MRBS and RSs associated with it transmit the same signal simultaneously in time and frequency, thereby creating diversity. *Cooperative transmit diversity* utilizes space-time codes distributed across MRBS and its cooperating RS antennas to generate diversity gain. *Cooperative hybrid diversity* is a combination of the two methods just outlined [33].

8.6.2.5 Reliable Coverage for In-Building and In-Door Scenarios
Recognizing that signal is significantly attenuated when it crosses outer and inner walls of a building (the so-called *penetration loss*), the wireless link connecting an outer BS to an inside-building MS/SS may provide QoS that is below the acceptable level. This limits the data rate in this link while consuming excessive time-frequency resource from the BS, which is usually much more than resources required for providing the same service to a MS/SS located at the same distance from the BS but located outside of the building. Additionally, there are other coverage problems for buildings if the link is associated with an outside BS. For instance, SSs residing in high floors of a high-rise complex, in many cases, receive signal from a multiple of outside BSs. As a result, two problems may occur. First, signals originated from multiple BSs may interfere with each other, hence degrading the SINR of the SS. Second, the SS may experience an undesired number of repeated handover process with the BSs from which they receive signal. This situation may result in unnecessary radio power consumption, squandering of backhaul and computational resources, and might eventually cause the termination of the call.

There are several methods for providing links with adequate QoS to indoor SSs. The selection of a specific method is based on the type of the building (residential, commercial, industrial, government), technical requirements, and cost. In general, depending on the size and the type of the building, one or more RSs may be used for providing coverage to SSs inside the building. One method uses fixed transparent RSs mounted in a way that their antenna maintains good link quality simultaneously with the MRBS and SSs, thereby providing a direct connection with the MRBS. A second method calls for the usage of lightweight nomadic RSs that are very similar to a Wi-Fi router. Such a unit is placed where its antenna

---

1 In general, cooperative diversity is generated in a wireless network when a set of nodes relay messages for each other to produce redundant signal copies over multiple paths in the network. This redundancy allows the ultimate receivers to essentially average channel variations resulting from fading, shadowing, and so on.

maintains a good link with the MRBS and at the same time provides good links to the SSs at the vicinity of the RS. When large area floors need to be covered, and a single RS may not be sufficient, multiple Rss can be distributed over the floor connected to each other using the multihop capability of the RS. The internal Rss can be chained to a RS mounted at the edge of the floor with a good link to the MRBS. In some cases, it may be possible to provide in-building coverage with external Rss that are "illuminating" the building from outside [33].

8.6.2.6 The Mobile Relays

Mobile Rss may be used to provide service to SSs located in a moving vehicle such as a bus, a train, or an aircraft. The RS is normally mounted on the roof of the vehicle and connects to a BS with a LOS or a near-LOS link. In this application, the RS is mobile and therefore will have to repeatedly perform handovers with the network. Clearly, multihop capability is required for supporting this scenario. Since the NTRS delays the messages related to channel estimation, only NTR is suitable for this scenario.

8.6.2.7 The Temporary Relay Stations

When a heavy load of traffic is expected in a certain location, or when, for instance, an incident has occurred on the airport surface that requires a wireless link for a limited period of time, Rss may be deployed temporarily to provide coverage or additional capacity. In this case, it may be possible to use a transparent RS that will add capacity by improving SS link throughput due to link improvement to and from the SSs. The advantage of using a transparent RS is the prevention of handover between the coverage area of the RS to its surrounding base stations. If, however, high capacity is required, one or more non-transparent Rss may be added.

8.7 IEEE 802.16j-Based Relays for AeroMACS Networks

Section 8.6 addressed potential applications and general usage scenarios for IEEE 802.16j-based technology. The current section is devoted to discussion of particular radio coverage situations on airport surface for which IEEE 802.16-2009-WiMAX system either fails to offer a viable solution or the resolution it provides is inefficient, costly, or excessively power consuming. In all these cases, it is argued that IEEE 802.16j technology offers a much better alternative with the application of multihop relays. In addition, a strong case is made in favor of adoption of multihop relays into AeroMACS networks.

8.7.1 Airport Surface Radio Coverage Situations for which IEEE 802.16j Offers a Preferred Alternative

There are a number of situations on an airport surface that call for relay-fortified network solution; a list of selected few is provided here.

1) When a portion of an airport is significantly shadowed by a new obstacle, such as a building constructed for a new terminal in an airport, an IEEE 806.16j-based transparent or non-transparent relay can be added to the airport network to provide higher capacity and acceptable QoS to the shadowed area. Adding a relay to an already established network does not require network reconfiguration and radio resource reallocation. The alternatives that IEEE 802.16-2009 offers are either the addition of another BS to the network, which requires network redesign and entails reallocation of resources, or an increase in the output power of the other BSs, which may or may not resolve the problem while increases the total AeroMACS output power.

2) When a heavy load of traffic is expected in a certain location, or when an incident has occurred that requires a wireless link over a limited period of time, RSs may be deployed temporarily to provide coverage or additional capacity. In these cases, it may be possible to use a TRS that will add capacity by enhancing MS/SS link throughput due to link improvement to and from the MS/SSs. The advantage of using a TRS is that no handover between the coverage area of the RS to its surrounding is needed. Mobile relays may be most suitable for these cases.

3) If a station is located outside of the airport perimeter but needs connection to the AeroMACS network, an RS (as opposed to a BS) can be used to establish the connection. An example of this sort is a wake vortex detection sensor that may need to be located a mile or more from the end of a runway, which can place it well outside of the airport boundary.

4) Coverage to single point assets on the airport surface that are outside of the BS coverage area can be readily rendered by an RS. This may be particularly suitable for airport security equipment such as cameras. Relays may also be used to provide coverage to other airport assets such as lighting system, navigational aid, and weather sensors.

5) Spectrum efficiency and/or network capacity may be improved by application of spatial diversity. This is feasible with IEEE 802.16j with RS.

6) Cooperative relaying enables improvement in spectrum efficiency and enhancement in reliability and can substantially increase the through-put and provide cooperative diversity. Multihop relays and the MRBS can function together to provide cooperative diversity. The IEEE 802.16j amendment allows for generation of cooperative diversity

through three mechanisms. In cooperative source diversity, the MRBS and RSs associated with it transmit the same signal simultaneously in time and frequency, thereby creating diversity. Cooperative transmit diversity utilizes space-time codes (see Chapter 5) distributed across MRBS and its cooperating RS antennas to generate diversity gain. Cooperative hybrid diversity is a combination of the two methods just outlined [34].

7) Relays can provide coverage to relatively small areas on the airport surface outside of the BS cell footprint, including permanently shadowed areas. Relays can also extend the radio outreach to airport surface locations outside of base station coverage domain, in which there are relatively few users.

In short, relays can be applied to extend radio outreach to various shadowed corners of an airport. They may be used for capacity and throughput enhancement, anywhere, anytime, temporary or permanent.

8.8 Radio Resource Management (RRM) for Relay-Fortified Wireless Networks

Radio resource management refers to measurement, exchange, and control of radio resource-related (e.g., the current subchannel allocations to service flows) indicators in a wireless network. *Measurement* refers to determining values of standardized radio resource indicators that measure or assist in estimation of available radio resources. *Exchange* refers to procedures and primitives between functional entities used for requesting and reporting such measurements or estimations [34]. Clearly, the control of radio resource assets is the central part of RRM. In this fashion, MS/SSs and MRBSs are controlled for the purpose of the optimization of the capacity of the whole network.

There are two primary tools that are used in RRM: power control and handoff. Through lowering the transmitter output power, one can reduce or control the interference to unintended receivers. Through handoff process, the network replaces the serving BS to the one which would allow lower transmit power, and therefore the interference to unintended receivers is reduced. The handoff process can play a multiple role in MRR, for example, consider the following two instances.

- A mobile station can be handed off from one BS to another on the basis of signal loss between the MS and the original BS in comparison with the signal loss between the MS and the target BS. That is to say that handoff would take place when the signal loss from the MS to the target BS is less than that of between the MS and the original BS.

- An MS might be handed over from a "crowded" BS to a less utilized BS, thus the underutilized BS becomes more utilized, leading to higher level of spectral efficiency. The overutilized BS now has room for new MS/SSs, and until they arrive, it can either increase bandwidth allocation for the remaining subscribers, or use more robust modulations to reduce transmit power.

When relays are introduced to the network, they complicate the RRM in several ways.

1) The existence of relays creates more opportunities for interference. The relays are new actors in the network that may cause interference to MS/SSs trying to reach other MSs and BSs.
2) The second important reason for which relays complicate RRM is the fact that each relayed SS/MS now takes part in more than a single wireless link, and RRM aims to optimize all these links simultaneously.
3) Deciding which RS will serve the MS is a function of load over the various air interfaces. Which RS is better suited to handle the MS depends on how many subscribers and how much traffic is relayed not only thorough candidate RSs, but also through the corresponding superordinate MRBSs.
4) Mobile RSs introduce another complication for RRM. A mobile RS can change its superordinate MRBS as it moves around. In terms of mobility, a mobile RS essentially functions as an MS. The network should take into account the special requirements of a mobile RS handoff.

The WiMAX networking architecture defines two profiles. Profile A defines distributed RRM, in which every BS has a radio resource agent (RRA) that collects radio utilization and interference metrics, and a radio resource controller (RRC) that communicates with the RRA and with other RRCs to initiate handovers. Profile C defines the RRC to reside in the ASN-GW. In Profile C, the RRC polls RRAs inside every BS to decide when a handover is needed.

When using network-based relay, together with an RRA and possibly an RRC, each RS is in essence a BS. If, for the initial stages of deployment, traffic demands are moderate, then using the existing WiMAX RRM, which does not take into account the load over the backhaul links, with little or no modifications, is probably sufficient for the initial stages of deployment.

Building RRM for an optimal network can be delayed to when the network is more mature. When using MAC-based relay, a RS does not inherently have RRM capabilities. It is therefore mandatory to implement RRM for the RS in the MRBS [22].

8.9 The Multihop Gain

In this section, we aim to quantify the positive effect of routing a signal through intermediate multihop relays, prior to its arrival at the destination node. The application of multihop relay enables a reduction in path loss, accordingly, from the perspective of link budget analysis, a "gain" is resulted. We designate this gain as *multihop gain*. Multihop gain can be realized in a variety of ways, we will consider the simplest possible case in our analysis. Once achieved, the multihop gain may be translated into one or more of the following system enhancements for AeroMACS.

- Radio outreach extension
- Improvement in throughput and network capacity
- Reduction in transmitted power

Reduction in the aggregate transmitted output power is of the primary concern in AeroMACS application and deployment. This is regarding the critical question of IAI, which will be addressed in this chapter.

8.9.1 Computation of Multihop Gain for the Simplest Case

Under the following assumptions, a simple analysis can be conducted for calculation of a raw measure of multihop gain, expressed in decibel.

1) RS and MS/SS receivers have the same sensitivity shown by S_p.
2) Propagation path loss between the MRBS and RS is represented by L_{BR} dB.
3) Propagation path loss between the RS and SS/MS is represented by L_{RS} dB.
4) Direct propagation path loss between the MRBS and SS/MS is represented by L_{BSS} dB.

The minimum required transmit power by the RS; P_{RS}, and the minimum required transmit power by the BS, for BS to RS transmission; P_{BR}, are given by Equations 8.1 and 8.2, respectively.

$$P_{RS} = S_p(10)^{L_{RS}/10} \tag{8.1}$$

$$P_{BR} = S_p(10)^{L_{BR}/10} \tag{8.2}$$

The minimum required power for signal transmission from the BS to the RS and then to the SS; P_{BRS}, is the sum of the powers given in Equations 8.1 and 8.2, expressed in Equation 8.3.

$$P_{BRS} = P_{BR} + P_{RS} = S_P\left[(10)^{L_{BR}/10} + (10)^{L_{RS}/10}\right] \tag{8.3}$$

The minimum required transmit power for direct transmission of signal from BS to SS; P_{BSS}, is determined by Equation 8.4.

$$P_{BSS} = S_p(10)^{L_{BSS}/10} \tag{8.4}$$

We define multihop gain as the ratio of Equation 8.4 to Equation 8.3. This gain, G_{MH}, in decibel, is then calculated by Equation 8.5.

$$G_{MH} = L_{BSS} - 10\log\left[(10)^{L_{BR}/10} + (10)^{L_{RS}/10}\right] \quad \text{dB} \tag{8.5}$$

This is intuitively satisfying and can readily be generalized for the case in which RS and MS/SS have unequal sensitivity values. Hence, although Equation 8.5 shows that multihop gain is independent from receiver sensitivities for this simplest case, in general multihop gain is a function of receiver sensitivities of MS and SS as well as other parameters involved in the equation. Equation 8.5 demonstrates, however, that multihop gain depends on the propagation path loss between various stations in the network (which in turn depends on positioning of the relay stations as well), in other words it varies from one propagation environment to the other.

The final conclusion from Equation 8.5 is that multihop gain is directly affected by the following three factors.

1) Relay stations positioning in the network
2) Propagation characteristics of the terrain through which signal travels
3) Transmit power settings and distribution

Path loss, as well as multihop gain, graphs can be plotted against various propagation loss models and parameters (see, Ref. [35] for an example). One such variable, which seems to be advantageous for the airport surface environment, is the factor α, expressed in Equation 8.6, and explained by Figure 8.11.

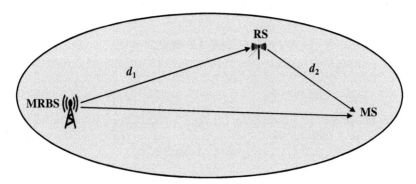

Figure 8.11 A simple scenario for multihop gain analysis.

$$d_0 = \alpha(d_1 + d_2) \quad \alpha \leq 1 \tag{8.6}$$

Regarding the generated multihop gain, expressed by Equation 8.5, the following important remarks can be made.

1) It is possible to achieve either of the following gains, vis-a-vis multihop gain, without increasing the total transmit power.
   a) Extend the radio coverage range for a required received signal strength (RSS)
   b) Enhance the RSS at a particular point in the network
2) Reduction in transmit power can be realized while maintaining the same RSS. This in turn reduces IAI, that is, the interference to coallocated applications. For AeroMACS this is highly desirable, since interference to mobile satellite system that shares the same spectrum will be reduced [9].

8.10 Interapplication Interference (IAI) in Relay-Fortified AeroMACS

A major concern about deployment of AeroMACS over the 5091–5150 MHz C-band is the IAI, for AeroMACS shares the C-band with other "coallocated" applications. Of particular importance is the case of commercial feeder links of nongeostationary satellite (low Earth orbit, LEO) systems in the mobile satellite service (MSS). The Globalstar Satellite Constellation is an example of an existing operational MSS system that operates feeder links in this band.[2] The potential for interference between AeroMACS and MSS feeder links limits the power levels that are allowed for AeroMACS networks [36].

Analytical methods and computer modeling have been employed to test and measure the level of interference posed by AeroMACS networks to co-allocated applications. At NASA Glenn Research Center, the software program Visualyse Professional [37] has been utilized to estimate the limitations of AeroMACS transmitter output power levels, vis-à-vis prevention of unacceptable levels of IAI into MSS feeder link signals [38]. A similar approach was previously adopted by MITRE Corporation [39]. In both of these models, a single BS per airport was assumed, and an airport was viewed as a power emitting point on the contiguous US global surface, with a total of 497 towered airports. In order to reflect a more realistic scenario, the antenna directivity pattern for each of the 497 towered airports was selected randomly.

---

2 http://www.globalstar.com/en/.

Table 8.3 Simulation scenarios for IAI measurement.

Simulation case	Mode	Number of BSs (or BS sectors)	Number of RSs	BS antenna relative beam orientation (degree)	RS antenna relative beam orientation (degree)
1	BS only	4	0	−90, 0, 90, 180	N/A
2	Mixed	3	2	0, 90, 180	−135, −45
3	Mixed	3	1	0, 90, 180	−90
4	BS only	3	0	−120, 0, 120	N/A
5	Mixed	2	2	0, 180	−90, 90
6	Mixed	2	1	0, 180	−90

To compare interference performance for IEEE 802.16-2009-based AeroMACS with IEEE 802.16j-based AeroMACS, six model cases were created and simulated at NASA GRC. Two of these cases represent all-BS airport networks, while the other cases are associated with mixed BS–RS networks. The output power of a BS antenna (or a sector of a BS antenna) is assumed to be 100 mW, and the output power of an RS antenna is assumed to be 3 dB lower, that is, 50 mW. Table 8.3 provides various information about these simulation cases, including data about relative directions of the BS and RS antenna beams.

For each of the six cases, shown in Table 8.3, all of the 497 towered airports in the contiguous United States were assumed to have the given antenna gain pattern, but the direction corresponding to 0° was randomized from airport to airport. Ten runs each with a different randomization were generated for each case. An example of the resulting interference power profile at LEO is shown in Figure 8.12 for one of the runs with Case 1. The colors are associated with the intensity of interference. As Figure 8.12 shows, in some areas in Northern Canada and Central America, interference has its highest intensity. For each run, the position and value of the maximum aggregate interference power was recorded [3]. Figure 8.13, adopted from Ref. [3], illustrates the simulation results and Table 8.4 shows the corresponding average maximum interference power and standard deviations.

In Case 1, four 100 mW beams spaced apart by 90° were used. This closely approximates an omnidirectional gain pattern with 400 mW radiation power. Thus, the different randomization runs have little variation in maximum interference power as shown in Figure 8.13, and reflected by the small standard deviation in Table 8.4. In Case 2, the total radiated power

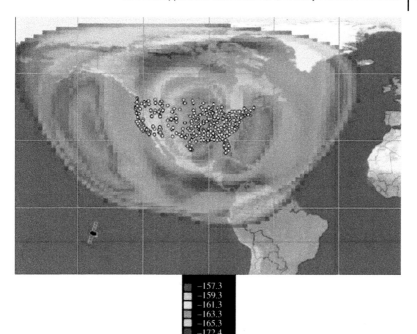

```
    -157.3
    -159.3
    -161.3
    -163.3
    -165.3
    -172.4
```

Figure 8.12 Aggregate interference power (dBW) at LEO from 497 airports with antenna randomized orientation gain pattern for Case 1. (Adopted from Ref. [3] with permission of IEEE.)

is still 400 mW, but one of the BS beams has been replaced by two RS beams. Figure 8.13 shows that the average maximum interference power is almost identical to that of Case 1, but Table 8.4 indicates that there is a larger spread among the ten randomized runs. This is due to the asymmetrical nature of the gain pattern. Both Cases 1 and 2 generate interference power higher than the threshold of −157.3 dBW established so as to limit the increase in the MSS feeder link satellite receiver's noise temperature to less than 2% [40].

In Case 3, one of the BS beams is replaced with a single RS beam, reducing the total radiated power from 400 to 350 mW. It is seen in Figure 8.13 and Table 8.4 that this is enough to reduce the maximum interference power below the threshold value [3].

In Cases 4 and 5, the total radiated power is further reduced to 300 mW. Case 4 has three BS beams, while Case 5 has two BS beams and two RS beams. The maximum interference power decreases as expected and as in the comparison between Cases 1 and 2, there is not much difference between the results of Cases 3 and 4, although there is somewhat more variation among the randomized runs in Case 4.

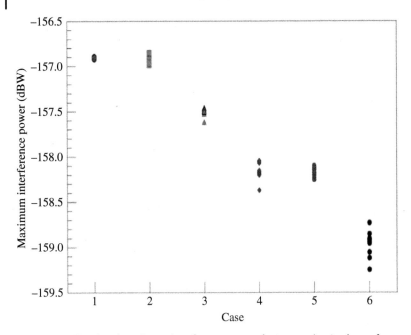

Figure 8.13 Simulated maximum interference power for ten randomized runs for the six antenna configuration cases described in Table 8.3. (Adopted from Ref. [3] with permission of IEEE.)

Case 6 radiates 250 mW with just two BS beams and one RS beam. The maximum interference power decreases again as expected and the spread among the runs is higher than in any of the other cases because of the increased asymmetry.

These results reveal that under equal output power transmission for each airport and each configuration, there is no additional interference

Table 8.4 Aggregate IAI power posed by all 497 contiguous US towered airports into MSS feeder link for configurations given in Table 8.3.

Simulation case	Total output power (mW)	Average maximum interference power (dBW)	Standard deviation of interference power (dBW)
1	400	−156.91	0.01
2	400	−156.92	0.05
3	350	−157.50	0.04
4	300	−158.16	0.09
5	300	−158.17	0.05
6	250	−158.96	0.15

into MSS feeder link from the IEEE 802.16j-based AeroMACS as compared to that of all-BS AeroMACS network. It is the total power that is radiated from each airport that is most important; the distribution is only of secondary importance.

In summary, we make the following important observations that are the key conclusions of this preliminary simulation study.

- No additional interference to the MSS feeder link is caused by deployment of IEEE 802.16j-based AeroMACS.
- The total antenna output power level has the dominant effect on AeroMACS interference to coallocated applications. Antenna orientation and directivity, and whether the network employs all-BS configuration or BS–RS architecture, play an insignificant role in this regard [3].

8.11 Making the Case for IEEE 802.16j-Based AeroMACS

Early on in this chapter, it was pointed out that, recognizing that practical shortfalls arising from early implementation of IEEE 802.16e-based WiMAX networks, some modifications in the Standard was deemed necessary. Even with the application of all advanced signal-processing techniques that are available in WiMAX technology, it was shown that the projected data rates require SINR levels at the front end of the receivers that are difficult to obtain at the WiMAX cell boundaries or in shadowed areas [3].

On the other hand, it has also been widely accepted that using an all-BS WiMAX network to cover a large area might be economically infeasible. Infrastructure deployment and maintenance for an all-BS WiMAX network may be too costly. This was evident from the beginning that the deployment of an all-BS WiMAX system in competition with the existing 3G network may be a losing battle, at least economically [23]. These issues shaped the key incentives behind the development of IEEE 802.16j amendment. In the following paragraphs, we present strong arguments in favor of IEEE 802.16j-based AeroMACS network, for this Standard allows the application of the multihop relay as a design option.

8.11.1 The Main Arguments

The main argument in favor of application of IEEE 802.16j technology in AeroMACS is the flexible and cost-effective extension of radio outreach inside and outside of the airport real estate, which can be provided by IEEE 802.16j networks, with virtually no increase in the power requirements and no additional interapplication interference (as shown in the previous section).

By flexible radio outreach extension we mean adding relays to the network as needed, like when a new runway, or a new terminal, or a new parking deck, is added to the airport. A basic IEEE 802.16j-based WiMAX cellular network can be rolled out for an airport, and as the network expands, relays can be added to meet the new coverage and transmission requirements.

8.11.1.1 Supporting and Drawback Instants

The following comments are in order at this point

- In most cases the use of an RS can increase throughput per MS/SS which can be translated into system capacity enhancement or QoS improvement.
- A single link between the BS and MS with high SINR requirement can be replaced with multiple BS to MS links, through relays, that require low SINR; the result is an increase in spectral efficiency.
- Further spectral efficiency may be realized through cooperative communications and diversity gain.
- Latency increases, particularly when NTRSs are employed for radio outreach extension.
- More signal overhead may be needed, the mode of relay and the number of hops will have direct effect on additional signal overhead.
- The MRBS will be more complex at both physical and MAC layers, and it will become more complex as the number of hops is increased.
- The MS/SS systems need no upgrade, that is, they can be based on IEEE 802.16e.
- IEEE 802.16j standard is written under a mandatory constraint that the MS apparatus conforms to the original IEEE 802.16e standard and has no part that is designed based on the IEEE 802.16j amendment. This implies that the MS will not be aware of the existence of the relays as intermediate nodes (even if NTRS is being used), this is a hindrance for this novel technology [19], although it keeps the legacy MS/SS intact.

8.11.2 The Second Argument

The next argument in support of IEEE 802.16j-based technology is the capability of increasing throughput and capacity, anywhere and at any time, temporary or permanent, in the IEEE 802.16j-based AeroMACS networks.

A BS to MS link with low SINR may be replaced with a BS to RS to MS link in which both BS to RS and RS to MS links can enjoy high SINR, particularly in C-band frequencies. Clearly, the WiMAX AMC protocols will interface these links to a modulation/coding combination with higher level of modulation and greater coding rate, and therefore the data rate for the MS, and as a result the overall throughput of the network will be effectively boosted and a higher spectral efficiency is achieved. For this sort of enhancement, TRS may be more appropriate.

The IEEE 802.16j amendment has specified a wide variety of options. It is necessary to consider which subset of these options is the most appropriate for AeroMACS application. The most recent version of AeroMACS profile, authored by RTCA, is described in some detail in Chapter 6. This profile provides a guide for AeroMACS network design and implementation. one has to note that, should the IEEE 802.16j-based WiMAX be adopted for AeroMACS, a number of technical issues remain to be addressed.

8.11.3 How to Select a Relay Configuration

Finally, regarding the selection of a proper relay configuration for a given applications, the following observations are made.

- For radio coverage extension NTRSs are required, exclusively.
- In order to accommodate and support mobile relays, NTRSs with distributed scheduling are required.
- NTRSs can provide a modest throughput/capacity improvement whereas TRS are incapable of supporting mobile relays or providing radio coverage extension.
- Conceptually, there can be cases in AeroMACS networks that call for simultaneous usage of TRSs and NTRSs in an airport. The IEEE 802.16j standard provides very little detail of how this can be handled/realized [1].

Thus, it is apparent that in case it is desired or necessary to exclusively employ one type of relay mode for all applications throughout an AeroMACS network, the proper selection is the NTRS.

8.11.4 A Note on Cell Footprint Extension

Coverage at a given point in the network is determined by by the output power of the transmitter and the noise figure of the receiver at that point. Use of directional antennas and increase in transmitter power generally expands the coverage area. However, in many scenarios, including in many parts of airport surface, cell coverage is significantly affected by obstructions such as building or topography. In such cases, transmitter power rise enlarges the IAI and raises the cost of the system electronics, while it has an insignificant impact on the coverage area. For instance, in highly obstructed links with 40–50 dB path loss per decade of distance, doubling the transmitter power extends the range of the cell footprint by less than 20% [5]. The other alternative is to create a new cell with its own BS, which requires reconfiguration and redesign of the whole cellular network. This increases the cost of the system considerably and raises the network output power significantly. The use of a relay seems to be the optimum choice, particularly in light of the fact that a relay can be deployed in places with a LOS or near-LOS link to the MRBS. When the mobile station moves away from the area covered by the MRBS, and into the extended area covered exclusively by the NTRS, a handover process is required.

8.12 Summary

In this chapter, the main focus is on AeroMACS networks that are based on IEEE 802.16j amendment. This amendment enables the network designer to use the multihop relay as yet another design option in her design arsenal set.

The crux of the idea of the IEEE 802.16j amendment is expressed in the Standard's abstract and opening statements. *"This amendment specifies OFDMA physical layer and medium access control layer enhancements to IEEE Std 802.16 for licensed bands to enable the operation of relay stations."* *"This amendment updates and expands IEEE Std 802.16, specifying OFDMA physical layer and medium access control layer enhancements to IEEE Std 802.16 for licensed bands to enable the operation of relay stations. Subscriber station specifications are not changed. As of the publication date, the current applicable version of IEEE Std 802.16 is IEEE Std 802.16-2009, as amended by IEEE 802.16j-2009"* [5].

The chapter defines and classifies IEEE 802.16j-based relays in a variety of ways. The main division is in relation to the mode of operation and functionality of a relay in the network. A nontransparent relay station (NTRS) may operate as a "mini base station," and therefore has the authority to manage network resources. NTRS can be used for radio outreach extension into severely shadowed areas in an airport. The TRS, on the other hand, acts essentially like a relay, receiving traffic data from the MRBS or another RS and passes it on to the next node. TRS is normally used for capacity enhancement and throughput improvement. Frame structure and MAC layer characterization of several types of relays are discussed.

The chapter contains a great deal of information regarding the applications and usage scenarios for multihop relays in wireless networks, in general, and AeroMACS network, in particular. IAI is an important issue, it has been shown through a preliminary simulation study that deployment of IEEE 802.16j AeroMACS poses no additional IAI to coallocated applications. At the end, a strong case is made in favor of IEEE 802.16j-based AeroMACS.

In regard to four key aspects of wireless networks, namely; throughput, radio coverage extent, signaling overhead/latency, and bandwidth efficiency, for different classes of IEEE 802.16j-supported relays, the following applications and usage scenarios are deemed feasible.

1) TRSs may be used for low-cost intracell throughput improvements but not for coverage extension.
2) NTRSs in centralized scheduling provides throughput improvement and coverage extension; particularly in low-mobility scenarios.
3) NTRSs with distributed scheduling mode present versatile relays for multipurpose applications. They may be used for simultaneous throughput improvement and radio outreach extension in various environments [14].
4) NTRS with distributed scheduling and security, as Ref. [14] argues and demonstrates; through system level simulation, achieves the best performance, at least in regard to intracell coverage and capacity.

It is plausible that the maximum number of hops in a IEEE 802.16j-based AeroMACS network should be limited to two. That would make all technical issues at PHY and MAC layer much more manageable. Hence, it would make sense to keep the usage of relays, at least as an option, in AeroMACS networks. Also, it should be mentioned that it is always possible to incorporate IEEE 802.16j standards into AeroMACS networks, even if the network is rolled out as an 802.16-2009-based network originally.

References

1 V. Gene, S. Murphy, Y. Yu, and J. Murphy, "IEEE 802.16j Relay Based Wireless Access Network: An Overview," *IEEE Wireless Communications*, 15(5), 56–63, 2008.
2 IEEE, 802.16's Relay Task Group, "IEEE 802.16 Relay TG Minutes of Session# 61," June 2009, http://www.ieee802.org/16/relay/.
3 B. Kamali, J. D. Wilson, and R. J. Kerczewski,"Application of Multihop Relay for Performance Enhancement of AeroMACS Networks," *Proceedings of IEEE ICNS Conference*, pp. G2-1–G2-11, 2012.
4 REWIND Project document, "REWIND WP4D4.2" Available at http://rewind-project.eu/deliverables, 2009.
5 IEEE, IEEE Std. 802.16j TM-2009, Part 16: Interface for Broadband Wireless Access Systems, Amendment1: Multihop Relay Specification, 2009.
6 B. Kamali, R. J. Kerczewski,"On Selection of Proper IEEE 802.16-Based Standard for Aeronautical Mobile Airport Surface Communications (AeroMACS) Application", *Proceedings of IEEE ICNS Conference*, pp. G3-1–G3-8, 2011.

7 C.-Y. Chang, C.-T. Chang, M.-H. Li, and C.-H. Chang,"A Novel Relay Placement Mechanism for Capacity Enhancement in IEEE 802j WiMAX Networks," *Proceedings of IEEE ICC*, 2009.

8 G. Shen, et al., "Multi-Hop Relay Operation Modes," *IEEE802.16 Broadband Wireless Access Working Group*, C802.16m/1429, 2008.

9 B. Kamali and R. J. Kerczewski,"IEEE 802.16j Multihop Relays for AeroMACS Networks and the Concept of Multihop Gain," *Proceedings of IEEE ICNS 2013*, Herndon VA, April 2013.

10 WiMAX Forum, "WiMAX Forum® Network Architecture: Architecture Tenets, Reference Model and Reference Points Base SpecificationWMF-T32-001-R021v01WMF, December 2012.

11 K. Loa, C. Wu, S. Sheu, Y. Yuan, M. Chio, and D. Huo,"IMT-Advanced Relay Standard," IEEE Communications Magazine, August 2010.

12 V. Genc,"Performance Analysis of Transparent Mode IEEE 802.16j Relay-based WiMAX Systems," PhD. Thesis, University College Dublin, March 2010.

13 S.-O. Seo, S. Kim, Y. Kim, H. Lee, and H. Yyu, "Relay Performance Analysis of TTR and STR Relay Modes in IEEE 802.16j MMR Systems," *EITR Journal*, 32(2), 203–240, 2010.

14 M. Okuda, C. Zhu, and D. Viorel, "Multihop Relay Extension for WiMAX Networks: Overview and Benefits of IEEE 802.16j Standard," *FUJITSU Scientific and Technical Journal*, 44, 292–302, 2008.

15 D. Soldani and S. Dixit,"Wireless Relays for Broadband Access," *IEEE Communication Magazine*, pp. 58–66, March 2008.

16 A. Sendonaris, E. Erkip, and B. Aazhang, "Cooperation Diversity: Part I – System Description," *IEEE Transactions on Communications*, 51(11), 1927–1938, 2003.

17 A. Sendonaris, E. Erkip, B. Aazhang, "User Cooperation Diversity. Part II. Implementation Aspects and Performance Analysis," *IEEE Transactions on Communications*, 51(11), 1939–1948, 2003.

18 IEEE, IEEE 802.16–2009 Standard for Local and Metropolitan Area, Part 16: Air Interface for Broadband Wireless Access Systems, May 2009.

19 REWIND, project document: "4.1 Summary of Network Architecture Analysis and Selected Network Architecture," January 2008.

20 REWIND project document: "System Specification, Architecture Definition, D 4.3 Relay Station specification," November 2009.

21 REWIND Project Document http://rewind-project.eu/deliverable/REWIND_WP4_D4.2.

22 REWIND project document: "System Specification, Architecture Definition D 4.2 Interface definition," May 2009.

23 B. Li, Y. Qin, C.P. Low, and C. L. Gwee,"A Survey on Mobile WiMAX", *IEEE Communications Magazine*, December 2007.

24 A. F. Bayan and T.-C. Wan, "A Review on WiMAX Multihop Relay Technology for Practical Broadband Wireless Access Network Design," *Journal of Convergence Information technology*, 1(9), 363–372, 2011.

25 J. Cho and Z. J. Haas, "On the Throughput Enhancement of the Downstream Channel in Cellular Radio Networks through Multihop Relaying," *IEEE Journal on Selected Areas in Communications*, 22(7), 1206–1219, 2004.

26 W. Zou,"Capacity Analysis for Multihop WiMAX Networks," *Proceedings of Auswireless Conference*, 2006.

27 REWIND Project Document. Available at http://rewind-project.eu.

28 R. Fei, K. Yang, and S. Ou,"A QoS-Aware Dynamic Bandwidth Allocation Algorithm for Relay Stations in IEEE 802.16j-Based Vehicular Networks," *Proceedings of IEEE WCNC*, 2010.

29 M. I. M. N Nazirah, et al. "Cross-Layer Routing Approach in Highly Dynamic Networks", *Proceedings of ICMSAO* 2011.

30 J. Zhang, G. Liu, F. Zhang, and D. Zhang,"Emerging Challenges to IMT-Advanced Channel Model," *IEEE Vehicular Technology Magazine*, June 2011.

31 A. Adhikary, V. Ntranos, and G. Caire,"Cognitive Femtocells: Breaking the Spatial Reuse Barrier of Cellular Systems," *Proceedings of IEEE Information Theory and Application Workshop*, February 2011.

32 REWIND Project Document. Available at http://rewind-project.eu/deliverables/REWIND_WP4_D41_V03P.pdf.

33 S. W. Peters and R. W. Heath, "The Future of WiMAX: Multihop Relaying with IEEE 802.16j," *IEEE Communications Magazine*, 47(1), 104–111, 2009.

34 WiMAX Forum, "WiMAX Forum® Network Architecture: Architecture Tenets, Reference Model and Reference Points Base SpecificationWMF-T32-001-R021v01WMF, December 2012.

35 IEEE, IEEE 802.16j Tutorial: 802.16 Mobile Multihop Relay, March 2006.

36 B. Kamali and R. J. Kerczewski,"On Selection of Proper IEEE 802.16-Based Standard for Aeronautical Mobile Airport Surface Communications (AeroMACS) Application," *Proceedings of IEEE ICNS* 2011.

37 Transfinite Systems Limited, www.transfinite.com.

38 J. D. Wilson,"Modeling C-Band Co-Channel Interference from AeroMACS Omni-Directional Antennas to Mobile Satellite Service Feeder Uplinks" NASA/TM—2011-216938, NASA Glenn Research Center, 2011.

39 I. L. Gheorghisor, Y.-S. Hoh, and A. E. Leu,"Compatibility of Wireless Broadband Networks with Satellite Feeder Links in the 5091–5150 MHz Band," *Proceedings of ICNS*, 2010.

40 I. L. Gheorghisor, Y.-S. Hoh, and A. E. Leu,"Analysis of ANLE Compatibility with MSS Feeder Links," MITRE-CAASD Report MTR090458, 2009.

Index

AeroMACS: An IEEE 802.16 Standard-Based Technology for the Next Generation of Air Transportation Systems, First Edition. Behnam Kamali.
© 2019 the Institute of Electrical and Electronics Engineers, Inc. Published 2019 by John Wiley & Sons, Inc.